Advanced Analytical Methods for Climate Risk and ESG Risk Management

Advanced Analytical Methods for Climate Risk and ESG Risk Management

A Concrete Approach to Modeling

JORGE R SOBEHART

WILEY

This edition first published 2024

Registered Office(s)
John Wiley & Sons, Inc., 111 River Street, Hoboken, NJ 07030, USA
John Wiley & Sons Ltd, The Atrium, Southern Gate, Chichester, West Sussex, PO19 8SQ, UK

Editorial Office
The Atrium, Southern Gate, Chichester, West Sussex, PO19 8SQ, UK

For details of our global editorial offices, customer services, and more information about Wiley products visit us at www.wiley.com.

Library of Congress Cataloging-in-Publication Data is Available

ISBN 9781394220090 (Cloth)
ISBN 9781394220106 (ePDF)
ISBN 9781394220113 (ePub)

Cover Design: Wiley
Cover Image: © Ali Kahfi/Getty Images

SKY10078858_070324

This book is dedicated to Valeria,
Lionel, Nadia, Yurie, Michael, Leon, Ray and Gisella.

Contents

Introduction: Climate Risk and Environmental, Social and Governance Challenges

The implementation of safe and sound risk management practices is a vital concern for financial institutions. One of the most fundamental problems in risk management is the identification, classification, measurement, and monitoring of the risks faced by the institution.

The aim of this book is to outline a framework for modeling and quantifying different aspects of climate risk and Environmental, Social and Governance (ESG) issues and their impact on risk management practices. Credit risk, market risk, profit and loss risk, asset and liability management risk, reputation risk and conduct risk have historically been thought of as related but different areas. However, the impact of climate risk and ESG, which includes physical and transition (or policy-driven) risks and social issues affecting potentially all other risk types in the short and long term, suggests that financial risks need to be analyzed on a fully integrated basis. Furthermore, in today's highly competitive financial world, successful risk management and financial engineering call for increasing levels of quantitative expertise, including the ability to effectively apply advanced modeling techniques to complex areas such as climate risk and ESG.

Reliance on models to trade and manage risks carries its own risks. Models are susceptible to errors: from incorrect assumptions about the impact of climate risk and ESG to incorrect descriptions of market interactions and social behavior, to inaccurate descriptions of the evolution of companies in highly competitive environments. Risk models used for climate risk and ESG are often based on or are extensions of methodologies developed for other purposes, such as stress testing for market and credit risk. In practice, model imperfections lead to substantial differences between the way markets behave and the models' implications.

This book introduces basic concepts of risk management, provides a broad treatment of modeling theory and methods, and explores their application to risk management practices. It draws on diverse quantitative disciplines across climate-related sciences, mathematical modeling, statistics and statistical inference, econometrics, and behavioral finance. This book is directed, first and foremost, to professionals in the financial industry involved in the development, implementation, and stress testing of credit risk management analytics, tools, and systems supporting climate risk and ESG initiatives. Second, it is directed toward risk managers who must evaluate practices and set policies for risk management within their own institutions. Third, it is directed to professionals involved in supervising, and in developing regulatory policies for banks and financial institutions. Given the highly evolving nature of the field of climate risk globally, we deliberately sought to avoid focusing on specific regulatory rules, requirements, or regimes. However, we concentrate on general principles and modeling approaches that will be of interest to practitioners and regulators everywhere. Lastly, the topics we discuss may be of interest to academics, professional auditors, and consultants, who play an important role for financial institutions in reviewing methodologies and providing effective challenge and guidance on industry best practices.

CHAPTER 1

Introduction to Climate Risk

DIMENSIONS OF CLIMATE RISK

Climate change, environmental, social and governance (ESG) risks are no longer concepts limited to academic and scientific discussions. These terms have become common topics of conversation across the media, for politicians, policymakers, statesmen, corporations, governments, and regulatory agencies around the world. Currently, the international community is struggling to define appropriate approaches, policies, and treaties that can mitigate the challenges of climate change while keeping a level of healthy growth to support economic prosperity (NGFS, 2020; Smith, 2020; FSB, 2021; IEA, 2021; USDT, 2021).

There are four dimensions to understanding climate change (Table 1.1) and its impact:

1. the science of climate change
2. the politics of climate change
3. the economics of climate change
4. the social impact of climate change.

As a result of the increased social awareness on climate change issues and continuous media coverage, many people now believe that extreme weather events are occurring more frequently than in the past and that the pace of climate change is accelerating toward a catastrophic tipping point. This view also affects statesmen and policymakers, whose decisions can have significant social and economic consequences. The political, economic, and social consequences of perceived climate change are already tangible. Developed (wealthier) nations want less-developed (poorer) nations to delay or limit their path to industrialization by reducing their carbon footprint and greenhouse gas emissions. In turn, the less-developed nations want the developed nations to take the economic burden of fixing the problem they have created in the first place. Other nations want to take advantage of this situation by accelerating their

TABLE 1.1 Dimensions of climate change

Dimension	Description	Chapters in this Book
Science	Scientific evidence of natural and man-made impact on the environment and climate change (physical risk)	1, 2 and 3
Politics	Policies, regulations, and incentives for transitioning to a low-carbon economy, which can impact sectors, markets, and consumer behavior (transition risk)	4
Economics	Risks and opportunities derived from corporate and investment strategies, practices, and actions to mitigate physical and transition risks	6–13
Social impact	Social behavior, credibility, trust, social polarization, and their impact on the migration to a low-carbon economy	4, 5

development (resulting in increases in their levels of greenhouse gas emissions), while keeping their status of less-developed economies to avoid limiting their path to prosperity. These conflicting goals make it difficult for the international community to come to terms with climate change. The debate over these issues often breaks down into strongly polarized views: those who favor a man-made-only cause of the observed changes in climate in recent times, and those who favor natural causes due to solar activity, oceanic and atmospheric variability, tectonic and volcanic activity with an important but smaller human contribution.

Given the relevance of these issues, gaining a clear understanding of how and why climate changes, with or without human influence playing a role, is one of the most important issues for society today (Mayewski and White, 2002; Plimer, 2009).

The starting point for any meaningful discussion on the topic is to define the concepts clearly. For example, what is the difference between *weather* and *climate*? What does *climate change* really mean? Are the changes due to natural or man-made causes?

BASIC CONCEPTS OF CLIMATE

The primary distinctions between weather and climate are the period used to record observations and the geographic extent of these observations. Weather is the behavior of the atmospheric system in a specific region over a short period of time, ordinarily one week, a few days or even a few hours. In contrast, climate is the behavior of the weather over extended periods of time, ranging from seasons to years, decades, centuries, or even millennia, and across extended geographic regions such as continents, subcontinents, or other large regions (NOAA, 2024).

The sciences of meteorology, oceanography, geology, and physical geography have advanced dramatically in the past century, assisted by advances in the systematic collection of local measurements across multiple regions of the globe, satellite imaging, the digitalization of information, and increased computational power. Furthermore, the scientific community has learned a great deal in recent decades about the dynamics of atmospheric and oceanographic phenomena and has constructed (and continues to refine) numerous complex models to simulate the global effects of climate change and the consequences of extreme climate patterns.

These models provide valuable insights into possible climate scenarios, their drivers, and the speed of change. But they also require objective scientific validation through strong evidence that demonstrates that these predictions are reasonably accurate and that the physical models driving these predictions and the data used to calibrate these models are comprehensive and complete. This limitation is a significant challenge for the scientific community, given the synergies and non-linearities observed in atmospheric and oceanographic phenomena, including complex energy and mass flows, which are exacerbated for long-term forecasts. Also, many physical and chemical phenomena occurring in the atmosphere, oceans, and continents are poorly represented, and even ignored, in these models, due to their complexity or lack of detailed knowledge.

Furthermore, actual records of weather patterns and climate change recovered from instruments that measure temperature, wind speed, precipitation, atmospheric pressure, and sea levels only extend back into the nineteenth century for the northern hemisphere and cover even shorter periods for the rest of the world. Constructing historical records on climate change requires extensive detective work to stitch different pieces of information into a coherent picture of past events. For example, changes in temperature or the level of CO_2 and other gases in the atmosphere over thousands of years can be inferred from the analysis of ice cores drilled around the globe.

When snow falls through the atmosphere and is deposited and compacted into ice in extremely cold regions of the planet (such as Greenland or Antarctica), it traps gases, dissolved minerals and chemicals, dust particles and sediments that can reveal the composition of the atmosphere at that moment in time. If the snow that falls does not melt or evaporate during the year, next year's snow will fall and accumulate on top of the previous year's snow. Over time, the snow layers compress and recrystallize under their own weight, creating ice and trapping small pockets of air, dust, and sediments. Ice cores capture changes in the physical and chemical composition of ice, trapped air bubbles, dust, pollen, and sediments that provide a vivid view of past environments and events, even recording climate changes over relatively short periods of time in great detail. The analysis of ice cores also reveals evidence of past volcanic eruptions, forest wildfires, changes in air humidity and dryness reflected in the composition of the trapped dust particles.

With long climate change records of 800,000 years, 2.7 million years or more available from ice cores and other paleoclimate sources, scientists are beginning to describe the natural processes of climate change, which can provide the theoretical foundations for assessing the extent of human influence on the climate (EPICA, 2004; Plimer, 2009; Voosen, 2017).

The term "climate change" is often misused as a substitute for any global shift in climate patterns that could result in a catastrophic anomaly leading to extreme weather events, such as severe hurricanes, floods, and droughts, or sea levels rising at abnormal rates. However, the climate has changed in the past in response to natural phenomena, including dramatic and sudden shifts. Ice core records confirm that the climate has changed significantly in the past and will likely change in the future. The Earth is a highly complex dynamical system, driven by internal factors (such as ocean-atmosphere interactions, heat redistribution between the Earth's core and mantle, and tectonic and volcanic activity) and external factors (such as the Sun's electromagnetic radiation, or the gravitational pull from the Sun, the Moon, and other celestial bodies). Furthermore, while natural climate changes were once viewed as moderate shifts against a backdrop of ever-changing short-term weather patterns, we now know that there have been several rapid climate change events (RCCE) in the past. These rapid changes shifted the regional and global climate very quickly. The ice core records have demonstrated that RCCEs occurred far more frequently than previously thought. RCCEs are massive reorganizations of the Earth's climate system and were more extreme during glacial times (10,000–70,000 years ago), when the ice sheets in the northern hemisphere provided a positive-feedback mechanism to amplify the colder climate of these periods. These events do not lessen the impact of human activity on climate but highlight that a simplistic view of greenhouse gas emissions as the single driver of climate change cannot provide accurate predictions and can lead to incorrect economic incentives and policies, with detrimental consequences for society.

Although the impact of human activity on climate is very important, an equally relevant question is how climate shapes the development of human civilizations. Human activity can alter naturally occurring climate changes, and these climate changes can in turn affect human activities and economic prosperity. Climate might be one of the most influential factors shaping human history. Furthermore, climate changes have played a critical role, and even triggered, major breakdowns in societies and civilizations around the world for millennia. The Akkadians, the Egyptians, the Mayans, the Norse, and many other peoples were impacted by climate shifts, rising and falling through history, driven by the whims of Mother Nature.

CARBON DIOXIDE, GREENHOUSE GASES (GHG), AND AIR POLLUTION

Sunlight is primarily radiation in the visible range of the electromagnetic spectrum. Greenhouse gases are transparent to sunlight, which passes through the atmosphere heating the Earth's surface. As the Earth's surface warms, it radiates part of the energy back to the atmosphere in the form of infrared radiation (heat). Greenhouse gases absorb a portion of it. This absorption reduces the amount of energy that escapes back into space, trapping energy in the atmosphere that blankets the Earth's surface. If there was no natural air convection to redistribute the excess energy, the trapped energy (heat) would result in an increase in the air temperature, much like in a typical garden glass greenhouse.

Although most discussions on climate change focus on carbon dioxide (CO_2) and other greenhouse gases, water vapor in the atmosphere and clouds are the biggest contributors to the greenhouse effect (representing about 50% and 25%, respectively). Water molecules are far more efficient as greenhouse gases than CO_2. However, their concentration is a function of the atmosphere's temperature, and therefore they are not considered external forcings of climate sensitivity. In contrast, gases such as CO_2 (which represents about 20% of the greenhouse contributions), nitrous oxides, methane, tropospheric ozone, and CFCs can be added to or removed from the atmosphere independently from temperature and, therefore, are considered *external forcings* of global temperatures.

Before the Industrial Revolution, water vapor and greenhouse gases in the atmosphere caused the air near the Earth's surface to be warmer (about 14°C) than it would have been in their absence (about -19°C) (IPCC, 2007). Since the Industrial Revolution, human activity related to extracting and burning fossil fuels (coal, oil, and natural gas) has increased the amount of greenhouse gases in the atmosphere. This leads to a radiative imbalance that can result in global warming.

Man-made greenhouse gas emissions are roughly equivalent to 60 billion tons of CO_2 (IPCC, 2022c). These emissions include roughly 75% from CO_2, 18% from methane, 4% from nitrous oxide, and 2% from CFC gases. Emissions of CO_2 come from burning fossil fuels for energy generation to support transportation, manufacturing, heating, and electricity. Additional contributions are from deforestation and industrial processes with CO_2 as a byproduct of chemical reactions (e.g. manufacturing cement, steel, aluminum, or fertilizers). Methane emissions come primarily from livestock, crop cultivation, landfills, wastewater, coal, oil, and gas extraction. Nitrous oxide emissions largely come from decomposition of fertilizers.

The Earth's surface absorbs a significant amount of CO_2 as part of the natural carbon cycle. CO_2 is absorbed by plants, and a fraction is released back when biological matter is digested, burned, or decays. Carbon fixation in photosynthesis or in soil formation (land-surface carbon sinks) removes about 30% of annual CO_2 emissions. The oceans absorbed another 20–30% of emitted CO_2 globally. Note, however, that CO_2 is only removed from the atmosphere when it is stored in the Earth's crust, the oceans, or the soil through physical or chemical reactions.

In addition to greenhouse gases, air pollution in the form of aerosols can scatter and absorb solar radiation, affecting climate on a large scale. Man-made contributions of aerosols come primarily from the combustion of fossil fuels and biofuels, and from dust generated from human activities. Aerosols have indirect effects on the Earth's energy budget, typically limiting global warming by reflecting sunlight. However, black carbon in soot or dark dust particles that fall on snow or ice can contribute to global warming by reducing the reflectivity of glaciers and ice sheets (reduced albedo), increasing the amount of the Sun's radiation that is absorbed.

THE SCIENCE OF CLIMATE CHANGE

Determining the human impact on climate requires understanding the natural processes leading to climate change. Without a sound baseline of comparison for natural processes, it is difficult to make definite statements about human influence and man-made effects. The Earth's climate changed significantly over time long before humans existed, and many of the mechanisms for change are still at play today.

Climate is driven (or *forced*) by a wide range of internal and external factors working on short- and long-term scales, usually compounding their effects non-linearly. For example, the inclination of the Earth's orbit around the Sun wobbles and the orbit itself shifts over time, causing changes in the radiation received by the planet from the Sun, changing the Earth's energy and heat balance, and leading to periodic climate changes over 23,000, 41,000, and 100,000 years. Short-term changes in the energy emitted by the Sun and its redistribution by atmospheric and oceanic circulation can also result in climate changes, as observed during the El Niño Southern Oscillations (ENSO) or during regular solar cycles (about every 11 years), marked by changes in sunspot activity and sporadic electromagnetic energy and mass ejection events. While changes in solar radiation during the regular solar cycles may be weak, changes to the physical and chemical aspects of the atmosphere could have measurable effects. Other short-term effects have been observed over periods of decades or even hundreds of years, which could be influenced by regular shifts in the position of the Sun as it is pulled by the planets in the solar system (barycenter shifts).

The Earth is a complex system with multiple interacting components (the atmosphere, the oceans, the mantle and land masses, the planet's core) and external driving forces (the Sun's radiation, the gravitational pull from the Sun, the planets, and celestial bodies). Climate is only one subsystem in the Earth's system. No single factor can account for the behavior of the Earth's climate over extended periods of time. Rapid climate change events may be set in motion by a variety of factors reinforcing themselves or canceling out each other's effects.

The Earth's rotation around its axis and its orbit around the Sun change over time due to gravitational interactions with other celestial bodies. In the 1860s, James Croll pioneered the view of an astronomical link between Ice Ages and changes in the Earth's orbit. In the 1920s, Milutin Milankovitch refined this view, introducing three key elements of planetary mechanics that resulted in changes to the magnitude and distribution of the solar radiation hitting the Earth (Strahler, 1973; Buis, 2020). These elements of the Earth's orbit are: (1) axial precession, (2) obliquity, and (3) eccentricity (Figure 1.1). The Earth's axis of rotation is not fixed but shifts over long periods of time (axial precession). Every 23,000 years, the Earth is closer to the Sun in an opposing hemisphere, which results in the precession of the equinoxes as the Earth's axis of rotation shifts direction. The tilt of the Earth's axis of rotation (obliquity) relative to the orbital plane also varies a few degrees (between about 22° and 24.5°) over 41,000 years. The shape of the Earth's orbit (eccentricity) is not fixed and changes from nearly circular to slightly elliptical over a period of about 100,000 years. These three effects impact the amount of solar radiation (*insolation*) reaching different latitudes on the Earth's surface and are major driving forces for climate change.

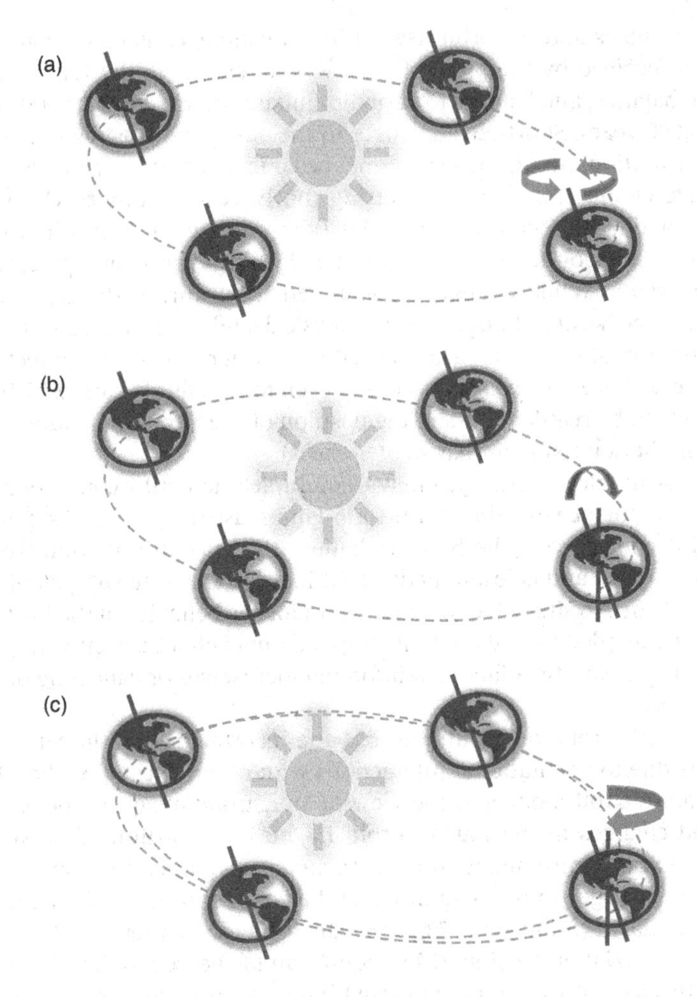

FIGURE 1.1 Drivers of the astronomical theory of climate change: (a) precession of the equinoxes (orientation of the Earth's axis of rotation); (b) obliquity (tilt in the Earth's axis of rotation); and (c) eccentricity (changes in the shape of the Earth's orbit).

The regular changes of the Earth's orbit and rotation axis impact solar insolation, which results in changes in atmospheric and oceanic circulation. Furthermore, they affect *polar* atmospheric and oceanic circulation, as reflected in ice core records. When the ice sheets in the northern hemisphere expand, the polar atmospheric and oceanic circulation is energized, and wind speed increases, as indicated by large changes in dust and sea salt deposits found in ice cores. The polar atmospheric and oceanic circulation cycles seem related to the amount of energy received from the Sun and have been found in marine sediment records and other paleoclimate records. The changes in the Earth's orbit are also associated with changes in the sea level, which is largely controlled by how much water is tied up in ice sheets over land masses. Larger ice sheets result in a lower sea level (Berner and Berner, 1987; Macdougall, 2004).

The differences between the Earth's orbit and its rotation changes and the observed polar circulation cycles from ice cores reflect the time lag effects in the complex climate system, as it takes time for ice sheets to adjust to changes in solar radiation. Also, shorter cycle events about 6,000 years apart (called Heinrich events) reflect massive discharges of (fresh water) icebergs into the North Atlantic, which affect sea level, the relative salinity of the ocean, and oceanic circulation.

Ocean water sinks when it is dense; and the saltier and colder, the denser it becomes. Water in the North Atlantic is not only cold but also salty (saline), resulting in significant amounts of dense water sinking to great depths in the ocean, pushing warmer and less salty water out of its way. As winds blowing from east to west across the equator transport humidity, they make the Atlantic more saline. In addition, waters flowing from the Mediterranean into the Atlantic are also more saline. The stream of cold and salty water sinking in the North Atlantic acts as a "conveyor belt" that drives ocean circulation. The water stream eventually warms up and surfaces from the ocean depths in the long return flow, sending heat from the tropics up along the east coast of North America toward Europe and higher latitudes (the Gulf Stream), closing the circulation loop. The Atlantic *thermohaline* circulation (conveyor belt) drives climate in the northern hemisphere, and did shut off in the past, leading to dramatic changes in global climate (Berner and Berner, 1987).

The mechanisms for climate forcing described above become even more complex when we introduce the impact of air pollutants (dust and particles in the atmosphere) and greenhouse gases (water vapor in clouds, carbon dioxide (CO_2), methane, and other gases) that can change the energy balance of the atmosphere.

Although significant progress has been made in recent decades in our understanding of climate, scientists are still debating the fundamental mechanisms for global atmospheric and oceanic circulation and the non-linearities and delayed effects in the complex climate system. Table 1.2 summarizes the drivers of observed climate changes.

TABLE 1.2 Driving factors for understanding climate change

Contribution	Driving Factor
Physical effects	Earth's orbital cycles ■ Precession ■ Obliquity ■ Eccentricity Changes to ice sheets Oscillations in atmospheric and oceanic circulation Variations in the radiation received from the Sun Changes to the carbon cycle between oceans, atmosphere, and land (soil) Changes in heat, energy, and mass transfer to the oceans and atmosphere due to tectonic and volcanic activity
Aerosols and atmospheric pollutants	Changes in the concentration of dust, sea salt, acids, aerosols, and other pollutants in the atmosphere ■ Natural and man-made
Greenhouse gases	Changes in greenhouse gases in the atmosphere (including water vapor in clouds, carbon dioxide (CO_2), methane, nitrous oxide, etc.) ■ Natural and man-made

CLIMATE CHANGE, THE INTERGOVERNMENTAL PANEL ON CLIMATE CHANGE (IPCC) REPORTS, AND SOCIAL CHANGE

Although the debate on climate change is primarily driven by statesmen, politicians, and policymakers, the scientific community has provided evidence and arguments to clarify erroneous concepts on climate change and help support adequate policy decisions.

Since 1988, the Intergovernmental Panel on Climate Change (IPCC) has taken on the role of evaluating the technical and scientific merits of climate change information and providing regular assessments. The IPCC is an intergovernmental body of the United Nations, established by the World Meteorological Organization (WMO) and the United Nations Environment Program (UNEP). The member states elect a bureau of scientists to govern the IPCC.

The IPCC does not conduct original research on climate change. It prepares reports on special topics and produces methodologies that can help countries estimate their greenhouse gas emissions. The IPCC has released six assessments and numerous reports reflecting growing evidence for climate change and its relation to human activity (IPCC, 2007, 2018, 2020, 2021a, 2021b, 2021c, 2022a, 2022b, 2022c). The IPCC includes working groups of scientific and technical experts from different fields, and their assessments and reports are expected to reflect an objective and dispassionate view, obtained by consensus of its participants (Mayewski and White, 2002).

The IPCC's sixth assessment report (AR6) (IPCC, 2022c) includes scientific, technical, and socio-economic information concerning climate change. The latest assessment shares many of the arguments of earlier assessments. According to the assessment, the main source of the increase in global warming is primarily due to the increase in CO_2 emissions that "likely" or "very likely" will exceed 1.5°C under higher emission scenarios. Table 1.3 shows the estimated increases in global temperatures for the period 2040–2060 for different shared socio-economic pathways (SSPs), which are climate change scenarios projected up to 2100 as defined in AR6. SSPs are used to derive greenhouse gas emissions scenarios with different climate policies.

Some headline statements derived from AR6 include:

- Human activities (through greenhouse gas emissions) have unequivocally caused global warming, with global surface temperature reaching 1.1°C above the 1850–1900 average temperature.
- Global greenhouse gas emissions have continued to increase.
- Continued greenhouse gas emissions will lead to increasing global warming.
- Deep, rapid, and sustained reductions in greenhouse gas emissions would lead to a discernible slowdown in global warming within around two decades.
- Climate change is a threat to human well-being and planetary health.

Notice that atmospheric concentrations of carbon dioxide, methane, nitrous oxide, and other gases have increased significantly since pre-industrial

TABLE 1.3 Global temperature increases for selected shared socio-economic pathways (SSPs)

SSP	Scenario	Estimated Warming (2040–2060)
SSP1 1.9	Very low GHG emissions: CO_2 emissions cut to net zero around 2050	1.6°C
SSP2 4.5	Intermediate GHG emissions: CO_2 emissions around current levels until 2050, then falling	2.0°C
SSP5 8.5	Very high GHG emissions: CO_2 emissions triple by 2075	2.4°C

times. Many greenhouse gases have a positive radiative forcing on climate (warming effect) and can remain in the atmosphere for decades. In contrast, man-made aerosols from fossil fuels and forest fire burning can have a negative radiative forcing (cooling effect) but they do not remain long in the atmosphere.

The IPCC report acknowledges that there is significant annual variability in weather patterns and severity but argues that aggregated temperature records since the 1850s (reflecting a range of data quality and measurement precision issues) show an overall systemic increase in global temperatures. Also, variable natural climate events, such as solar variability or volcanic activity, can make the detection of human influence more difficult to estimate. Regardless of these effects, models suggest climate will continue to change. Overall warming will be the result of an increase in the frequency of hot days and a decrease in the frequency of cold days. The increase in global temperature will result in the thermal expansion of the oceans and melting of ice sheets and glaciers, and sea levels will rise. Early model simulations also suggested a reduction in the strength of the Atlantic thermohaline circulation (conveyor belt), which will have the opposite effect, leading to colder weather in the northern hemisphere.

Although existing models are very valuable to gain insight into possible scenarios for climate change, there is still significant uncertainty in these models to provide the level of comfort required for taking drastic decisions to reduce carbon emissions that can severely affect a country's economic stability and society's wellbeing. For example, capturing the long-term effects of solar radiation variability, clouds, sea ice and ice sheet dynamics, vegetation, carbon sequestration in the deep ocean and land (i.e., seashell and soil formation) and non-linear interactions between the atmosphere, oceans and underwater volcanic activity will require further investigation. True scientific discourse demands a dose of healthy skepticism and strong debate on the merits of model assumptions and economic decisions that can affect society.

REFERENCES

Berner, E.K. and Berner, R.A. (1987). *The Global Water Cycle*, Prentice Hall, Englewood Cliffs, NJ, pp. 1–43.

Buis, A. (2020). Milankovitch (Orbital) Cycles and Their Role in Earth's Climate, Global Climate Change. *NASA News*, February 27.

EPICA (European Project for Ice Coring in Antarctica) (2004). Eight Glacial Cycles from an Antarctic Ice Core. *Nature* 429 (June): 423–429.

FSB (Financial Stability Board) (2021). *FSB Roadmap for Addressing Climate-Related Financial Risks*. Financial Stability Board, Basel, pp. 1–29.

IEA (International Energy Agency) (2021). Net Zero by 2050; A Roadmap for the Global Energy Sector. *IEA Special Report*. IEA, Paris.

IPCC (Intergovernmental Panel on Climate Change) (2007). AR4 WG1 Ch1 2007, FAQ1.1: To Emit 240 W m−2, a surface Would Have to Have a Temperature of Around -19 °C. This Is Much Colder Than the Conditions That Actually Exists at the Earth's Surface: the Global Mean Surface Temperature Is About 14 °C. https://www.ipcc.ch/site/assets/uploads/2018/02/ar4-wg1-spm-1.pdf

IPCC (Intergovernmental Panel on Climate Change) (2018). Global Warming of 1.5°C. IPCC Special Report .www.ipcc.ch/srl5/

IPCC (Intergovernmental Panel on Climate Change) (2020). The IPCC and the Sixth Assessment cycle. (Retrieved May 5, 2020).

IPCC (Intergovernmental Panel on Climate Change) (2021a). *Climate Change 2021: The Physical Science Basis. Contribution of Working Group I to the Sixth Assessment Report of the Intergovernmental Panel on Climate Change*. (Eds.) V. Masson-Delmotte, et al. Cambridge University Press, Cambridge. doi:10.1017/9781009157896.

IPCC (Intergovernmental Panel on Climate Change) (2021b). AR6 Climate Change 2021: The Physical Science Basis.. www.ipcc.ch. (Retrieved November 20, 2023).

IPCC (Intergovernmental Panel on Climate Change) (2021c). Sixth Assessment Report. (Retrieved April 19, 2021).

IPCC (Intergovernmental Panel on Climate Change) (2022a). Working Group II Contribution to the Sixth Assessment Report of the Intergovernmental Panel on Climate Change, Summary for Policymakers. www.ipcc.ch/report/ar6/wg2

IPCC (Intergovernmental Panel on Climate Change) (2022b). Intergovernmental Panel on Climate Change, United Nations Environmental Programme, World Meteorological Organization. (Retrieved March 2, 2022).

IPCC (Intergovernmental Panel on Climate Change) (2022c). AR6 WG3 Summary for Policymakers 2022, Figure SPM.1.

Macdougall, D. (2004). *Frozen Earth: The Once and Future Story of Ice Ages*. University of California Press, Berkeley, CA, pp. 75–85.

Mayewski, P., and White, F. (2002). *The Ice Chronicles: The Quest to Understand Global Climate Change*. University Press of New England, New Hampshire, pp. 1–108.

NGFS (Network for Greening the Financial System) (2020). NGFS Climate Scenarios for Central Banks and Supervisors. June, pp. 1–39. https://www.ngfs.net/sites/default/files/meelias/documents/820184 ngfs scenarios final version v6.pdf

NOAA (National Oceanic and Atmospheric Administration) (2024), Climate Data Primer. https://www.climate.gov/maps-data/climate-data-primer

Plimer, I. (2009). *Heaven and Earth: Global Warming – The Missing Science*. Taylor Trade Publishing, Lanham, MD.

Smith, P. (2020). The Climate Risk Landscape: A Comprehensive Overview of Climate Risk Assessment Methodologies. United Nations Environmental Programme Finance Initiative, pp. 1–48.

Strahler, A. (1973). *Introduction to Physical Geography*, John Wiley & Son, New York, pp. 23–113.

USDT (U.S. Department of the Treasury) (2021). Fossil Fuel Energy Guidance for Multilateral Development Banks. https://home.treasury.gov/system/files/136/Fossil-Fuel-Energy-Guidance-for-the-Multilateral Development-Banks.pdf

Voosen, P. (2017). Record-Shattering 2.7-Million-Year-Old Ice Core Reveals Start of the Ice Ages. *Science AAAS*, 14 August. https://www.science.org/content/article/record-shattering-27-million-year-old-ice-core-reveals-start-ice-ages

CHAPTER 2

Forces of Nature

THE ASTRONOMICAL THEORY OF CLIMATE CHANGE

Debate on climate change has renewed the interest in the field of environmental sciences, which focus on environmental problems caused primarily by man, and in earth sciences and physical geography, which focus on physical drivers of land, ocean, and atmospheric changes. The first step in understanding climate change is to learn more about the processes driving our physical environment. We start by reviewing key concepts in earth sciences and physical geography (Strahler, 1973; Millero, 1996; Prager and Earle, 2000; Mathez and Webster, 2004).

At first blush, the Earth can be viewed as a spherical planet, although its shape is more like an oblate ellipsoid compressed slightly along the polar axis and made to bulge along the equator. The oblateness of the Earth measures the flattening of the poles and is defined as the difference between Earth's equatorial radius and the polar radius (roughly 27 miles) divided by the equatorial radius (about 7,927 miles). That is, Earth's polar radius is roughly 1/300 (or 3.3%) shorter than the equatorial radius. Earth's oblateness is primarily attributed to the Earth's rotation and the somewhat plastic nature of the Earth.

The spinning of the Earth around its axis defines two nature points, the north, and south geographic poles. These points define a geographic grid of intersecting north-south lines or *meridians*. The Earth's spinning also defines a set of lines running parallel to the equator or *parallels*.

The location of any point on the Earth's surface can be measured along meridians and parallels. Taking a specific meridian as a reference line, meridian arcs can be measured eastwards (E) or westwards (W) of any point. More precisely, the *longitude* of an arbitrary point on the Earth's surface can be defined as the arc (measured in degrees) between the selected point and the prime meridian. The generally accepted prime meridian is the one that passes through the Royal Observatory of Greenwich in the United Kingdom (0° longitude).

Knowledge of the longitude of a point only cannot tell us its exact location because it applies to all points in the meridian. To determine the exact location of a point, a second measure is needed: *latitude*. The latitude of a point on the

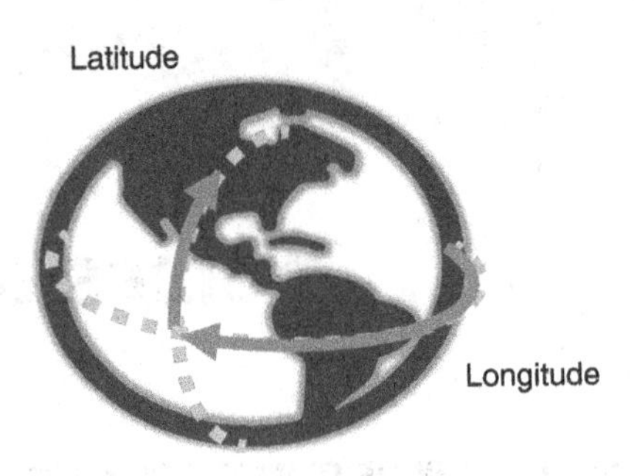

FIGURE 2.1 Longitude and latitude of a point.

Earth's surface can be measured northwards (N) or southwards (S) as the arc (also measured in degrees) of a meridian between that point and the equator, which is the natural prime parallel (0° latitude). Longitude and latitude allow us to locate any place on the surface of the earth with precision (Figure 2.1).

If we ignore the Earth's oblateness and consider Earth to be a perfect sphere, parallels can be assumed to be equidistantly spaced. For example, for every 1° of arc we would expect the same distance between parallels. The length of a 1° degree of latitude is roughly the same as that of a 1° degree of longitude at the equator (about 69 miles per degree). Note, however, that due to the Earth's oblateness, the surface curvature is less strong near the poles than at the equator. As a result, a 1° degree of latitude changes in length from the equator to the poles (the difference is about 0.7 miles). Thus, 1° degree of latitude at the poles is roughly 1% longer than at the equator. These small differences in the Earth's shape can affect atmospheric and ocean circulation patterns.

Let's now turn our attention to the relationship between the Earth and the Sun. The Earth is turning on its axis daily at the same time as it is circling the Sun. Furthermore, the Earth's rotation axis is tilted (*obliquity*) with respect to the plane of its orbit and wobbles over time (Figure 2.2). The angles at which the Sun's rays strike the Earth at different latitudes at different times of the day through the year are fundamental drivers of the ocean circulation and atmospheric temperatures, air pressure, winds, storms, and precipitation. These factors drive both weather and climate.

Earth's spin on its polar axis is called Earth rotation. The period of Earth rotation is the mean solar day, a 24-hour period that represents the average time required for the Earth to make a complete turn on its axis with respect to the Sun. The true direction of the Earth rotation is eastward, the opposite of the apparent westward motion of the Sun, Moon, and stars across the sky.

FIGURE 2.2 Earth's rotational axis.

Although the Earth spins daily on its axis roughly at a constant rate, the velocity at each point on its surface varies with latitude. At the equator (0° latitude), the velocity of a point on the Earth's surface is about 1,050 miles per hour. At the 60° parallel (close to the Arctic and Antarctic polar circles), the velocity reduces to half the amount, about 525 miles per hour. At the poles (90° latitude), the velocity is zero.

The change in velocity at different points on the Earth's surface has important implications for atmospheric and ocean circulation and, consequently, for climate. First, as the Earth rotates, objects on its surface moving at the Earth's surface velocity would fly off into space unimpeded if not physically restrained to the surface or pulled in by gravity. This effect due to Earth's rotation appears as a (fictitious) *centrifugal force* to an observer rotating with the Earth, and is stronger near the equator, where the velocity is the highest. Although the force of gravity is about 300 times stronger than the centrifugal force due to the Earth's rotation, this effect results in a small reduction of the weight measured for any object rotating with the Earth. Second, the decrease in rotational velocity with increasing latitude results in a slight deflection to the right or to the left for objects in motion relative to an observer moving with the Earth's surface (the Coriolis effect). These effects affect the water in the oceans and the air in the atmosphere, impacting their global circulation and dynamics in the southern and northern hemispheres, which drive both weather and climate.

The Earth also circles around the Sun counterclockwise, that is, in the same direction of turning as the Earth spins. The time it takes to complete one orbit around the Sun is the revolution period or year. When the revolution period measures the time required to return to a given fixed point in its orbit relative to the fixed stars, the period is called *sidereal year*. However, to reflect Earth-Sun effects, the *tropical year* is more relevant as it measures the time required from one vernal equinox to the next, which determines the regular seasons, weather, and climate patterns. The tropical year length is approximately 365 and ¼ days. Every four years the additional one-fourth days accumulates to nearly one whole day. We can align the calendar year to the tropical year by introducing one extra day in February (29th) every four years (a leap year).

The Earth's orbit around the Sun is not exactly a circle. It is an ellipse that differs slightly from a circle (low *eccentricity*), where the Sun occupies one focus (Figure 2.3). Every ellipse has two foci that lie on the line with the maximum diameter of the ellipse (major axis). The shortest diameter of the ellipse is at a right angle to the major axis (minor axis). On about July 4th, the Earth is at its farthest distance from the Sun at about 94.5 million miles (orbit *aphelion* along the major axis). On about January 3rd, the Earth is at its nearest distance from the Sun (*perihelion*) at about 91.5 million miles. This difference in distance to the Sun is about 3 million miles, about 3.5 times the diameter of the Sun (850,000 miles).

Although the change in the distance to the Sun is not the cause of the seasons, it does impact the amount of solar energy received by Earth. Summers and winters are slightly milder in the northern hemisphere and slightly more

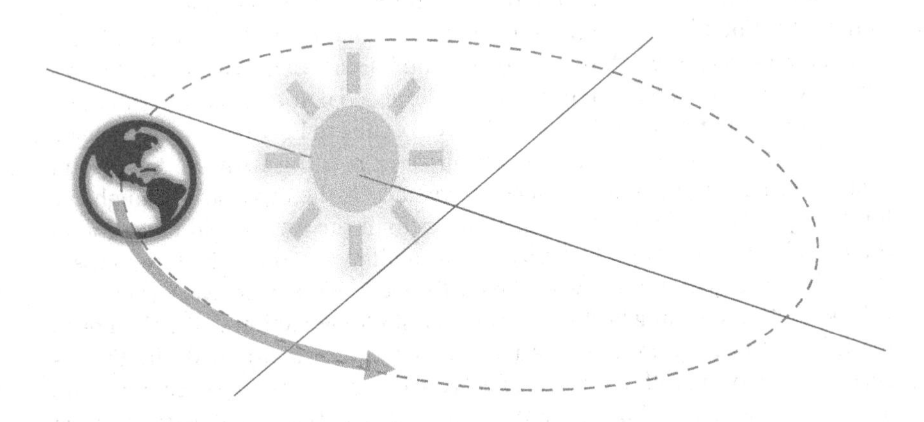

FIGURE 2.3 The Earth's orbit.

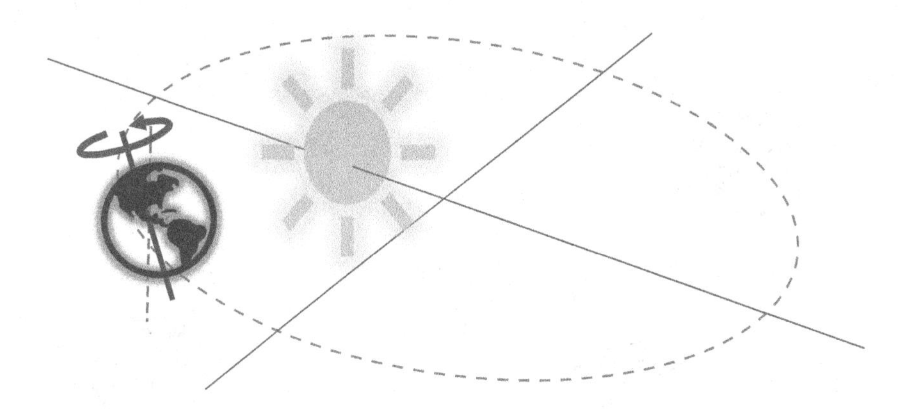

FIGURE 2.4 The Earth's rotation.

intense in the southern hemisphere due to the impact of reaching the orbit perihelion and aphelion during the summer and winter seasons.

Another consequential driver of weather and climate is the inclination of the Earth's axis of rotation, which is currently tilted 23.5° from a perpendicular line to the orbit plane. Therefore, the plane of the equator makes an angle of 66.5° (90°–23.5°) with the plane of the Earth's orbit (Figure 2.4). As the Earth circles the Sun, the Earth's axis maintains a nearly fixed orientation with respect to the stars, acting as a giant gyroscope. At one point in the orbit, the Earth's axis leans toward the Sun, and at an opposite point in the orbit the axis leans away from the Sun. At the two intermediate points in the orbit, the axis leans neither toward nor away from the Sun (Figure 2.5).

Note, however, that the inclination of the Earth's rotation axis is not completely fixed. It shifts and wobbles over a 23,000-year cycle (*axis precession*), leading to significant changes in climate due to variations in the amount of solar energy received by the Earth.

On June 21 or 22 each year, the Earth's axis leans at a 23.5° angle toward the Sun. The northern hemisphere points toward the Sun while the southern hemisphere points away from it. This is the *summer solstice* in the northern hemisphere. Half the way around the orbit, on December 21 or 22, the Earth is at the opposite point, where the southern hemisphere points toward the Sun while the northern hemisphere points away from it at a 23.5° angle. This is the *winter solstice* in the northern hemisphere.

Midway between the summer and winter solstices, on March 20 or 21 and September 22 or 23, the Earth's axis makes a right angle with a line drawn to the Sun. At these two points (the *vernal* and *autumnal equinoxes*), neither the south hemisphere nor the north hemisphere has any inclination toward the Sun. At the equinoxes, the boundary between the sunlit and shadowed halves

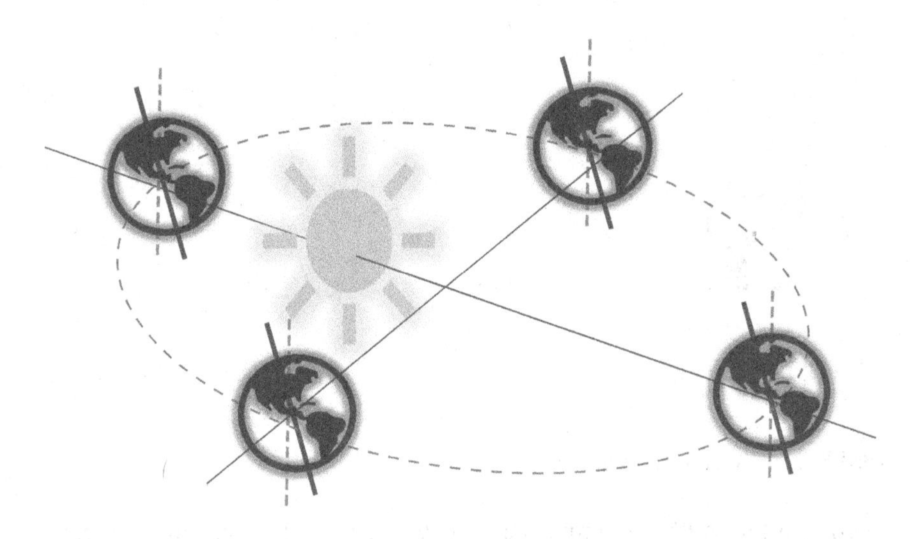

FIGURE 2.5 The Earth's rotation and tilt.

of the Earth (or circle of illumination) passes through the poles and aligns with the meridians. Day and night last 12 hours each, that is, equal length at all latitudes. At the equator, the Sun at noon is at an altitude of 90°. For other latitudes, the Sun's noon altitude is 90° minus the latitude (or colatitude).

During the summer solstice, day and night are unequal in length over most of the globe. Night is longer than day in the southern hemisphere, while day is longer than night in the northern hemisphere. The difference between day and night increases from the equator to the poles. The relative lengths of day and night are exactly the opposite for the same latitude north and south of the equator. Between the Arctic Circle (66.5 °N) and the North Pole, day lasts 24 hours as the entire polar area lies on the sunlit side of the circle of illumination. The reverse situation occurs in the Antarctic Circle (66.5 °S), where night lasts the entire 24 hours because the polar area lies on the shaded side of the circle of illumination.

At the winter solstice, the exact opposite conditions occur. The north polar end of the Earth axis leans away from the Sun, and the southern hemisphere experiences the same conditions of increased sunshine that the northern hemisphere enjoyed during the summer solstice. The determination of the summer and winter solstices and vernal and autumnal equinoxes were of great importance to earlier civilizations because of their alignment to the seasons and impact on agricultural activities.

Earth's rotation around its axis and its orbit around the Sun change over long periods of time due to gravitational interactions with the Sun, the

Moon, the planets, and other celestial bodies. These changes have profound implications for the Earth's climate. Many of the modern scientific discussions on climate started with the cause of the advance and retreat of glaciers and ice sheets (Macdougall, 2004, 2009). In the 1840s, Louis Agassiz, who pioneered the study of glaciers, suggested that there once were great ice sheets across Europe and North America. Soon after, Joseph Adhemar suggested that because the southern hemisphere had longer periods of darkness in winter, it must be cooling, and attributed ice sheets to this effect over long periods of time. In the 1860s, James Croll suggested an astronomical link between ice ages and changes in the Earth's orbit. In the 1920s, Milutin Milankovitch introduced a more refined theory of ice ages based on the amount of solar radiation hitting the Earth as result of changes in the Earth's orbit and its rotation. The modern version of the theory is summarized by three astronomical concepts: eccentricity, obliquity, and precession.

The Earth's orbit varies between nearly circular and slightly elliptical, which is measured by the orbit's eccentricity. A more elongated orbit results in more variation in the distance between the Earth and the Sun, impacting the amount of solar radiation received at different times during the year. In addition, changes in the tilt of the Earth's axis of rotation (obliquity), in the direction of the axis of rotation (axial precession) and in the direction of the elliptical orbit around the Sun (apsidal precession) make seasons more extreme.

Furthermore, over long periods of time, the gravitational pull from Jupiter, Saturn, and other planets causes the shape of the Earth's orbit (eccentricity e) to change from nearly circular ($e = 0.003$) to more elliptical ($e = 0.058$) over a period of about 100,000 years. Eccentricity is the reason for different lengths in seasons. In the northern hemisphere, summers last about 4.5 days longer than winters and springs about 3 days longer than autumns. As the eccentricity of the Earth's orbit decreases, the length of seasons evens out. Currently, the Earth's orbit eccentricity ($e = 0.017$) is decreasing slowly toward its least elliptic shape, that is, the Earth's orbit is becoming more circular (Buis, 2020).

The Earth's axis of rotation is tilted with respect to the Earth's orbit around the Sun. This tilt (obliquity) is the reason the Earth has seasons. As the Earth travels around the Sun, the tilt of the Earth's axis of rotation relative to the orbital plane also varies a few degrees (between 22.2° and 24.5°) over 41,000 years. This variation of about 2.3° in the angle of the Earth's axis of rotation alters the height of the Sun in the daytime sky. Even small shifts in the Sun's elevation impact the amount of solar radiation received at different latitudes. The greater the tilt angle, the more extreme the seasons are, as the hemispheres receive more solar radiation. Large tilt angles favor periods of climate warming (melting and retreat of glaciers and ice sheets). Currently, the Earth's axis of rotation is tilted about 23.5 degrees, roughly half the way between its minimum and maximum values. Its maximum was about 11,000 years ago (end of the last Ice Age), and its minimum will occur again roughly about 9,000–10,000 years from now, when seasons will be milder, with warmer winters and colder summers leading to a build-up of ice sheets and glaciers at higher latitudes. The increase in ice sheets will reflect more of the Sun's energy back into space, leading to further climate cooling.

The overall direction of the Earth's axis of rotation is not fixed but wobbles and shifts over long periods of time (axial precession). The wobble is due to tidal forces caused by the gravitational influence of the Sun and the Moon that makes the Earth bulge at the equator, affecting its rotation. The cycle of axial precession spans about 26,000 years and makes seasons more extreme in one hemisphere and more moderate in the other. The conditions revert every 13,000 years. Currently, the perihelion (closest distance to the Sun) occurs during summer in the southern hemisphere. This makes summers hotter in the southern hemisphere and less extreme in the northern hemisphere.

The Earth's axis of rotation not only wobbles but the orbit around the Sun also wobbles, primarily due to the gravitational pull of Jupiter and Saturn (*apsidal* precession). This orbital precession changes the orientation of the Earth's orbit relative to the Sun and the other planets. The combined effects of axial and apsidal precession result in an overall precession with a cycle of approximately 23,000 years. Every 23,000 years, the Earth is closer to the Sun in an opposing hemisphere, which results in the precession of the equinoxes as the Earth's axis of rotation shifts direction.

The combination of these three effects (changes in eccentricity, obliquity, and precession) impacts the amount of solar radiation (*insolation*) reaching different latitudes on the Earth's surface and are major natural driving forces for climate change (Figure 2.6).

The changes in the Earth's orbit and rotation axis impact solar insolation, which results in changes in atmospheric and oceanic circulation. Because of differences in the distribution of land and oceans, the impact of these changes is not the same everywhere. Oceans cover most of the southern hemisphere, which makes climate changes more moderate as they can store large amounts of solar energy. In contrast, the northern hemisphere has the largest land masses (North America, North Africa, Europe, and Asia), which are more reactive to climate changes. As solar radiation heats the land surface, heat can penetrate only a short distance into the ground, which can be heated and cooled to a far larger extent than oceans. Furthermore, land surfaces vary widely across flat and mountainous terrains, forests, deserts, glaciers, and ice sheets. Each terrain type responds differently to changes in solar radiation and atmospheric circulation, creating complex (short-term) weather and (long-term) climate patterns.

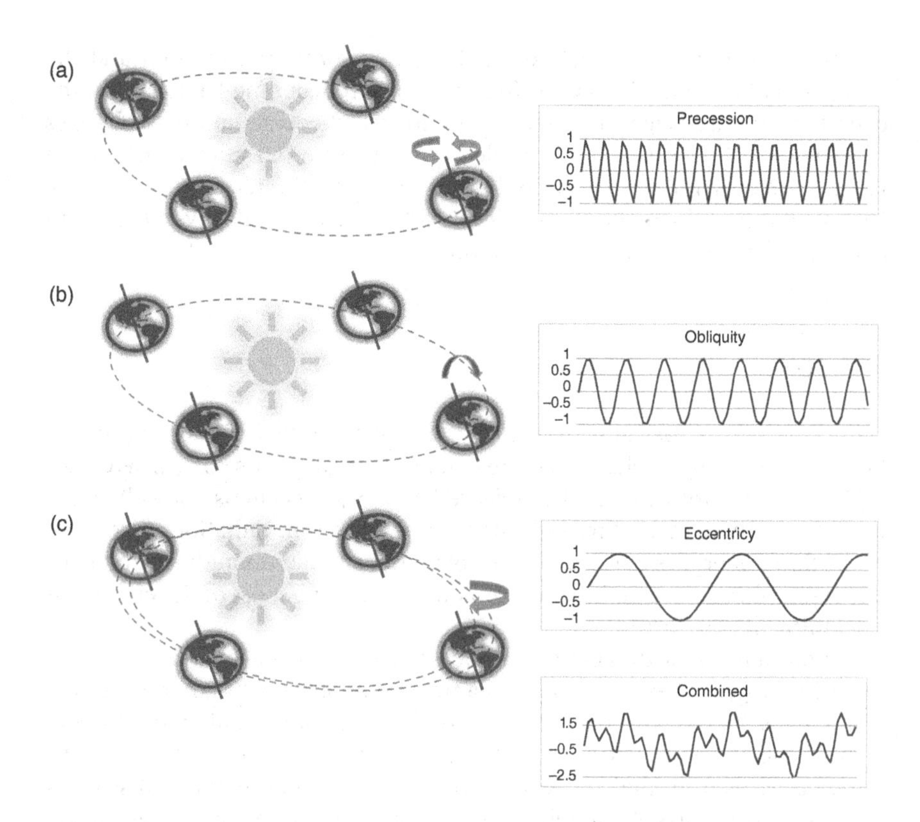

FIGURE 2.6 Drivers of the astronomical theory of climate change: (a) precession of the equinoxes (orientation of the Earth's axis of rotation); (b) obliquity (tilt in the Earth's axis of rotation); and (c) eccentricity (changes in the shape of the Earth's orbit).

The Earth's orbit and axis of rotation changes also affect *polar* atmospheric and oceanic circulation, as reflected in ice core records. When the ice sheets in the northern hemisphere expand, the polar atmospheric and oceanic circulation is energized, and wind speed increases, as indicated by large changes in dust and sea salt deposits found in ice cores. The polar atmospheric and oceanic circulation cycles seem related to the amount of energy received from the Sun and have been found in marine sediment records and other paleoclimate records. The changes in the Earth's orbit are also associated with changes in the sea level, which is largely controlled by how much water is tied up in ice sheets over land masses. Larger ice sheets result in a lower sea level.

Differences between changes in the Earth's orbit and rotation and the observed polar circulation cycles from ice cores reflect time-lag effects in the complex climate system as it takes time for ice sheets to adjust to changes in solar radiation. Also, shorter cycle events about 6,000 years apart (called Heinrich events) reflected massive discharges of (fresh water) icebergs into the North Atlantic, which affected sea level and the relative salinity of the ocean and, therefore, impact ocean circulation.

THE SUN

What is the Sun? Basically, the Sun is a massive, hot ball of plasma (ionized hot gas), heated by nuclear fusion reactions occurring at its core, where matter is converted into energy. The released energy heats atoms and subatomic particles in the Sun, and part of this energy escapes as radiation. A fraction of the Sun's energy is emitted as electromagnetic radiation in the form of visible light, ultraviolet and infrared radiation, providing most of the energy for sustaining life on Earth.

Viewed from Earth (about 93 million miles away), the Sun does not seem to be very big. However, the Sun's diameter is about one hundred times the diameter of the Earth. Imagine the Earth the size of a small coin. The Sun would be as tall as a major league basketball player, standing more than two hundred yards from the coin-size Earth. The Sun is not only big in size but has a very large mass. Although the Sun is only a mid-size average yellow star (G-type star), its mass is about 330,000 times the Earth's mass. There are scores of stars much bigger and massive than the Sun.

Every second, thermonuclear (proton-proton) chain reactions in the Sun's core fuse about 600 million tons of hydrogen into helium and convert about 4 million tons of matter directly into energy. Thermonuclear fusion energy is the source of the Sun's light and heat. The Sun has enough fuel to last for several billion years. Its core extends from the center to about 20–25% of the solar radius and has an average density of up to about 150 g/cm^3 (that is, about 150 times denser than water). The Sun's density decreases dramatically as we move from the core to the outer layers dominated by low density hot gases, plasma, and magnetic fields.

The temperature of the Sun varies widely. In its interior, where nuclear fusion reactions occur, the temperature can reach tens of millions of degrees. At the Sun's core, the temperature is about 16 million degrees (usually measured in Kelvin (K)). In contrast, the Sun's surface temperature is merely 5,800K. The Sun is surrounded by a region of extremely low-density, hot plasma with a temperature of roughly about 2 million degrees (*solar corona*).

The Sun may seem relatively quiet from Earth, but it is a very active star (Fan, 2009; Nandy, 2009). The surface of the Sun (*photosphere*) shows a granular cell structure of hot gases, plasma, and magnetic fields rising and sinking in the surrounding gas, much like convection cells in boiling water, although each cell is about 1,000 miles across. The Sun continuously releases streams of radiation, atoms, electrons, and other atomic particles, flying deep into space. The Sun also produces gigantic bursts of mass and energy in the form of solar storms, twisted flame-like electromagnetic and plasma prominences, and jets of gases and energy larger than Earth. The solar wind produced by these mass and energy streams travels to the very edge of the solar system.

The Sun's activity varies over short and long periods of time. There is a nearly periodic 11-year solar cycle of changes in solar magnetic activity with significant variations in the number of observed sunspots on the Sun's surface (regions of lower brightness relative to their surroundings). During the solar cycle, the number, size, and location of sunspots, the amount of solar radiation and number of solar flares, and the ejection of mass and energy into space exhibit significant variations from minimum solar activity to a period of a maximum solar activity, reverting to low activity as a new cycle begins (Hathaway et al., 1994, 1999; Miyahara, 2004; Wilson et al., 2008; Charbonneau, 2010; Hathaway, 2011). The north-south orientation of the Sun's magnetic field flips during each solar cycle. The magnetic field flips occur when the solar cycle is near its maximum, returning to its original state after two cycles (the Hale cycle). These short-term solar cycles have been observed for centuries. Other longer cycles (extended hundreds and thousands of years) have been inferred from a range of ice core records, marine sediments, and other paleoclimate records.

Although the Sun represents most of the mass in the solar system (over 99.8%), it is not located at the center of the solar system). The Sun orbits around the *barycenter* of the solar system (center of mass of the Sun and planets). The Sun is so massive that ordinarily the barycenter of the solar system is located within or very near the surface of the Sun (Jose, 1965; Charvátová, 2000; Hejda et al., 2008). However, the Sun's orbit around the barycenter shifts and wobbles primarily due to the gravitational pull of Jupiter, Saturn, and the other planets. At times, changes in the relative distance between the solar system's barycenter and the Sun can be as large as the diameter of the Sun itself, repeating the pattern over hundreds of years. Unlike the Earth's nearly circular orbit, the Sun's orbit is complex and can be characterized by a combination of ordered trefoil patterns reoccurring every 180 years and a more disordered orbital pattern between ordered orbits (Figure 2.7). The full orbital cycle is completed in about 2,400 years (Charvátová, 2000; Palus et al., 2007). Since the Sun does not move as a rigid body, the swings and wobbles of its orbit can impact its internal structure and solar activity.

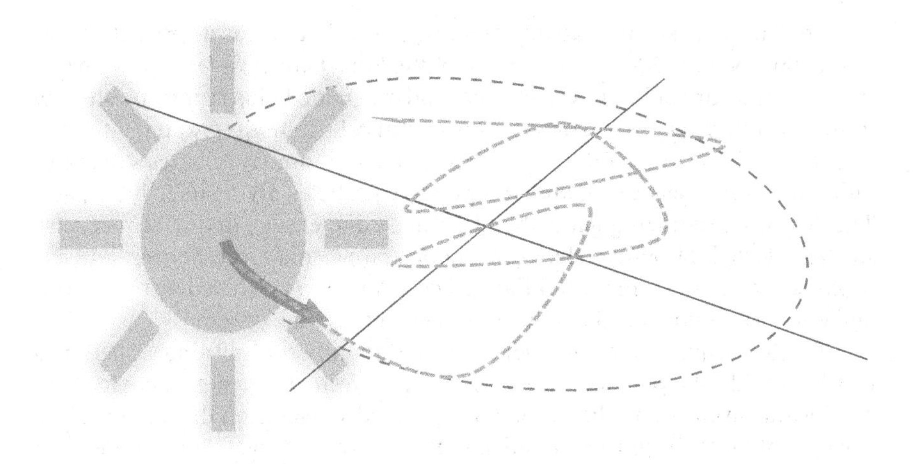

FIGURE 2.7 The Sun's orbit and variability over time.

Furthermore, there have been cycles of 200–300 years and 1,000–1,500 years recognized from fossil and ice core measurements that seem linked to solar cycles (Charvátová, 2000; Zhang et al., 2008; Plimer, 2009; Kern et al., 2013). There have been somewhat regular ice rafting and temperature swing events (Bond and Dansgaard–Oeschger (D-O) events) tentatively linked to climate fluctuations over a period of 1,000–1,500 years. Cooling and warming of the Arctic and Antarctic regions seem out of phase, as shown in ice cores. Temperatures in the Antarctic region increase while there is cooling in Greenland on millennial scales. Warming in Antarctica and the circumpolar current leads to changes in the North Atlantic about 1,500 years later. In the past, climate changes in Antarctica were about 1,000–2,000 years ahead of those in Greenland. The origin of this polar climate see-saw, which operates like a pendulum between the two poles, needs further research.

Solar activity due to solar cycles, orbital changes, and idiosyncratic events drive space weather in the solar system, which affects Earth's atmosphere, leading to climate fluctuations on scales of decades, centuries, and longer. Understanding and predicting changes in solar activity remain a key challenge for modeling climate change. Note, however, that neither direct measurements nor proxies of solar radiation variations seem to correlate well with the increases in global temperature and CO_2 concentration levels observed over the last 170 years, which are attributed primarily to human activities such as fossil fuel burning, deforestation, and industrial pollution.

THE EARTH

Seen from outer space, Earth looks like a blue and white marble whirling around the Sun. The Earth is accompanied by the Moon in its annual journey. From afar, it is difficult to note that our planet has just the right temperature and just the necessary atmosphere and water to foster an amazing diversity of life.

Earth was not always a comfortable place to live. Our present understanding of planetary formation suggests that the Earth is a highly evolved planet. Earlier views of the Earth's creation envisioned a gentle accumulation of dust, grains, small asteroids and comets, and other celestial bodies that shaped our planet. However, the Earth's evolution has not been so simple.

The impact of a large asteroid or comet can produce a huge amount of energy, enough to melt and vaporize the asteroid and part of the planet's surface. Our planet was likely hit by scores of asteroids and comets, melting, and separating different component materials. Heavy materials sank to the interior of the planet, creating the *core*, and lighter materials remained on its surface, creating the *crust*. All the intermediate materials were deposited in a region called the *mantle*, just between the planet's core and the crust (Figure 2.8). The core is mainly composed of iron and nickel and is partially molten. The mantle is composed mainly of magnesium, silicon, and oxygen. It contains other elements such as iron, calcium, and aluminum. The mantle is the source of magma and molten rocks ejected during volcanic eruptions and is the largest part of Earth's mass (Verhoogen et al., 1970; Press and Siever, 1974; Tarbuck and Lutgens, 1979). Earth's oceans and the atmosphere originated from volcanic activity and outgassing from the planet's interior. Over time, water vapor condensed into the oceans with additional contributions from asteroids, comets, and protoplanets.

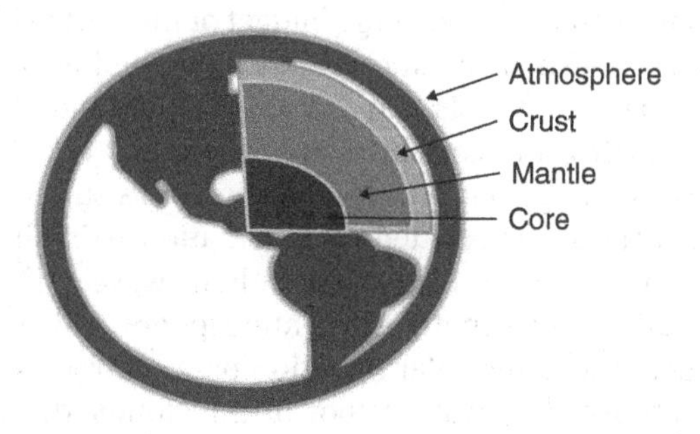

FIGURE 2.8 The Earth's core, mantle, crust, and atmosphere.

Encasing the mantle is the crust, composed of solidified molten lava, and solid rocks. There are giant cracks in the crust which divide it into pieces called *tectonic plates*. Continents are located on top of these plates, which are in motion with respect to another. Plates move and collide with each other, creating mountains and producing earthquakes (Verhoogen et al., 1970; Zumberge and Nelson, 1976; Tarbuck and Lutgens, 1979; Engdahl and Villasenor, 2002; Mathez and Webster, 2004). The continuous creation of new crust and the continuous motion of the plates made the Earth's face change dramatically over its history of 4.5 billion years. Over the last two hundred million years the Earth's surface changed from having a single supercontinent to having multiple continents, which are still moving today. These changes in the Earth's surface and its interior impact the position of land masses, the shape of the oceans and their current circulation, and the composition of the atmosphere, which drives climate.

Besides the wonders of the Earth's interior, it is the Earth's atmosphere which is amazing. Without it, life would not be possible. It not only provides us with oxygen for breathing, but it shields us from dangerous radiation coming from the Sun and outer space. It also shields us against small meteorites, which are heated and destroyed by friction on their way to the Earth's surface.

The most remarkable aspect of the atmosphere is that it acts as a thermal blanket to keep the surface of the planet warm enough for water to condense and life to occur. Earth's atmosphere is mainly composed of nitrogen and oxygen with smaller amounts of carbon dioxide and other gases. Water from oceans, lakes, and rivers also evaporates to form clouds in the atmosphere, giving our planet its distinctive blue and white colors.

Solar radiation is the most important source of energy for the Earth's energy balance, and it is the dominant influence on the ocean and atmospheric circulation. Solar radiation is received through the atmosphere and is reflected, absorbed, or converted to other forms of energy.

Figure 2.9 shows the overall energy budget of the Earth (Plimer, 2009). Of the about 340 W/m^2 of solar radiation received by the Earth, about 77 W/m^2 is reflected into space by clouds and the atmosphere and about 23 W/m^2 is reflected by the Earth's surface. This leaves about 240 W/m^2 of solar energy input to the Earth's energy budget. Because the Sun's surface temperature is around 5,800K, most of the emitted solar radiation is in short wavelengths (less than 4 μm), with a peak in the visible light wavelengths (0.4–0.7 μm). This visible radiation can penetrate the atmosphere and reach the ground mostly unimpeded. Incoming solar radiation passes through the atmosphere and is partially absorbed by water vapor in the clouds, dust, ozone, carbon dioxide (CO_2), and other greenhouse gases. A fraction of the incoming energy is reflected into space by clouds, ice sheets, land, and the oceans, and the rest is finally absorbed by the Earth's surface. The fraction that is

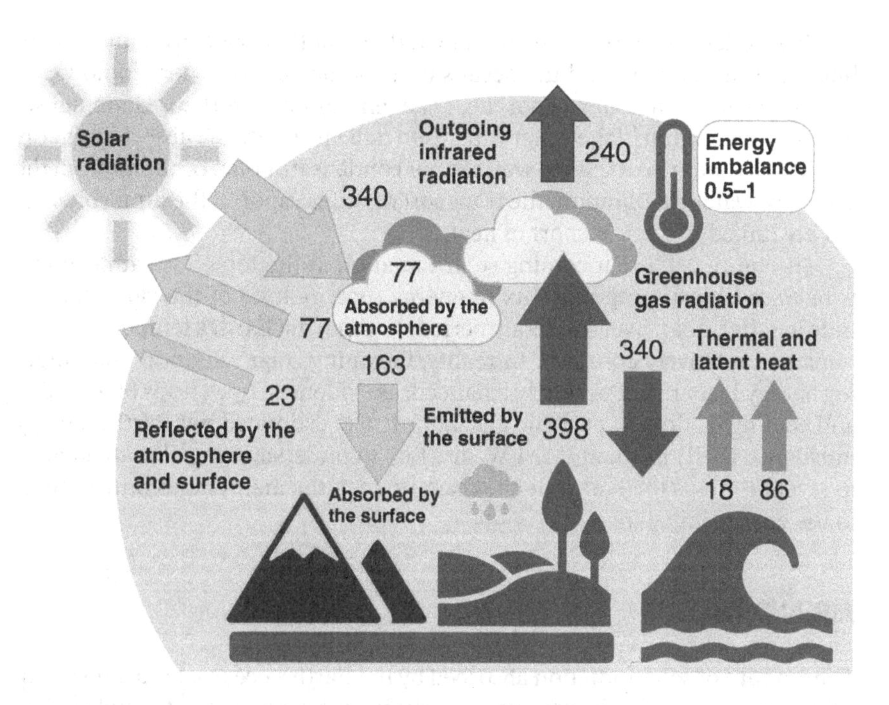

FIGURE 2.9 Radiation and heat energy budget of the Earth (in W/m^2). About 340 W/m^2 overall is received from the Sun.

reflected back is the Earth's *albedo*. The remaining energy is absorbed and reradiated back as infrared radiation (longer wavelengths; longer than 4 μm) by the ground, oceans, and the atmosphere. By contrast, almost all of the harmful solar radiation in the ultraviolet and shorter wavelengths (less than 0.4 μm) is absorbed in the upper layers of the atmosphere by oxygen (O_2) and ozone (O_3). Some of the incoming solar radiation in the infrared and longer wavelengths is also absorbed in the atmosphere by water vapor, CO_2, and water droplets in the clouds. Furthermore, atmospheric water vapor is a very effective absorber of energy in infrared wavelengths for both incoming and outgoing energy fluxes. Water vapor (and, to a lesser extent, CO_2) allow incoming short-wave radiation to pass through the atmosphere to heat the planet's surface, while absorbing and reradiating back to the surface the outgoing long-wave radiation. This process is referred to as the *atmospheric greenhouse effect* by analogy with a glass greenhouse. Note, however, that glass greenhouses are heated by trapping warm air, that is, by restricting air convection, which transfers heat and air.

Figure 2.9 also shows the latent heat flux and thermal convection contributions from the land and the oceans to the atmosphere. When liquid water evaporates to form water vapor, heat (i.e. energy) is absorbed. Upon subsequent condensation of the water vapor into rain and snow, the absorbed energy is released (latent heat). Since water vapor condensation can occur far from the source of water evaporation, the transport of water vapor in the atmosphere is closely linked to the transport of heat.

In Figure 2.9, the incoming solar radiation at the top of the atmosphere is balanced by an outgoing flux of energy (energy loss) of the same magnitude to reflect that, over moderate periods of time, the Earth's temperature has remained relatively constant. In reality, incoming solar radiation and outgoing energy loss are not perfectly balanced, as evidenced by periods of warming and cooling leading to periodic glaciations. The estimated net Earth's energy imbalance (EEI) is about 0.5–1 W/m^2. Key to understanding climate change in recent times is to determine the magnitude of the man-made contributions to the energy imbalance.

AIR AND WIND

The amount of solar radiation absorbed by the Earth's oceans, land, and atmosphere decreases with increasing latitude from the equator to the poles. This is due to two factors: (1) the angle at which sunlight hits the Earth's surface at different latitudes, and (2) the tilt in the Earth's axis of rotation, which determines the duration of daylight and seasons, that is, the amount of sunlight received through the year. The variation in solar radiation and Earth's heating drives the circulation of the atmosphere and oceans (Berner and Berner, 1987; Plimer, 2009).

The variation in the angle of the Sun's rays with latitude and the seasons results in a large regional variation in total solar radiation. By contrast, long wavelength (infrared) radiation from the Earth's surface varies little with latitude. This results in regional energy imbalances between incoming solar radiation and the outgoing energy loss. From about 40° north and south latitudes to the poles (90°), the energy radiated back into space is more than the energy received from the Sun (energy deficit). From 40° north and south latitudes to the equator (0°), more solar energy is received than energy is radiated back into space (energy surplus). These energy imbalances drive the ocean and atmospheric circulation, transporting heat from lower latitudes to higher latitudes to keep the equator and tropics from getting warmer and the poles from getting colder.

Heat is transported from the equator and tropics to the poles in three different ways:

1. by ocean currents carrying warm water
2. by atmospheric circulation carrying warm air
3. by atmospheric circulation carrying latent heat in the form of water vapor.

Latent heat is subsequently released by water vapor condensation, which warms the atmosphere.

Overall atmospheric circulation is driven by energy and heat imbalances for different latitudes. However, if atmospheric circulation were solely due to heating imbalances, hot air would rise at the equator and flow poleward at high levels, cooling and sinking at the poles, and returning to the equator as cold air. Such a circulation would produce two symmetrical closed cells (Hadley cells), one in the northern hemisphere and one in the southern hemisphere. In practice, winds do not blow along meridional lines in the north and south direction. Actual atmospheric circulation is broken up into several latitude zones due to the Earth's rotation, which deflects winds rightwards in the northern hemisphere and leftwards in the southern hemisphere through the Coriolis force. The north and south latitude zones can be divided into (a) low latitude Hadley cells, (b) mid-latitude Ferrel cells, and (c) high-latitude polar cells, each with distinct wind and atmospheric circulation patterns (Figure 2.10).

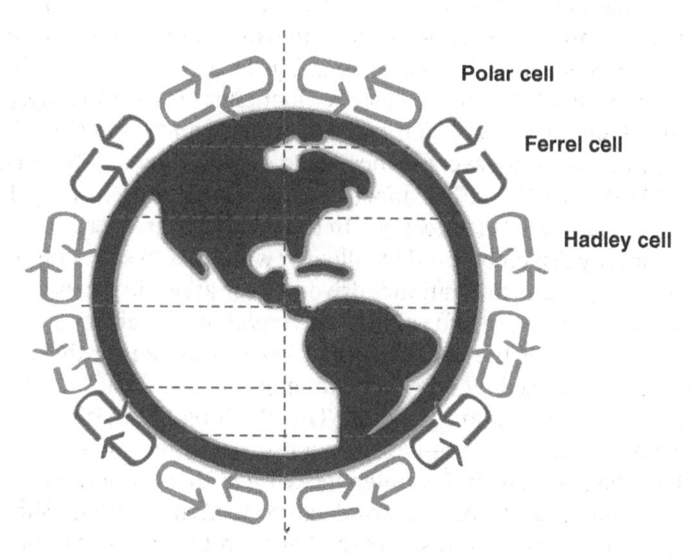

FIGURE 2.10 General circulation of the atmosphere.

WATER AND ICE

Water is essential for life on Earth and plays an important role in regulating weather and climate patterns (Press and Siever, 1974; Berner and Berner, 1987; Millero, 1996; Rahmstorf, 2003; Prager and Earle, 2004). Water impacts heat transfer through the movement of water vapor in the atmosphere, evaporation, and condensation, and through ocean circulation. Water also regulates heat in the atmosphere by reflecting incoming solar radiation and absorbing outgoing long wavelength radiation emitted by the atmosphere, oceans, and land.

Water is a remarkable substance with unique physical properties. Water ice should be denser than liquid water but is not. If water became denser when it froze, ice in lakes of fresh water would sink to the bottom and lakes would freeze solid in winter. To change states from ice to water or water to steam, a large amount of energy (latent heat) is needed. Water requires large amounts of energy to raise its temperature by one degree (heat capacity). Water does not behave like other liquids because it has an effective network of hydrogen bonds that hold the water molecules (H_2O) together. When water is heated, the energy is used primarily to break the hydrogen bonds without increasing the average energy of motion of its molecules (temperature). Because of the hydrogen bond breaking mechanism, water can absorb a lot of heat for a small change in temperature. However, sea water does not behave like pure water or fresh water because of its high salt content and relative high density.

Most of the Earth's water is stored as sea water in the oceans (roughly 97%). The remaining 3% is either in the atmosphere or on the continents in the form of lakes, glaciers, ice sheets, or subsurface ground water. The ocean structure can be divided into two basic components: (1) a *surface layer* (about 150–1,000 feet deep) stirred and mixed by surface winds, and (2) colder and *deeper water* divided into layers of increasing density (Ross, 1977; Berner and Berner, 1987; Millero, 1996). The upper layer of deep water has a steep decrease in temperature gradient (*thermocline*). In the surface layer, lateral ocean circulation is primarily driven by wind (shallow or wind-driven circulation). In the deep-water layer, ocean circulation is driven by variations in density reflecting differences in temperature and salinity (*thermohaline* circulation).

The circulation of the shallow ocean layer is driven by the prevailing winds, which are caused by uneven heating of the Earth's surface combined with atmospheric circulation patterns. The resulting ocean circulation patterns can be described by ring-like structures (*gyres*) that flow clockwise in the northern hemisphere and counterclockwise in the southern hemisphere. Each gyre exhibits a strong poleward current on its western side and a weaker current to the east as result of the Earth's rotation and the continental land masses that obstruct their circulation. Because of the Coriolis force (caused by the Earth's rotation), water in the surface layer in the northern hemisphere does not simply move downstream but moves to the right of the wind direction. This creates clockwise-flowing gyres. In the southern hemisphere, the Coriolis force is to the left of the direction of motion, creating counterclockwise-flowing gyres.

At depths below the surface layer, the ocean circulation is not affected by the winds. Circulation is driven by differences in water density. The density of sea water increases continuously with decreasing temperature (as opposed to fresh water, which has a maximum at 4 °C). Density also increases with increasing salt content (*salinity*). Because of the decrease in temperature with depth, the deep ocean is stratified into density layers (*water masses*), with the densest water mass at the bottom and the lightest water mass at the top. The density stratification inhibits vertical motion, making it easier to move water along regions of constant density. That is, deep water circulation is primarily horizontal.

The deep ocean density stratification can be affected by processes occurring in the ocean surface layer, such as heating and cooling of surface water, evaporation, formation of sea ice and large discharges of fresh water from glaciers and ice sheets. Thermohaline circulation involves the flow of warm surface water that cools and sinks through convection, creating denser water due to the decreased temperature and increased salinity (Figure 2.11). At high latitudes in the far North and South Atlantic Oceans, surface water becomes

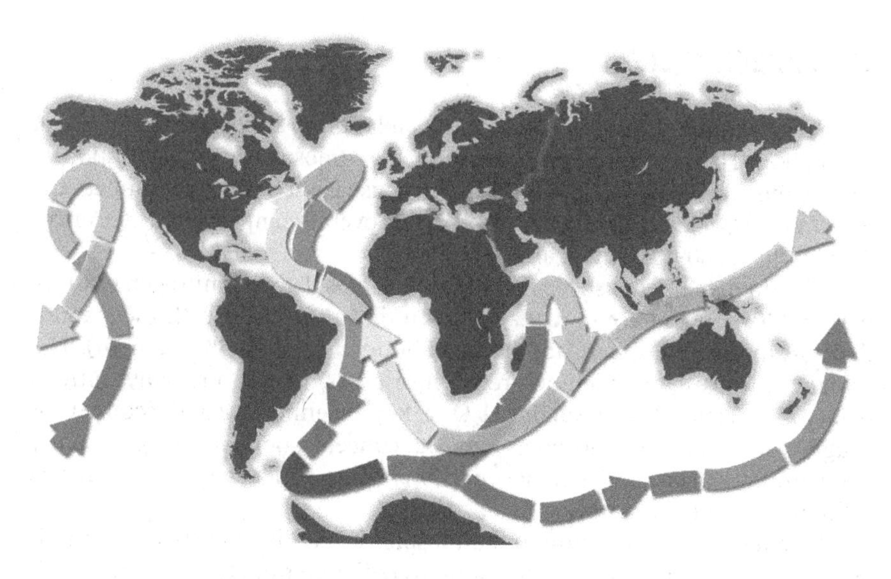

FIGURE 2.11 General ocean circulation of surface and deep water.

denser than surface water at lower latitudes due to cooling and sea ice formation. As sea water freezes, dissolved salts are left out of the ice, increasing the salinity (and density) of the remaining sea water. The increase in density can be so great that the denser surface water sinks downwards through convection to replace the underlying (lighter) water, creating new dense deep water that is replenished from above. Once this denser deep water reaches the appropriate stratification depth, it flows laterally throughout the oceans, creating deep water circulation. As the deep water travels through the oceans, there is a deep water upwelling that slowly diffuses into the surface layer, supplemented by a more intense coastal upwelling, closing the deep water circulation loop.

Deep water circulation is dominated by two cold water masses: the North Atlantic Deep Water (NADW) and the Antarctic Bottom Water (AABW). NADW involves the conversion of warm salty northward-flowing surface waters (the Gulf Stream) to cold dense deep waters in the North Atlantic Ocean behind the Greenland-Iceland-Scotland Ridge. In contrast, AABW involves the conversion of southward-flowing surface waters to cold dense deep water in the South Atlantic Ocean surrounding Antarctica, around the Weddell and Ross Seas. AABW is the densest water mass in the oceans and creates the lower branch of the world's large-scale thermohaline circulation. The shutdown or disruption of the global thermohaline circulation due to weakening the replenishment of dense deep water could have dramatic effects on climate.

THE CARBON CYCLE

We have reviewed the energy cycle and its relation to atmospheric and ocean circulation. Here we focus on the rise of the atmospheric concentration of carbon dioxide (CO_2) gas and other greenhouse gases, which can impact the Earth's energy balance and can raise Earth's overall temperature (McCarthy, 2009; Plimer, 2009; Tripati et al., 2009).

The three major constituents of Earth's atmosphere are nitrogen (78.1%), oxygen (20.9%), and argon (0.9%). Carbon dioxide is the fourth most abundant gas in the atmosphere and its contribution is only 0.04% (420ppm). The remaining trace gases include greenhouse gases, mainly methane, nitrous oxide, and ozone. Water vapor (H_2O, a key greenhouse gas) accounts for roughly 0.25% of the atmosphere, and its concentration varies significantly from around 10ppm in the coldest portions of the atmosphere to as much as 5% in hot, humid air masses.

Despite its low concentration, atmospheric CO_2 is important because it strongly absorbs infrared (low wavelength) radiation from the Earth and radiates it back, reducing energy loss to space that helps maintain the Earth's temperature (atmospheric greenhouse effect). Atmospheric CO_2 concentration, measured at Mauna Loa in Hawaii, shows a persistent increase of about 1.6ppm/year (in the period 1958–2023). The records also show an annual oscillation around 6ppm related to biological cycling that represents the uptake of CO_2 by plants for photosynthesis in spring and summer, and its release due to plant respiration and decomposition during autumn and winter.

The persistent rise in atmospheric CO_2 has been attributed to the burning of fossil fuels (coal, oil, and natural gas), which releases CO_2 into the atmosphere. The concentration of fossil fuel-related CO_2 has risen over the last 170 years in response to an increasing demand for energy for industrial development and better standard of living. Increased levels of CO_2 can also result from deforestation due to burning or breaking down carbon-containing plant material.

Understanding the relative importance of natural and man-made CO_2 contributions to the atmosphere requires understanding the carbon cycle, which covers how much CO_2 is produced, how much is consumed, and how much is stored (Berner and Berner, 1987; Millero, 1996). The atmosphere stores only about 720Gt (gigatons of carbon). There are three other carbon reservoirs that are much larger than the atmosphere: (1) the terrestrial biosphere and soil (2,000Gt), (2) the oceans (about 38,000Gt), and (3) carbonate rocks[1] and buried organic matter (>15,000,000Gt).

The terrestrial biosphere, forest, and soil exchange CO_2 with the atmosphere rapidly as shown by the annual oscillations in CO_2 discussed above and can increase its CO_2 storage capacity by increasing the size of the biosphere (e.g. more plant coverage as CO_2 increases). The oceans reflect the fastest exchange of CO_2 with the atmosphere since the surface layer of the oceans are well mixed with the atmosphere. Carbon is present in the oceans as inorganic carbon, in the form of dissolved bicarbonate ions and carbonate ions. The carbonate rock reservoir is the largest one but acts over very long periods of time by sequestering carbon in rocks and in buried organic matter.

Contrary to common belief, it is the water cycle that controls climate as opposed to the carbon cycle. Water molecules are far more efficient as greenhouse gases than CO_2. Water vapor and clouds are the main contributors to the atmospheric greenhouse effect (representing about 50% and 25% respectively). Water vapor fluxes represent one of the largest movements of mass and energy in the Earth's climate system (atmosphere, oceans, and land), yet the relative contributions and dynamics of biological and non-biological water vapor fluxes are still not well known. However, water vapor concentration is a function of the atmosphere's temperature, and therefore water vapor is usually not considered an external forcing of climate sensitivity. By contrast, gases such as CO_2 (representing about 20% of the greenhouse contributions), nitrous oxides, methane, tropospheric ozone, and CFCs are considered external forcings of the climate system because they can be added to or removed from the atmosphere independently from the air temperature.

Significant progress has been made in recent decades in our understanding of climate. However, scientists are still debating key mechanisms for global atmospheric and oceanic circulation. In order to make more accurate predictions of future rises in global temperature due to CO_2 from fossil fuel burning and other human activities, more refined models of the carbon cycle, carbon sources, and sinks are required. Other key factors need to be taken into consideration. The level of cloudiness in the atmosphere can result in either positive or negative effects (Berner and Berner, 1987). High-level clouds can absorb low wavelength radiation, which reduces the radiation outflow into space, resulting in warming (atmospheric greenhouse effect). In contrast, low-level clouds can reflect more incoming solar radiation back into space, resulting in cooling (increased albedo). Separately, the oceans can slow down or delay any rise in the Earth's temperature because of their large capacity to absorb and store heat.

NOTE

1. Carbon storage in the lithosphere in sedimentary carbonates can reflect additional capacity but these carbon sequestration processes extend over geological time scales.

REFERENCES

Berner, E.K. and Berner, R.A. (1987). *The Global Water Cycle*. Prentice Hall, Englewood Cliffs, NJ, pp. 1–43.

Buis, A. (2020). Milankovitch (Orbital) Cycles and Their Role in Earth's Climate, Global Climate Change, *NASA News*, February 27.

Charbonneau, P. (2010). Dynamo Models of the Solar Cycle. *Living Reviews in Solar Physics* 7: 3–91.

Charvátová, I. (2000). Can the Origin of the 2400-Year Cycle of solar Activity Be Caused by Solar Inertial Motion? *Annales Geophysicae* 18: 399–405.

Engdahl, E.R., and Villasenor, A. (2002). *Global Seismicity: 1900–1999*, University of Colorado, Boulder, CO.

Fan, Y. (2009). Magnetic Fields in the Solar Convection Zone. *Living Reviews in Solar Physics* 6: 4–96.

Hathaway, D.H. (2011). A Standard Law for the Equatorward Drift of the Sunspot Zones. *Solar Physics* 273: 221–230.

Hathaway, D.H., Wilson, R.M., and Reichmann, E.J. (1994). The Shape of the Sunspot Cycle. *Solar Physics* 151: 177–190.

Hathaway, D.H., Wilson, R.M., and Reichmann E.J. (1999). A Synthesis of Solar Cycle Prediction Techniques. *Journal of Geophysical Research* 104(A10): 22375–22388.

Hejda, P., Charvátová, I., and Střeštík, J. (2008). On the Long-Term Variability and Predictability of Geomagnetic Activity. XIIIth IAGA Workshop on Geomagnetic Observatory Instruments, Data Acquisition and Processing. Golden and Boulder, Colorado, United States, June 9–18.

Jose, P. (1965). Sun's Motion and Sunspots. *The Astronomical Journal* 70(3): 193–200.

Kern, A.K., Harzhauser, M., Soliman, A., Piller, W.E., and Oleg Mandic, O. (2013). High-Resolution Analysis of Upper-Miocene Lake Deposits: Evidence for the Influence of Gleissberg-Band Solar Forcing. *Palaeogeography, Palaeoclimatology, Palaeoecology* 370: 167–183.

Macdougall, D. (2004). *Frozen Earth: The Once and Future Story of Ice Ages*. University of California Press, Berkeley, CA, pp. 75–85.

Macdougall, D. (2009). *Nature's Clocks: How Scientists Measure the Age of Almost Everything*. University of California Press, Berkeley, CA.

Mathez, E.A. and Webster, J.D. (2004). *Machine*. Columbia University Press, New York.

McCarthy, J.J. (2009). Reflections on Our Planet, Its Life, Origins, and Futures. *Science* 326 (December): 1646–1655.

Millero, F. (1996). *Chemical Oceanography*. CRC Press, Boca Raton, FL, pp. 1–96.

Miyahara, H., Sokoloff, D., and Usoskin, I.G. (2006). The Solar Cycle at the Maunder Minimum Epoch. *Advances in Geosciences* 2: 1–20.

Nandy, D. (2009). Outstanding Issues in Solar Dynamo Theory. In: *Magnetic Coupling between the Interior and the Atmosphere of the Sun*, Eds. S.S. Hasan and R.J. Rutten. Springer-Verlag, Heidelberg.

Palus, M., Kurths, J., Schwarz, U., Seehafer, N., Novotna, D., and Chárvátova, I. (2007). The Solar Activity Cycle Is Weakly Synchronized with the Solar Inertial Motion. *Physics Letters A* 365: 421–428.

Plimer, I. (2009). *Heaven and Earth: Global Warming, the Missing Science*. Taylor Trade Publishing, Lanham, MD.

Prager, E.J., and Earle, S.A. (2000). *The Oceans*. McGraw-Hill, New York.

Press. F., and Siever, R. (1974). *Earth*. W.H. Freeman and Company, San Francisco, CA, pp. 395–487.

Rahmstorf, S. (2003). The Current Climate. *Nature* 421 (February 13): 699.

Ross, D.A. (1977). *Introduction to Oceanography*. Prentice-Hall, London, pp. 192–236.

Strahler, A. (1973). *Introduction to Physical Geography*. John Wiley & Son, New York, pp. 23–113.

Tarbuck, E.J., and Lutgens, F.K. (1979). *Earth Science*. Charles E. Merrill Publishing Company, Columbus, OH, pp. 115–142.

Tripati, A., Roberts, C.D., and Eagle, R.A. (2009). Coupling of CO_2 and Ice Sheet Stability Over Major Climate Transitions of the Last 20 Million Years. *Science* 326(December): 1394–1397.

Verhoogen, J., Turner, F.J., Weiss, L.E., Wahrhaftig C., and Fyfe, W.S., (1970). *The Earth: An Introduction to Physical Geology*. Holt, Rinehart and Winston, New York, pp. 598–636.

Wilson, I.R.G., Carter, B.D., and Waite, I.A. (2008). Does a Spin–Orbit Coupling Between the Sun and the Jovian Planets Govern the Solar Cycle? *Astronomical Society of Australia* 25: 85–93.

Zhang, P., Cheng, H., Edwards, R.L., Chen, F., Wang, Y., Yang, X., Liu, J., Tan, M., and Johnson K. (2008). A Test of Climate, Sun, and Culture Relationships from an 1810-Year Chinese Cave Record. *Science* 322(November): 940–942.

Zumberge, J.H., and Nelson, C.A. (1976). *Elements of Physical Geology*. John Wiley & Sons, New York, pp. 125–177.

CHAPTER 3

A Brief History of Climate Change

NATURAL DRIVERS OF CLIMATE CHANGE OVER THE AGES

Fossil records of human evolution begin around seven million years ago with the discovery of an early hominid specimen (*Sahelanthropus tchadensis*) in Chad, north central Africa, followed by other hominid species that used stone tools around 4.5 million years ago. Around 2 million years ago, the first human-like species appeared and had different evolutionary descendants, leading to the human lineage.

Modern humans (*homo sapiens*) evolved from this lineage in Africa around 300,000 years ago and initially lived as hunter-gatherers in grassland or sparsely wooded areas. Around 100,000 years ago, modern humans had rituals, traces of art and tools, and were already using basic personal artifacts such as jewelry or ornaments. Around 50,000 years ago, they performed burials of the dead, used tools and weapons and, most importantly, improved their ability to communicate with each other as reflected in outstanding cave paintings and sculptures made from ivory, stone, and bone (Braidwood, 1959; Starr, 1991). They migrated out of Africa in waves through the continents and larger islands, reaching North Africa and the Middle East around 100,000–120,000 years ago, Australia around 50,000–65,000 years ago, Europe 45,000 years ago, and the Americas around 21,000 years ago. By the time the last Ice Age ended around 12,000 years ago, modern humans had populated most of the habitable regions of the planet.

Human history can be described as the collection of cumulative development, activities, knowledge, and culture. Human history (as defined above) goes back roughly to 50,000 to 100,000 years ago, although human development has a far longer history (Mithen, 2003). For a significant fraction of human history, people lived as hunter-gatherers, living in small communities or temporary settlements, continuously adapting to the environment and climate at the time.

About 15,000 years ago, there was a significant change in human activity, leading to the beginning of farming, organized settlements, and civilization.

By 5,000 BC, the foundation of our modern world was laid. Regional civilizations around the world rose, flourished, and fell as they faced their own social, environmental and climate challenges. This remarkable change in human history occurred roughly around the end of the last Ice Age, which peaked around 20,000 BC. Before that time, humans struggled with deteriorating and frigid climate conditions.

The world of the last Ice Age peak (maximum) at 20,000 BC was cold, dry, windy, and inhospitable. Changes in the Earth's orbit around the Sun and in the tilt of its rotation axis caused ice sheets to expand dramatically across much of North America, northern Europe, and Asia. Sea levels fell, exposing coastal plains and joining land masses, and droughts covered vast regions of the planet. Human communities were forced to move from regions that were inhabited before the last Ice Age. Other regions remained unoccupied because routes were blocked by impenetrable walls of ice and dry deserts.

This was not the first time our ancestors lived through periods of dramatic climate change and hardship conditions. Nor will it be the last. Our planet seesaws between Ice Ages (glacial periods) roughly every 100,000 years, driven by small changes in the shape (eccentricity) of its orbit around the Sun from roughly circular to slightly elliptical. The northern hemisphere develops greater seasonality, while the opposite happens in the southern hemisphere. This triggers the growth of northern ice sheets. When the orbit returns to a nearly circular shape, the north-south seasonality contrast is reduced, leading to global warming and the melting of the ice sheets. Changes in the Earth's tilt during its orbit (*obliquity*) also result in climate changes. Every 41,000 years, the inclination of the Earth's axis of rotation changes from roughly 22.2 to 24.5 degrees and back. As the tilt angle increases, seasons become more extreme: colder winters and hotter summers. The Earth's axis of rotation also wobbles and shifts every 23,000 years (axial precession). This influences the point in the Earth's orbit at which the northern hemisphere is tilted toward the Sun. If this happens when the Earth is close to the Sun, winters in the northern hemisphere will be warm and short. If the Earth is farther away, winters will be colder and longer.

FROZEN EARTH

Paleoclimate records show that glacial advances and retreats have been occurring regularly. The most recent glacial advance ended about 20,000 years ago, followed by a glacial retreat and warming about 12,000 years ago. However, more prolonged, and severe Ice Ages occurred in the Earth's distant past. Some of the early Ice Age events stretched over several hundred million years, including periods of global glacial conditions such as the

"Snowball Earth" period, about 650 million years ago, or the more recent Permo-Carboniferous Ice Age period, 340 million years ago when the Earth was cold for about 80 million years (Mcdougall, 2004).

During the Snowball Earth event, the planet's surface became entirely or nearly entirely frozen with no large liquid water surface exposed to the atmosphere. The snowball event could be initiated by an increase in the Earth's coverage of snow and ice. This would in turn increase Earth's albedo (reflectivity), which would create positive feedback for further cooling, resulting in runaway cooling if enough snow and ice accumulated. This positive feedback mechanism could be facilitated by the equatorial distribution of land masses at that time. This would allow ice to accumulate in the regions closer to the equator, where solar radiation is higher. The additional snow and ice would reflect more solar energy back to space, limiting heat and mass transfer from the equator to the poles, further cooling Earth, and further increasing the surface area covered by snow and ice.

The large accumulations of CO_2 in the atmosphere over millions of years, emitted primarily by volcanic activity during this period, could have resulted in global warming that triggered the melting of the Snowball Earth. As snow and ice melted, the Earth's surface reflectivity decreased, the absorption of solar radiation increased, creating a positive feedback mechanism for further melting of snow and ice and accelerating the deglaciation process.

During the Permo-Carboniferous Ice Age (340 to 260 million years ago), there were many intermediate cycles of glacial advances and retreats. At the time of this global glaciation, the Earth's land masses were joined in a supercontinent (Pangea). Since then, continents have slowly drifted apart to their current positions, changing the global ocean and atmospheric circulation, and affecting climate patterns.

WARMING AND FREEZING CYCLES AND PERIODIC GLACIATIONS

The regular cycle of Ice Ages seems to have started about 1.5 million years ago, with periodic swings between warming and cooling. For 500,000 out of 780,000 years in paleoclimate records, climate cycled from cooling to warming to cooling again. Glacial periods lasted longer than warming ones. Ice cores for the past 420,000 years from Greenland and Antarctica show four distinct climate switching periods from cold to warmer conditions around 335,000, 245,000, 135,000, and 18,000 years ago (about 100,000 years apart).

After 20,000 BC, when the last glacial period reached its maximum, a period of slow and uneven global warming began. By 15,000 BC, the extensive ice sheets had begun to melt, and, by 12,000 BC, the global climate exhibited fluctuating surges of warmth and rain, followed by periods of cold and drought. Following this period, soon after 10,000 BC, there was a spurt of global warming that brough the last Ice Age to its close. This led to the (interglacial) Holocene period, in which we live today. The last 10,000 years of climate change and global warming have shaped human history.

Earth's climate has been significantly colder and warmer than today for most of human history (Brooks, 1970; Mithen, 2003; Plimer, 2009). There have been periods of gradual climate changes and periods of very rapid change. Furthermore, global climate rapidly plunged into very cold periods from 12,900 to 11,500 years ago and from 8,500 to 8,000 years ago. These rapid climate changes to cold conditions led to the expansion of ice sheets and alpine valley glaciers, stressing human populations, and changing the distribution of plants and animals. In contrast, during a warmer period 6,000 years ago, sea level was about 2 meters higher than at the present time (Plimer, 2009). Temperatures were also higher, but the environment was far from a civilization-ending catastrophe.

Many of these changes happened suddenly, taking only a few years to decades to change from a relatively warm climate to a very cold one. Fluctuating warm and cold conditions continued during a period of thousands of years characterized by overall climate warmth of the Holocene.

THE LAST ICE AGE, YOUNGER DRYAS, AND CLIMATE CYCLES

About 15,000 years ago, the global climate began to shift from the cold conditions of the last glacial maximum (20,000 to 27,000 years ago) to a warmer interglacial state. During the transition to a warmer climate, the northern hemisphere suddenly returned to nearly glacial conditions around 12,000 to 13,000 years ago. The change was relatively sudden and resulted in a significant reduction in the temperature in Greenland by 4–10 °C. The event ended abruptly about 11,500 years ago as reflected in a 10 °C temperature increase in Greenland and similar changes in other locations. This period is known as the Younger Dryas, named after the *Dryas octopetala* flower that grows in cold conditions. The Younger Dryas event was the most severe of several interruptions of the climate warming occurring between 15,000 and 13,000 years ago and was likely due to a weakening of the Atlantic meridional overturning circulation (the thermohaline circulation/conveyor). There are several theories on the reason for the weakening of the ocean circulation, ranging from a sudden influx of fresh water into the North Atlantic due to deglaciation in North America, to volcanic activity that interrupted the thermohaline circulation temporarily. Eventually, the thermohaline circulation strengthened again, heat and mass transport resumed, and climate recovered.

CLIMATE CHANGE IN ANCIENT TIMES

About 8,200 years ago (6200 BC), there was a sudden decrease in global temperatures that lasted for two to four centuries. This cooling event was significantly less pronounced than the Younger Dryas cold period. The event may have been caused by a large discharge of meltwater into the North Atlantic from the final collapse of the Laurentide ice sheet, which was a continental ice sheet covering a vast portion of North America during the last Ice Age. The meltwater discharge may have affected the North Atlantic thermohaline circulation, reducing northward heat transport and causing significant cooling. The melting of large icebergs can also cause vast quantities of fresh water to be added to the North Atlantic, which can alter the thermohaline circulation in the oceans. In the past, large groups of icebergs have broken off from the Laurentide ice sheet and moved into the North Atlantic (Heinrich events).

Over the millennia, climate changes have impacted the development of human settlements and activities. Some of these changes have been gradual, while others were sudden or catastrophic (Braidwood, 1959; Pritchard, 1959; Starr, 1991; Flanagan, 1998; Mithen, 2003; Fagan, 2004; Diamond, 2005; Plimer, 2009).

The period 6200 to 5800 BC was a catastrophic mini-Ice Age for communities between the ancient Euxine Lake (the Black Sea today) and the Euphrates River. Cold and dry conditions forced local communities to move from farming to sheep farming as drought increased. By 5800 BC, the North Atlantic circulation strengthened again, and moisture-laden Mediterranean westerlies resumed, leading to better farming in the Fertile Crescent between the Tigris and Euphrates rivers (Mesopotamia). The small settlements of 5800 BC slowly became the earliest cities (Ur, Nippur, and Uruk in modern Iraq) that eventually led to the Mesopotamian civilizations.

Around 5600 BC, Lake Euxine was flooded with water from the Mediterranean Sea, suddenly changing the content of the lake from fresh water to salty water. The flood pushed the population into the plains of Hungary, to the Danube River and Dnieper River to the west of Europe. Until 4000 BC, agriculture was simple, supplemented with hunting-gathering. By 3500 BC, farms in Europe had become closer, with more established agriculture. As the population grew, farms filled the gaps between the hunter-gatherers' territories.

During this mini-Ice Age period, conditions over North Africa and throughout southwestern Asia became drier. Deserts, including the Sahara, expanded, streams and springs dried up and grasslands withered. The Sahara grew slightly wetter again after 5000 BC with the increase in rainfall through North Africa and southwestern Asia. By 4000 BC, intense drying conditions had settled again over the Sahara, impacting vegetation and animal populations that once enjoyed wetter conditions.

Around 2200 BC (late Bronze Age), human civilization's vulnerability to climate change was tested once again, especially in Egypt and Mesopotamia. For nearly a thousand years (from 3100 BC to 2160 BC), the Egyptian civilization in the early dynasties and Old Kingdom prospered, as reflected by an era of pyramid and temple building. Widespread disruption resulted from a 300-year drought cycle that led to the collapse of several empires and depopulation. The Nile floods, critical to the crop season, faltered, leading to political and social unrest. The collapse of the Old Kingdom was followed by decades of famine and strife. As the drought cycle ended, a new period of prosperity followed in the Middle Kingdom, interrupted by periods of low floods in the Nile but not to the extent observed earlier (Grimal, 1991; Fagan, 2004).

A new drought cycle descended over the Mediterranean, North Africa, and the Near East (Egypt, Libya, the Balkans, the Aegean, and Anatolia) by 1200 BC. This time, prosperous and powerful civilizations like the Hittites, Myceneans, Assyrians, and Babylonians were affected, as the delicate balance of power and trade fell apart (Starr, 1994; Fagan, 2004). The Mycenean civilization imploded, the Hittite state collapsed, and the Assyrians and Babylonians lived through hard times in deteriorating conditions. The Egyptian state weakened but survived. The Late Bronze Age collapse was a time of widespread social disruption and violence associated with environmental change, invasions and mass migration, and the destruction of civilizations.

ROMAN WARM PERIOD

The Roman Warm Period was a period of unusually warm climate in Europe and the North Atlantic that ran from approximately 250 BC to AD 450 (Fagan, 2004; Ruddiman, 2005; Plimer, 2009). Literary fragments describing tree and plant growth and other activities from that time indicated that summer temperatures in the fourth and fifth centuries BC were about the same as current temperatures. Populations increased, there was excess wealth, forests expanded, and agriculture could be undertaken in areas at much higher latitudes and altitudes than today due to the warm climate. This period also brought an unparallel level of engineering technology and general prosperity, with roads, aqueducts, buildings, monuments, and sculptures that have perdured for almost two millennia.

THE CLIMATE IN THE DARK AGES

Deterioration of the climate conditions toward the end of the Roman Warm period, social unrest, and wars may have played a role in the migration and settlement of various tribes (the Franks, the Goths, the Alemanni, the Alans, the

Huns, the Slavs, the Avars, the Bulgars, and the Magyars) within the territories of the Roman Empire and throughout Europe. The migration period began around AD 300–375 and ended well after the collapse of the Roman Empire in AD 450. The Roman Warm period was followed by the Dark Ages.

The Dark Ages (AD 450–900) were bitterly cold, with frequent crop failures, famine, disease, war, depopulation, expansion of the ice sheets, and increased wind. During the Dark Ages, there was great social disruption and climate refugees wandered Europe, desperately looking for food. This was exacerbated by a plague of unprecedented intensity in AD 540–542 (Plague of Justinian) that swept through the Middle East, North Africa, and the Mediterranean, spreading into southern and western Europe.

Severe droughts hit China, Mongolia, and Siberia around AD 536–538. Tree ring analysis has revealed some of the coldest conditions (Fagan, 2004). In the Americas, civilizations, such as the Mayans, collapsed during the ninth century. The central Maya region suffered a major political decline, marked by the abandonment of cities and the ending of dynasties, likely caused by endemic warfare, over-population, severe environmental degradation, and climate change-related droughts. Although significantly reduced, the Maya presence continued after the abandonment of the major cities until its final collapse following the Spanish conquest in the mid-1500s (Fagan, 2004; Plimer, 2009).

THE MEDIEVAL WARM PERIOD

The Medieval Warm Period was a time of warm climate in the North Atlantic region that lasted from about AD 900 to about AD 1300 during the European Middle Ages. This warmer period was a better time for economic development, when the ice sheets and the sea ice contracted, enabling sea exploration and settlement at high latitudes. Grain crops, cattle, sheep, farms, and villages were established in Greenland by the Norse (the Vikings), which was warmer than today. Although the Medieval Warm period included a short cold period, there was enough food to feed an increasing population, and crop failures and famine were rare.

During the Medieval Warm Period, the population of Europe exploded and reached levels that, in some places, were not matched again until centuries later. The crop yield ratio (the number of seeds one could harvest and consume per seed planted) significantly increased after favorable harvests during this warm period. That is, for every seed planted, the number of seeds harvested was more than the one needed for next year's crops. The excess number of seeds could be used for food or sold at a profit. Excess wealth created over generations was used to build cathedrals, monasteries, and universities (Fagan; 2004; Ruddiman, 2005; Plimer, 2009).

THE LITTLE ICE AGE

The Medieval Warm Period was followed by the Little Ice Age, a cooler period, which was more pronounced in the northern hemisphere. The period extends from about 1300 to about 1850. The Little Ice Age started in the late thirteenth century following a decrease in solar activity (Fagan, 2004; Ruddiman, 2005; Plimer, 2009). This period was characterized by cold and fluctuating climate during periods of reduced solar activity (1280–1340, 1450–1540, 1645–1715, and 1795–1825).

The Little Ice Age brought colder winters to Europe and North America. In high latitudes sea-ice packs surrounded seaports, forcing ships to stay in port. Many regions where sea ice was rare in the period 1000–1200 had increasingly abundant sea ice between 1300 and the late 1800s. Sea ice became rare again in the late 1900s (Ruddiman, 2005). The Baltic Sea froze over twice, in 1303 and 1306–1307. In the mid-seventeenth century, farms and villages in the Swiss Alps were destroyed by advancing glaciers. In Great Britain and the Netherlands, canals and rivers frequently were frozen. In 1658, with the winters in Scandinavia being exceptionally frigid, the Swedish army marched across frozen areas of the Great Belt to Denmark to attack Copenhagen.

The Great Frost (or Great Winter) in 1708–1709 was an extraordinarily cold winter in Europe, with a subsequent famine estimated to have caused hundreds of thousands of deaths by the end of 1710. The winter of 1794–1795 was also particularly harsh. The French invasion army marched on the frozen rivers of the Netherlands, while the Dutch fleet remained locked in the ice.

There was crop failures, famine, disease, social disruption, wars, and depopulation globally. The coldest decades in the Little Ice Age impacted farmers growing frost-sensitive crops at high latitudes or at altitudes in the valleys and mountains where such crops were barely possible. Crops were often destroyed by unanticipated freezes, or the harvest was delayed by summers with extended rain or cold periods. As the Little Ice Age cold intensified, regions of land under cultivation were progressively abandoned. The Norse settlements in Greenland died out.

The Great Famine of 1315–1317 was the first of several crises that struck Europe. The Great Famine, primarily restricted to northern Europe, caused significant numbers of deaths, and marked a clear end to the prosperity of the Medieval Warm Period. During the early 1300s, Europe suffered some of the worst and most sustained periods of bad weather, characterized by severe winters and rainy and cold summers, leaving little margin for error for food production. The Great Famine started with bad weather in 1315, followed by crop failures through 1316 and 1317. In the spring of 1315, unusually heavy rain began in much of Europe and temperatures remained relatively cool. Grain could not ripen, leading to crop failures, and straw and hay for the animals could not be cured, so livestock could not be fed, and disease spread. Europe did not fully recover until 1322. Famine periods were marked by extreme levels of crime, disease, and mass death.

Famines were followed by devastating plague pandemics. The Black Death pandemic (bubonic plague) from 1346 to 1353 was one of the most fatal pandemics in human history. The Black Death originated in Asia and swept through Europe unimpeded, devastating its populations. In the sixteenth, seventeenth, and eighteenth centuries, there was a series of European plague outbreaks. Other plague outbreaks were reported in Asia and North Africa.

Famine in Europe killed millions between 1690 and 1700, followed by famines in 1710, 1725, and 1816. In North America, tree-ring records and other paleoclimate records indicate intense drought cycles from 1650 to 1750 (Fagan, 2004). With the catastrophic eruption of the Tambora volcano in Indonesia in 1815, the situation was only exacerbated (Plimer, 2009). The top of the volcano blasted into the air, launching more than seven times the number of ash particles into the atmosphere than the Krakatoa eruption of 1883. The Indonesian islands plunged into darkness temporarily. Most crops were destroyed by the ash fall and tsunamis with significant loss of life in the eruption and as a result of the epidemics and famine that followed the eruption. Sunlight was reduced as far as Hainan Island, 1,200 miles north of Tambora. As the volcanic ash and dust dispersed through the atmosphere, low temperatures, excessive rainfall, and unseasonal frosts dramatically impacted subsistence farming. Europe and Asia experienced an exceptionally cold and stormy winter in 1816–1817 with disastrous crop failures across the northern hemisphere.

The year 1816 is known as the *year without a summer*. In the summer of 1816 in Geneva, poor weather conditions, akin to winter, forced Lord Byron, Mary Shelley, and other guests of Lord Byron to stay indoors. To help pass the time while stuck indoors, Byron suggested they have a competition to write the best ghost story. Mary Shelley, who was just 18 years old, won the contest with her creation "Frankenstein," which vividly describes the harsh conditions of the time.

The social cost of the Little Ice Age was reflected in endless wars, social disruption, civil unrest, bloody revolts, and revolutions, and the fight for independence and self-determination across many countries, colonies, and territories in Europe, the Americas, and elsewhere.

The Little Ice Age end year is defined as 1850. The end of this period roughly coincides with the beginning of large-scale industrial mechanization, at the end of the first wave of the Industrial Revolution. The 1850–1900 period is used by the IPCC as a reference to measure global temperature increases.

Tables 3.1–3.3 present climate history from 18000 BC to AD 2024, following closely the description in Fagan (2004).

THE INDUSTRIAL REVOLUTION AND MAN-MADE EFFECTS

The selection of the year 1850 as the end of the Little Ice Age roughly agrees with the end of the Industrial Revolution transition period from manual to mechanized manufacturing processes occurring from around 1760 to about 1820–1840. The mechanization of processes spread out rapidly from Great Britain to Europe and the United States, and to other countries.

TABLE 3.1 Climate history (18000 BC to 10000 BC)

Period (BC)	Climate Events	Human Events	Climate Conditions
18000	Ice Age (cold)	Cro-Magnons (little outside activity). Hunting when possible.	Cold most of the year (near freezing)
16000	Late Ice Age	Cro-Magnons in Europe, 40,000 years ago	Rapid retreat of ice sheets
15000	Some warming Variable temperature	Climate improvement in Eurasia	
14000		Final Ice Age in Europe	Rapid sea level rise
13000	Heinrich type I event ends	First settlements of Northeastern Siberia	Rapid warming
12000	Late Glacial Interstadial warming (Bolling-Allerod event)	Monte Verde, first settlement in Americas Niaux, France, paintings	Spread of forests in Europe
11000	(cold) Younger Dryas	Clovis in North America	Cold in Europe Atlantic circulation shuts down
10000	Younger Dryas	Farming begins in Southeastern Asia	Drought in Europe/Asia

TABLE 3.2 Climate history (9,000 to 250 BC)

Period (BC)	Climate Event	Human Event	Climate Conditions
9000	Pre-Boreal (warming)	Farming spreads in Asia	Atlantic circulation resumes Moist conditions
8000	Boreal	Farming spreads in Southwestern Asia Hunter-gatherers in Northern Europe	
7000	Mini-Ice Age (cold, dry)		
6000	Mini-Ice Age (cold, dry)	Linearbandkeramic farmers in Central Europe Settlements in Southern Mesopotamia Farmers in the Balkans	(~5600 BC) Euxine Lake flooded Sea-level rise (~5800 BC) Laurentide ice sheet collapses Atlantic circulation slows (~5200 BC)
5000	Change in the Atlantic	Ertebølle culture in Scandinavia	
4000	Sub-Boreal	Cattle herds in Sahara Rise of cities. Uruk, Gerzean	Warm, moist conditions in Europe
3000	Sub-Boreal	Sumerian civilization Old Kingdom, Egypt Unification of Egypt, towns appear in Egypt; Mesopotamia, Ur	Major aridification of Sahara and Egypt
2000	Drought event in Eastern Mediterranean	Old Kingdom, Egypt, ends in crisis Egypt reunifies (2046 BC) Akkadian empire	Major El Niño event 300-year drought in Eastern Mediterranean by 2200 BC
1000	Drought event in Eastern Mediterranean	Collapse of Hittites Collapse of Myceneans Phoenicians, Arameans Uluburun shipwreck (Turkey)	Major drought episode
250		Hellenistic Age	

TABLE 3.3 Climate history (250 BC to AD 2024)

Year	Climate Event	Human Event	Climate Conditions
250 BC–AD 450	Roman Warm Period Sub-Atlantic (wetter/cooler)	Ceasar conquers Gaul Celtic migration Avar Empire in Europe Decline of Rome	Droughts in eastern steppes Sudden cooling (AD 850) Major volcanic event causing cooling
450–900	Dark Ages		
900–1300	Medieval Warm Period	Ancestral Pueblo dispersal Collapse of Tiwanaku Collapse of Maya civilization	
1300–1850	Little Ice Age	Beginning of the Industrial Revolution, 1760–1850	Cooler, more volatile climate Major droughts in Western North and South America Warming after 1860
1850–2024	Global Industrial Revolution	Industrial Revolution in full progress. Accelerated development in developed and undeveloped economies Use of fossil fuels and release of CO_2 and other greenhouse gases into the atmosphere	

The Industrial Revolution in the late eighteenth century and early nineteenth century helped transition manual production methods to mechanized and reproducible processes driven by machines: using new chemical manufacturing and iron production processes and increasing use of water and steam power. Production, output, and profits greatly increased, while production costs decreased. The textile industry was one of the first to use modern production methods, and textiles became the dominant industry, followed by other industries.

The desire to leverage mechanization to increase production and profits across different industries and activities led to an ever-increasing demand for energy to power bigger, better, and more diverse machines. Transportation and trade also changed significantly with the development of trains and railroads and steam-powered ships. Coal and fossil fuels were cheap and efficient sources of energy. The increasing availability of coal and petroleum products further widened the potential for industrialization. Over the following decades, the emergence of large factories and the immense growth in coal and fossil fuel consumption gave rise to an unprecedented level of air pollution and emission of CO_2 and other greenhouse gases in industrial centers, and large volumes of industrial chemical discharges as byproducts of industrial activities.

Mechanization in the Industrial Revolution gradually grew to include iron (and later steel), chemicals, petroleum (refining and distribution of fuel and byproducts), mining, agriculture, industry, and consumer electrification and, later in the twentieth century, the automotive industry, electric and electronics goods, and a wide range of high and low carbon-emission manufacturing processes.

The rise of the atmospheric concentration of CO_2 and other greenhouse gases can impact the Earth's energy balance and can raise Earth's overall temperature. Carbon dioxide is the fourth most abundant gas in the atmosphere at only 0.04% relative to nitrogen (78.1%), oxygen (20.9%), and argon (0.9%). The remaining greenhouse trace gases include methane, nitrous oxide, and ozone. Despite its low concentration, atmospheric CO_2 is important because it strongly absorbs infrared (low wavelength) radiation from the Earth and radiates it back, reducing energy loss to space that helps maintain the Earth's temperature (atmospheric greenhouse effect). Atmospheric CO_2 concentration records showed a persistent increase in the CO_2 level in the period 1850–2024, with small annual oscillation related to biological cycling of CO_2 from plant photosynthesis, respiration, and decomposition. Over the last 170 years, CO_2 has accumulated in the atmosphere and its current concentration (420 ppm) is about 50% higher than pre-industrial levels.

The persistent rise in atmospheric CO_2 has been attributed primarily to the burning of fossil fuels (coal, oil, and natural gas), which releases CO_2 and other byproducts into the atmosphere. The concentration of fossil fuel-related CO_2 has risen over the last 170 years in response to an increasing demand for energy for sustaining industrial development and for improving the population's standard of living across the world. Increased levels of CO_2 can also result from deforestation due to burning or breaking down carbon-containing plant material.

Understanding the natural and man-made contributions to atmospheric CO_2 is critical for evaluating their impact on climate. This can lead to both actual physical risk of more frequent or extreme weather events and transition risk related to policies, regulations, and incentives for transitioning to a low-carbon economy.

TODAY AND TOMORROW

Significant progress has been made in recent decades in our understanding of climate. Although scientists continuously debate the merits and limitations of different mechanisms of climate change, the accumulated scientific evidence shows that we currently live in an interglacial and variable climate period exposed to significant swings in temperature. These changes are driven by complex processes involving the atmosphere, oceans and land, solar radiation, and changes in the Earth's orbit and rotation (McCarthy, 2009; Plimer, 2009).

The persistent rise in atmospheric CO_2 over the last 170 years seems primarily due to the burning of fossil fuels and other human activities in response to an increasing demand for energy globally to drive social prosperity. The increase in the emission of greenhouse gases could contribute to a destabilizing effect, leading to climate changes beyond the natural climate variability, which can impact society. Potential solutions to this dilemma include reduction of greenhouse gases, but also adapting to climate change that cannot be mitigated, and shifting priorities toward climate-stabilizing technologies that can protect the environment while allowing for economic growth and social prosperity (Mayewski and White, 2002).

Climate-related policies focusing only on greenhouse gas emissions, the reduction of fossil fuel consumption and their impact on global warming may look good for political posturing and virtuous signaling, but they distort the picture of actual climate change, which is natural but accelerated by human activities performed in the pursuit of social prosperity. Policies must consider both the long-term and short-term impact on climate and the environment, and their effect on society, which these policies are supposedly aiming to protect.

REFERENCES

Braidwood, R. (1959). *Prehistoric Men*. Anthropology 37. Chicago Natural History Museum, Chicago, pp. 7–91.

Brooks, C.E.P. (1970). *Climate Through the Ages*. Dover Press, New York.

Diamond, J. (2005). *Collapse*. Penguin Books, New York.

Fagan, B. (2004). *The Long Summer: How Climate Changed Civilization*. Basic Books, New York.

Flanagan, L. (1998). *Ancient Ireland: Life Before the Celts.* Gill & Macmillan, Dublin, pp. 1–44.
Grimal, N. (2000). *A History of Ancient Egypt.* Blackwell, Oxford, pp. 17–59.
Macdougall, D. (2004). *Frozen Earth: The Once and Future Story of Ice Ages.* University of California Press, Berkeley, CA.
Mayewski, P.A. and White, F. (2002). *The Ice Chronicles: The Quest to Understand Global Climate Change.* University Press of New England, Hanover, NE.
McCarthy, J.J. (2009). Reflections on Our Planet, Its Life, Origins, and Futures. *Science* 326(December): 1646–1655.
Mithen, S. (2003). *After the Ice: A Global Human History.* Harvard University Press, Cambridge, MA.
Plimer, I. (2009). *Heaven and Earth: Global Warming – the Missing Science.* Taylor Trade, Lanham, MD.
Pritchard, J.B. (1958). *The Ancient Near East: An Anthology of Texts and Pictures.* Princeton University Press, Princeton, NJ, pp. 28–118.
Ruddiman, W.F. (2005). *Plows, Plagues and Petroleum.* Princeton University Press, Princeton, NJ.

CHAPTER 4

Science, Politics, and Public Policy

SCIENCE, FACTS, PERCEPTION, SOCIAL INFLUENCE, MISINFORMATION, AND FEAR

Climate change, which is often used (and misused) as a reference to the impact of greenhouse gas emissions from human activity, is one of the most polarized topics of discussion today. The assessment and mitigation of the risks associated with climate change have direct implications on many sectors of the economy, including fossil fuel extraction, energy generation, transportation, manufacturing, and even agriculture and food production (Ruddiman, 2005; Plimer, 2009). Furthermore, inconsistent domestic policies and unilateral restrictions on high-carbon emission activities can be taken advantage of by countries with much larger high-carbon emission contributions but with fewer or no restrictions on manufacturing or energy generation activities. This creates an unfair competitive advantage that can result in a net transfer of wealth from countries that share the economic and social burden of climate risk mitigation to those countries polluting the most and doing the least to mitigate climate risk while developing a path to their own economic prosperity.

Discussions on climate change tend to be polarized around two extreme views: those concerned solely about damage to the environment and the increase in the emission of greenhouse gases caused by the large-scale use of fossil fuels, and those who oppose mitigating the effects of climate change because it could be damaging for the economy and social prosperity. For some, climate change is an inconvenient truth, while for others, it is a convenient lie. Those who have a more dispassionate and objective approach to listening to both sides of the argument tend to have a more conciliatory and moderate view.

Distortions on the soundness of arguments come from both extremes. Environmental extremists are prone to alarmist exaggerations and are eager to accept anecdotal cases or questionable information as strong support for their arguments. On the other hand, pro-industry extremists often deny basic knowledge from mainstream science or exercise unhealthy skepticism on sound information. These extreme views can affect the integrity of a healthy debate on climate change.

Unfortunately, the public often hears from people toward the extremes of the climate change debate issue because the media prefers controversial sound bites that align with a selected narrative. Supporters of extreme views usually cite selected results from scientific or technical work that supports their arguments, ignoring contradictory or critical information, or simply reporting distorted facts. People have a variety of degrees of trust in the information they receive and how they react to social influence, which affects the probability they assign to uncertain events and how they perceive credibility and reputation. Here we focus on social influence, conduct risk, credibility, disinformation, and conformity, using a probabilistic behavioral approach that quantifies investors and customers' trust. This is important for climate risk and ESG because for most institutions it takes years to acquire a good reputation and recognition, but it could take a single adverse event (either actual or perceived) to destroy an institution's reputation and franchise value completely and put the institution out of business.

BEHAVIORAL ASPECTS OF RISK TAKING AND DECISION MAKING

For decades, academics and practitioners have been debating the impact that investors' behavior and imperfect information have on asset prices and financial markets and, similarly, the impact that misinformation and social influence have on consumer behavior (Thaler, 1999; Shiller, 2000; Hirshleifer, 2001; Sobehart, 2003; Elan, 2010). Although there is extensive literature on the topic, there is still a need to understand the behavioral aspects of information filtering and social influence in risk taking and decision making, which are important during market booms, economic downturns, and financial crises and are particularly relevant in the context of climate risk and ESG. Any meaningful framework for understanding risk taking, decision-making and market behavior must be able to accommodate both the uncertainty of economic environments and the whims of human psychology.

The dominant academic view on risk taking and decision making is that market participants are presumed to make rational decisions and will quickly learn and correct their behavior, motivated primarily by self-interest. In the idealized rational view of decision making and risk taking, price changes are only the result of rational market participants responding to a flow of information and exogenous shocks. As a result, investors' predictions of the future value of economically relevant market variables are not systematically wrong in that all errors are completely random (Fama, 1965, 1970, 1998; Bernstein, 1992, 1998; O'Hara, 1998; Shleifer, 2000; Malkiel, 2003; Sobehart and Farengo, 2003; Sobehart and Keenan, 2005). In practice, persistent cognitive biases and behaviors can be at play during regular market upswings and downswings, as suggested by reoccurring asset bubbles and market crises (Damasio, 1994; Ariely, 2008; Coates et al., 2008; Lerner, 2009; Mlodinow, 2009; Coates et al., 2010; Calvin, 2011; Sobehart, 2012).

The elegant framework of rational behavior that underlies neoclassical economics cannot reconcile itself to actual human behavior. The problem with this framework is that it does not reflect how people behave in the real world. Our actions are often far from purely rational. Evolution did not make us closer to the idealized rational agents of economic theory, but, to the contrary, allocated significant parts of our brains to processing, interpreting, and censoring information that does not conform to our views, and to dealing with emotions and with other people, which strongly influence our judgment and actions (Tversky and Kahneman, 1974; Dorner, 1996; Gray, 2007; Gigerenzer, 2000; Czerner, 2010; Buonomano, 2011; Kahneman, 2011; Ariely, 2013; Sobehart, 2014). This in turn helped us make sense of the world around us and deal with events even in situations with limited, uncertain, or contradictory information. This also makes us prone to be influenced by social pressure, misinformation, deception, and manipulation.

People have different ways of forming opinions and making decisions, given the contextual information available to them and the way in which they make inferences about the plausibility of different outcomes. In an ideal rational world, objectivity would demand that judgment about the likelihood of occurrence of events needed to be based on all the available evidence, not just some arbitrary subset of it. Any such a choice would amount either to ignoring information that is available or presuming information that is not. In practice, our minds constantly filter information that does not conform to our perception of the world, or it is influenced by others' views, sometimes leading to biased inference and erroneous or unsupported conclusions, even when they may seem perfectly logical to us.

Behavioral studies have found that when a person is confronted with an uncertain situation when making a decision, the individual usually does not necessarily evaluate all the information or compute probabilities for the different outcomes carefully (Dorner, 1996; Pinker, 1997; Gigerenzer, 2000; Evans and Feeney, 2004; Ariely, 2008; Kahneman, 2011). Instead, the decision often relies on emotions and inferences (or heuristics). Furthermore, these studies showed that the pain of a loss seems about twice as potent as the pleasure of generating a gain. This effect is known as *loss aversion*. Everyone who experiences emotions is vulnerable to the loss aversion effect, which is part of a larger psychological phenomenon known as negativity bias (for most people, their emotional response to a *bad* outcome is stronger than to a *good* outcome).

In most day-to-day decisions, people use primarily loose associations, analogies, heuristics, and *plausible reasoning*, which may lead to erroneous conclusions based on preconceptions, biased contextual information, social pressure, and manipulation. Behavioral studies have also revealed that human emotions are critical in the decision-making process. Loose associations, emotions and feelings help us navigate in an uncertain environment and plan our actions accordingly without having to analyze each action rationally. In some pathological situations, when people are cut off from their feelings due to illness or injury, even the simplest decisions can be a struggle. A brain that cannot feel or connect with emotions cannot make up its mind correctly. The fact that people cannot make everyday decisions without feelings and emotions contradicts the conventional view of humans as rational agents. Of course, the discussion above does not mean that emotions should dominate every aspect of decision making. There is a range of emotional responses and cognitive biases affecting risk taking and decision making, which need to be better understood.

People's behavior can also translate into negative publicity regarding corporate practices and actions that can cause the loss of customers, costly litigation, or revenue reduction, which could affect a company's reputation as a good lender, borrower, or counterparty in financial transactions. This can also result in liquidity constraints and significant depreciation in market capitalization.

For most institutions it takes years to acquire a sound reputation and recognition, but it could take a single adverse event driven by misperception, social pressure, manipulation, or plain incompetence to destroy a company's reputation completely and put the institution out of business. In a world of compulsive communications through scores of daily phone calls, emails, and social media communications, the threat to a company's reputation can be sometimes only one phone call, one keystroke, or a mouse-click away. This is particularly relevant in situations where social pressure is motivated by altruistic goals (such reducing the impact of climate change) driving people with good intentions but with limited competence and incorrect or misleading information, who may have few qualms about pursuing their goals since they find their actions justifiable and usually supported by peers or social groups. As a result, social pressure combined with incompetence and misinformation can be dangerous as good intentions may override the natural inhibitions from self-reflection, leading to poor decision making. This can impact economic, social or policy decisions.

PERCEPTION AND PLAUSIBILITY OF EVENTS

Here we discuss the plausibility assigned to events from the behavioral point of view, focusing primarily on perception and credibility under uncertainty or limited knowledge. This is relevant for gaining insight into how divergence of opinions can occur and for understanding the impact of social pressure and manipulation on credibility.

In the plausibility-interpretation of probability (often referred to as *subjective* or perceived probability), even though people reason in a consistent and rational way and are exposed to the same information on a given event, they can assign completely different probabilities to the same event based on differences in their background information, preconceptions, or behavioral processes for making inference. This does not mean that the event is actually more or less likely to occur. It only means that it is more or less likely to occur to people who may disagree on how to interpret the limited information available to them.

Let's assume that after a reputation crisis of "Any-company" on the impact of its carbon footprint, its senior officers issue the following statement about their commitment to low-carbon emission targets:

$$A = \{\text{“}Any - company\text{”}\ \textit{is fully committed to low carbon emissions}\} \quad (4.1)$$

Judgment about the likely truth or falsity of statement A above is generally based on the perceived probability $P(A|BCD\ldots)$ that A is true, conditional on the available evidence B, C, D, etc. For example, B = {*There is a new lawsuit against "Any-company" for deceptive practices*} and C = {*"Any-company" has previous lawsuits for questionable ESG practices*}, etc. In our example, the company officers may be telling the truth about their company's commitment (statement A) and not be believed by the public as result of the background evidence and information (B, C, D, etc.), even though customers and investors are reasoning in a consistent and rational way and get truthful statements from the company officers.

In a strictly technical sense, there is no such a thing as an unconditional probability that statement A in Equation (4.1) is true. All probabilities are conditional, at least, on the background and contextual information available to form opinions. Also, background information and preconceptions are always available to us to form opinions on the validity of any given statement, which impacts the credibility of the source and the plausibility assigned to the event.

More precisely, given statement A in Equation (4.1), some new information B and background information C, the probability P that A and B are both true, given C is:

$$P(AB|C) = P(A|BC)\,P(B|C) = P(B|AC)\,P(A|C) \tag{4.2}$$

From Equation (4.2) we obtain the probability that the statement A is true, given the additional evidence and the background information C:

$$P(A|BC) = \frac{P(B|AC)}{P(B|C)}\,P(A|C) \tag{4.3}$$

Equation (4.3) (Bayes rule) indicates that introducing the additional information B, in addition to the background information C, can make statement A more or less likely to occur to the observer depending on whether B supports or contradicts statement A.

Notice that when $P(A|BC)$ is very close to one, we may conclude that A is likely to be true, given B and C and form an opinion accordingly. In contrast, when $P(A|BC)$ is very close to zero, we may conclude that A is likely to be false and form a different opinion. When $P(A|BC)$ is not far from 1/2 (that is, both outcomes are equally likely), we need to get additional evidence because the available information is not sufficient to support the likely truth or falsity of statement A.

TRUST, DECEPTION, CREDIBILITY, AND FAKE NEWS

Trust and perceived deception play a key role in determining credibility and plausibility and, therefore, the convergence or divergence of opinions (Jaynes, 2003).

Following closely early analysis on the topic (Sobehart, 2014), let's assume that after several disappointing quarters "Any-company" releases a plan to regain its reputation. This is summarized in statement A in Equation (4.1). Now suppose that two people, "Mr. Belief" (b) and "Mr. Doubt" (d) have different views concerning the truth or falsity of statement A as result of their different contextual and background information: C_b and C_d. Also suppose that both of them have access to the same additional information B that supports statement A (e.g. a new third-party research report on "Any-company").

Let Mr. Belief be a believer in the validity of statement A, that is, $P(A|C_b) \approx 1$. Also let Mr. Doubt be a doubter, that is, $P(A|C_d) \approx 0$. Since information B supports statement A, and Mr. Belief considers statement A

almost certainly true, we have $P(B|AC_b) \approx P(B|C_b)$, and from Equation (4.3) we have $P(A|BC_b) \approx P(A|C_b)$. That is, the new information B does not have any significant effect on Mr. Belief's positive opinion. In contrast, Mr. Doubt (who also reasons in a consistent and rational way) recognizes that $P(B|AC_d) \geq P(B|C_d)$ and, therefore, $P(A|BC_d) \geq P(A|C_d)$. Thus, Mr. Doubt's opinion changes in the direction of Mr. Belief's positive view. In contrast, if the information B refuted statement A, Mr. Belief's opinion would move in the direction of Mr. Doubt's negative view.

It may seem logical that by providing Mr. Belief and Mr. Doubt with additional information B that supports statement A, one could expect their opinions on statement A to converge (e.g. by measuring the difference in probabilities: $|P(A|BC_b) - P(A|BC_d)| \ll |P(A|C_b) - P(A|C_d)|$. However, providing the same additional information to all parties does not necessarily tend to bring people's opinions closer. In fact, additional information may cause a divergence of opinions as often observed in discussion of social, economic, political, legal, or environmental issues.

If both Mr. Belief and Mr. Doubt behave rationally in evaluating the same information B and C, how can this be possible? In the example above, the convergence of opinion occurs only because both Mr. Belief and Mr. Doubt use the same rules of logic and perceived probability (e.g. Bayesian statistical inference), and because they agree that information B *supports* statement A in Equation (4.1). That is, they believe that information B is *credible*. But if for one of them the probability of being deceived is high, getting additional information from somebody he distrusts may have the opposite effect from what the claimant intended. That is, as more information is introduced into the discussion, the perception of being deceived increases and the polarization of opinions grows stronger. This can be exacerbated by social pressure, misinformation, and manipulation, as commonly seen when biased media and news narratives on climate change motivated by partisan politics, economics, or self-interest influence the perceived credibility of the source and veracity of the information. In many cases, poorly selected "experts" can misinterpret or misrepresent the credibility of available information due to their own biases and desire to conform to a particular view.

Let's assume that Mr. Doubt believes that he is being purposely deceived by the new information B. That is, Mr. Doubt's perceived deception is described by: $P(B|AC_d) \ll 1$ and $P(B|C_d) \ll 1$. Then, from Equation (4.3), we have $P(A|BC_d) \approx (P(B|AC_d)/P(B|C_d))\, P(A|C_d)$, which depends on the perceived credibility of statement B given by the ratio $P(B|AC_d)/P(B|C_d)$. Mr. Doubt's opinion can easily change in either direction depending on how he perceives being deceived by the additional information B.

Now suppose that Mr. Belief and Mr. Doubt do not agree on their judgment about the additional information B because they have different contextual and background information: C_b and C_d.

$$P(B|AC_b) = p_b \qquad P(B|AC_d) = p_d$$
$$P(B|\overline{A}C_b) = q_b \qquad P(B|\overline{A}C_d) = p_d \tag{4.4}$$

Here $\overline{A}$ (not A) denotes the complement of A in Equation (4.1). More precisely, $\overline{A}$= {*"Any-company" is not committed to low-carbon emissions*}.

Now let Mr. Belief and Mr. Doubt have different initial views on the soundness of "Any-company" (that is, the veracity of statement A in Equation (4.1)). Then,

$$P(A|C_b) = f_b \qquad P(A|C_d) = f_d \tag{4.5}$$

From Equations (4.3), (4.4), and (4.5), Mr. Belief's and Mr. Doubt's probabilities of the veracity of statement A conditional on the available information are:

$$P(A|BC_k) = \frac{f_k}{f_k + \left(\frac{q_k}{p_k}\right)(1 - f_k)} \quad k = b, d \tag{4.6}$$

From Equation (4.6) and the condition $P(\overline{A}|BC_k) = 1 - P(A|BC_k)$, we obtain:

$$\log(\textit{Odds that A is true}) = \log\left(\frac{P(A|BC_k)}{P(\overline{A}|BC_k)}\right)$$
$$= \log\left(\frac{F_k}{1 - F_k}\right) = \log\left(\frac{f_k}{1 - f_k}\right) + \log\left(\frac{p_k}{q_k}\right) \quad k = b, d \tag{4.7}$$

Here $F_k = P(A\ |BC_k)$ is the perceived probability that A is true given information B and C_k $(k = b, d)$.

Equation (4.7) indicates that information B causes the opinions of Mr. Belief and Mr. Doubt to change in the same or opposite direction driven by the ratio p_k/q_k in Equation (4.7). The probabilities (or odds) they assign to the veracity of statement A can differ greatly if their information priors f_b and f_d are very different.

From the examples above, the posterior probability depends on the likelihood that statement A is true, the prior information C and the credibility of the source for information B. Discrepancies in prior information may cause a difference in people's opinions, which could have a dramatic impact on credibility.

The examples above are important because, in general, people do not exactly use pure deductive reasoning and Bayesian inference in their day-to-day decisions and opinions. To the contrary, people often use plausible reasoning based on analogies, heuristics, inference, preconceptions, and background information, which may result in biases and misperception leading to more noticeable effects than the rational view examples discussed here.

Given the fast pace of information flowing to us daily, we are frequently presented with limited, edited, or censored versions of the facts, and rarely have access to irrefutable evidence to corroborate these facts. Furthermore, listening to or reading about events in the news, social media, documentaries or even from professional, academic, and official sources sometimes cannot convince us of their veracity after finding that some of these sources had willingly and systematically reported distorted facts and fabricated stories motivated by self-interest, ideology, or simply to boost their profits at the expense of factual accuracy or other people's credibility and reputation.

SOCIAL PRESSURE, CONFORMITY, AND MEDIA BIAS

The issues discussed in previous sections suggest that the analysis of people's perceptions and opinions could help companies determine effective responses to credibility and reputation crises when they occur.

The impact of climate risk and ESG issues on a company's reputation could be significant. People's opinions are influenced strongly by the social situation in which they occur and by social norms, such as examples set by peers, expectations of those in authority positions, the social identities of actors, and the emotions that cause people behave as they do (Gray, 2007; Ariely, 2013).

Here we discuss a model of social influence and its impact on people's actions and their opinions. Key to this approach is the concept of *social pressure*, which can be defined as the psychological forces exerted on each of us by others through their examples, judgments, demands, and expectations, whether real or perceived. We are most influenced by people who are psychologically or physically closest to us, although the introduction of social media has created new channels for influence. If peers signal explicitly or implicitly through their actions that certain behavior is acceptable and another is not, we may feel pressured to follow the acceptable behavior to conform with the group (*normative compliance*). If the acceptable behavior fills a gap in our information about the situation, we may accept the behavior as a natural consequence of improving our knowledge (*informational compliance*).

Social pressure arises from the ways we collect, interpret, and respond to the situations around us and our desire to conform to a social group, promoting social acceptance by others and creating predictability in social interactions. However, it can also lead to unquestioned conformity and compliance, herd mentality, and morally unacceptable behavior.

The simplest form of social pressure is the mere presence of other people who can observe us perform, changing the way we would behave when we are alone. This can lead to social facilitation, when our performance improves in the presence of others, and social interference (or social inhibition), when our performance declines when people are present.

The facilitation or inhibition response seems related to the complexity of the tasks to be performed, the performer's ability and confidence to complete the tasks, and the performer's concern about being evaluated. Social facilitation and interference increase when the audience has high status or expertise. Social pressure can worsen performance by creating distracting thoughts about being evaluated, the difficulty of the tasks, the consequences of failure and related issues. This can lead to freezing under pressure and the inability to cope with problems when the demands on the performer's working memory and ability to execute are overtaxed, leading to poor decision making.

Besides influencing people's ability to perform, social pressure influences behavior and choices of what to say or do in front of others as people strive to influence their thoughts or conform to the group. People behave differently when others are present than when they are alone, either consciously or unconsciously.

People's behavior is not only influenced by the presence of others but also through the examples they set and the desire to be accepted by them. Social influence that works through providing clues about situations or events, perceived as gaining information, is referred to as *informational* influence. Another reason for conforming to a group is to promote acceptance and closeness to others. Conformity allows people to act as a coordinated unit as opposed to separate individuals. Social influence that works through the desire to belong to a group or be approved of by others (by choice or due to fear of the consequences) is called *normative* influence. People tend to conform to others' judgment when the objective evidence against the judgment is ambiguous, but they are expected not to conform when the evidence against the judgment is clear-cut. However, behavioral studies demonstrated consistently that the tendency to conform with others can lead people to ignore objective evidence in favor of conformity. Compliance and conformity have more to do with a desire to be accepted than a desire to be right. However, dissenting views from the majority not only can reduce normative pressure to conform and provide alternative information on the situation but also can shake people out of the complacent view that the majority must be right.

In many situations, the social context consists not only of what we hear from others or see them do, but also the signs that inform us on what behaviors are acceptable and which ones are not. Social influence is more effective when it emphasizes that most people behave in the desired way and explicitly or implicitly portray the undesirable behavior as abnormal (Gray, 2007).

In the next section, we discuss how both informational and normative influence can affect people's individual views or collective/social views, leading to polarization of opinions, which can impact decision making.

SOCIAL POLARIZATION

When people share information or make decisions, they can influence one another through normative social pressure by simply expressing their opinion or by taking a position on the issue in front of another. This results in unstated social pressure to agree.

When the group's opinions are evenly divided on an issue, the result of sharing credible information can result in a compromise where each side partially convinces the other, so the group majority adopts a more moderate/centered view or consensus (convergence of opinions). However, group participation can make attitudes more extreme when the opinions are not evenly split or when there is a credibility issue with the information shared and people believe they are being deceived. When a significant part of the group adopts the same side of an issue, discussions can push the group toward a more extreme view than the original one, leading to social polarization.

Social polarization can be *informational* or *normative.* Informational polarization occurs when people reinforce arguments that favor their side and tend to withhold arguments that favor the other side, as often seen in political discussions, social media, and biased arguments on climate risk and ESG presented by news organizations motivated by economic or political self-interest. This confirmation bias results in people hearing a disproportionate number of arguments aligned to their own view, which may persuade them to lean further in that direction. Listening to others repeat one's own arguments can have a validation effect where people become more convinced of their own views.

In contrast, normative polarization arises from people's concerns about being accepted or approved by others in the group. Normative influence can cause the opinions within a group to become more similar across the group. However, as adoption of the group's view increases and people are more sensitive to the group's view, people can become more vigorous supporters of the position that that majority favors, pushing the group toward an increasingly more extreme position. Other explanations may also apply to describe this behavior.

Polarization can have important social consequences that can be exacerbated by misleading information from social media and biased new organizations. To the degree that one group acts based on sharing the best unbiased available information, their decisions as a group could be better than decisions made by the individuals working alone. However, to the degree that the group acts based on misinformation, selectively withholding information on the less-favored side, and participants attempt to please, impress, or deceive one another rather than to arrive at the best objective view, it is likely that group decisions will be worse than the decision that group members would have made alone. This type of *group-think* self-deception arises when group members are more concerned with upholding group cohesion, unanimity, and conformity than having a critical appraisal of approaches to solve a problem, often suppressing criticism, and favoring a selected view, as often seen in partisan discussion. When this misguided behavior occurs at high levels of business or government, it can lead to poor decisions and disastrous social, economic, or environmental policies.

At first blush, it could appear extremely difficult to be able to model people's opinions on complex issues such as climate change, which may be driven by unknown causes and idiosyncratic behavior. Here, however, we model the simplest case of social pressure on individual opinions on the veracity of statement A in Equation (4.1). Our model of opinions closely follows early work on cooperative phenomena in interacting social groups (Sobehart, 2014).

Intuitively, people's opinions seem to be driven by cooperative effects. That is, the formation of the opinion of an individual is often influenced by the presence of groups of people with the same or opposite opinion. More precisely, we assume that the rate of change of positive (+) opinions to negative (−) ones ($\gamma_{+-} = n_+ w_{+-}$) can increase or decrease due to the size of the group with the same positive opinion n_+ or negative opinion n_- as reflected in the probability of a change in opinion per unit time $w_{+-}(n_+, n_-)$. Similarly, the rate of change of negative opinions to positive ones ($\gamma_{-+} = n_- w_{-+}$) can also increase or decrease due to the size of the group with the same or opposite opinion as reflected in the probability of a change in opinion per unit time $w_{-+}(n_+, n_-)$.

We now turn to the technical aspect of social polarization. Using quasi-equilibrium assumptions (Jaynes, 1958; Haken, 1975; Gardiner, 1985) about transitions between positive and negative opinions, the transition probabilities per unit time $w_{+-}(n_+, n_-)$ and $w_{-+}(n_+, n_-)$ can be expressed as a combination of (a) a systemic alignment of all individuals to a global preference (for the population as a whole), and (b) an individual's alignment to a specific positive or negative population (cooperative effect). Technically, the transition rates can be expressed as follows:

$$w_{+-}(n_+, n_-) = \nu e^{-(\theta(n_+ - n_-)+\delta)/\sigma}$$
$$w_{-+}(n_+, n_-) = \nu e^{(\theta(n_+ - n_-)+\delta)/\sigma} \quad (4.8)$$

Here ν is the average frequency of changes in opinion, θ measures the strength of adaptation (conformity) to other people's opinions, δ is the systemic opinion preference ($\delta > 0$ indicates that a positive view is preferred on aggregate for the whole population, while $\delta < 0$ indicates that a negative view is preferred on aggregate), and σ measures the characteristic uncertainty of change in opinion (opinion volatility that reflects the dispersion of individual opinions). The systemic opinion preference δ in Equation (4.8) partially captures the credibility of the common information $\log(F/(1 - F))$ discussed in Equation (4.7).

Notice that, in principle, the transition rates in Equation (4.8) could have different parameters for positive and negative opinions, creating asymmetric responses to changes in the populations. The competing effects in Equation (4.8) reflect the individuals' struggle to align their own individual views to the opinion of a large group (systemic preference for the whole population) or conform to the opinion of the groups with positive and negative views.

OPINION POLARIZATION UNDER SOCIAL PRESSURE AND MEDIA BIAS

Let's assume there is a single group of people with different positive or negative opinions about the veracity of statement A in Equation (4.1).

The probabilities $P(n_+, n_-, t)$ of the number of people with positive opinion n_+ or negative opinion n_- are driven by the following equation:

$$\frac{\partial P}{\partial t} = \gamma_{+-}(n_+ + 1, n_- - 1)P(n_+ + 1, n_- - 1)$$
$$+\gamma_{-+}(n_+ - 1, n_- + 1)P(n_+ - 1, n_- + 1)$$
$$-[\gamma_{+-}(n_+ n_-) + \gamma_{-+}(n_+, n_-)]P(n_+, n_-) \quad (4.9)$$

Let's introduce the excess number of people with a positive opinion relative to those with a negative opinion: $x = (n_+ - n_-)/n$, where $n = (n_+ + n_-)$ is the total number of people with different views (assumed constant for simplicity). To make the problem analytically tractable, we expand Equation (4.9) in terms of the excess variable x. In the limit of small excess x, Equation (4.9) can be approximated to second order by the following diffusion equation:

$$\frac{\partial P}{\partial t} = \frac{\partial}{\partial x}\left(\frac{2}{n}(\gamma_{+-} - \gamma_{-+})P\right) + \frac{\partial^2}{\partial x^2}\left(\frac{4}{n^2}(\gamma_{+-} + \gamma_{-+})P\right) \tag{4.10}$$

The stationary solution to Equation (4.10), when the positive and negative populations reach their equilibrium, has the form:

$$P(x) = \left(\frac{nC_0}{\gamma_{+-} + \gamma_{-+}}\right) exp\left[-n\int_{-1}^{x}\left(\frac{\gamma_{+-} - \gamma_{-+}}{\gamma_{+-} + \gamma_{-+}}\right)ds\right] \tag{4.11}$$

Here C_0 is a normalization constant, and

$$\begin{aligned} \gamma_{+-} - \gamma_{-+} &= -\nu n[\sinh(z) - x\cosh(z)] \\ \gamma_{+-} + \gamma_{-+} &= \nu n[\cosh(z) - x\sinh(z)] \end{aligned} \tag{4.12}$$

and

$$z = \alpha x + \beta, \alpha = n\theta/\sigma, \beta = \delta/\sigma \tag{4.13}$$

Equation (4.11) has three distinct regimes: (a) a bell-shape distribution of positive and negative views; (b) a transition regime where there is significant dispersion of views; and (c) a polarized regime with two distinct populations for positive and negative views.

Figure 4.1 illustrates the three regimes described above for the positive and negative populations when there are no systemic opinion preferences ($\beta = 0$). Figure 4.1(a) shows the distribution of opinions when there are frequent changes of opinion because uncertainty is relatively high compared to the strength of the coupling to other people's opinions ($\alpha = 0$). In contrast, if the uncertainty of opinion changes σ declines significantly, or if the coupling strength between individuals θ increases significantly, two very distinct groups of opinions develop as shown in Figure 4.1(c) ($\alpha = 1.3$). These two different segments can be identified as a *polarization of opinions*, with significant consequences for credibility, the assessment of event plausibility, and subsequent decision making. In this limit, individuals align to adopt to others' opinions creating stubborn bias toward positive or negative views ("polarized views"). Figure 4.1(b) shows an intermediate state ($\alpha = 1$) where the coupling strength between individuals reaches a critical transition level, leading to a wide range of outcomes covering people who remain undecided and those who have different degrees of positive or negative views as they are affected by social pressure to conform to a particular view.

A wide range of distributions of opinions can be generated using different combinations of parameters α and β (that is, the relative contributions of systemic views and cooperative effects), ranging from regimes where individual opinions are dominant to regimes with highly polarized views.

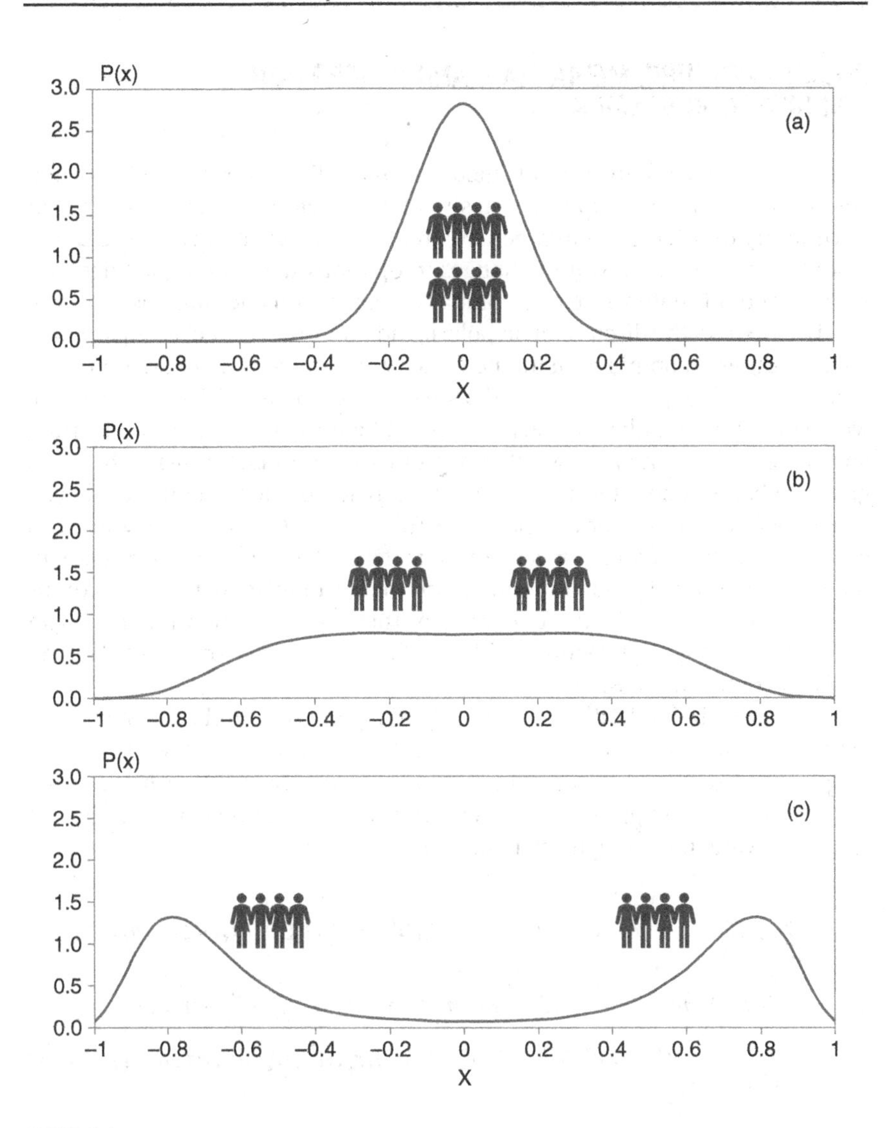

FIGURE 4.1 (a) Population distribution for frequent changes of opinion ($\alpha = 0$ and $\beta = 0$); (b) transition regime between frequent changes and polarized views ($\alpha = 1$ and $\beta = 0$); and (c) polarization of opinions ($\alpha = 1.3$ and $\beta = 0$).

SOCIAL PRESSURE, MEDIA BIAS, AND PERCEPTION FOR DIFFERENT GROUPS

Although this is a relatively crude model of credibility, cooperative effects, and conformity, it allows us to describe situations where changes in the level of uncertainty or interaction between different groups, social pressure, and perceived deception can result in changes in opinions that can impact decision making and ultimately impact a company's reputation and franchise.

Let's extend the discussion on polarization to three distinct groups of people with different opinions about the veracity of statement A in Equation (4.1): doubters (D), independents (I), and believers (B). For simplicity of exposition, we assume that members of each group can be influenced by social pressure and perceived deception from different groups, but they do not switch between groups. That is, a doubter from group D can react to information provided by believers from group B or independents from group I, and become more (+) or less (−) convinced of the veracity of statement A but the individual will remain a member of group D. Similar arguments apply to members of groups B and I. The analysis of the dynamics of these groups when members can switch between groups requires additional mathematical complexity beyond the scope of this discussion.

The probability $P(n_+^D, n_-^D, n_+^I, n_-^I, n_+^B, n_-^B, t)$ of the number of people in group D with positive opinion n_+^D or negative opinion n_-^D, the number of people in group I with positive opinion n_+^I or negative opinion n_-^I, and the number of people in group B with positive opinion n_+^B or negative opinion n_-^B is driven by the following master equation:

$$\begin{aligned}\frac{\partial P}{\partial t} = &\sum_{\{ijk\}} \gamma_{\{ijk\}}(n_+^D + m_i, n_-^D - m_i, n_+^I + m_j, n_-^I - m_j, n_+^B + m_k, n_-^B - m_k, t)\\ &\times P(n_+^D + m_i, n_-^D - m_i, n_+^I + m_j, n_-^I - m_j, n_+^B + m_k, n_-^B - m_k, t)\\ &-\sum_{\{ijk\}} \gamma_{\{ijk\}}(n_+^D, n_-^D, n_+^I, n_-^I, n_+^B, n_-^B, t) \times P(n_+^D, n_-^D, n_+^I, n_-^I, n_+^B, n_-^B, t) \quad (4.14)\end{aligned}$$

Here $\{ijk\}$ describes symbolically all combinations of positive (+) and negative (−) transitions for groups D, I, and B, and m_i, m_j, m_k represent the number of people in each group switching between positive and negative opinions driven by the transition rates $\gamma_{\{ijk\}}$.

Solving Equation (4.14) is beyond the scope of our discussion. To gain insight into the impact of social influence and polarization, we focus on a simplified version of Equation (4.14) when the D, I, and B groups are driven primarily by common information available to all groups (with differences in the credibility of the source for different groups) and when social pressure on individuals results from the interaction within each group separately ($m_i, m_j, m_k = \pm 1$). More precisely, the transition rates between positive and negative views for each individual group depend only on the common information and the interaction within each group.

Under this simplifying assumption, Equation (4.14) can be separated into three independent equations where the probability $P_D(n_+^D, n_-^D, t)$ of the number of doubters in group D with positive opinion n_+^D or negative opinion n_-^D is driven by the following equation:

$$\begin{aligned}\frac{\partial P_D}{\partial t} &= \gamma_{+-}^D(n_+^D+1, n_-^D-1)P_D(n_+^D+1, n_-^D-1)\\ &+ \gamma_{+-}^D(n_+^D-1, n_-^D+1)P_D(n_+^D-1, n_-^D+1)\\ &- [\gamma_{+-}^D(n_+^D n_-^D) + \gamma_{+-}^D(n_+^D, n_-^D)]P_D(n_+^D, n_-^D)\end{aligned} \tag{4.15}$$

The probability $P_I(n_+^I, n_-^I, t)$ of the number of independents in group I with positive opinion n_+^I or negative opinion n_-^I is driven by the following equation:

$$\begin{aligned}\frac{\partial P_I}{\partial t} &= \gamma_{+-}^I(n_+^I+1, n_-^I-1)P_I(n_+^I+1, n_-^I-1)\\ &+ \gamma_{-+}(n_+^I-1, n_-^I+1)P_I(n_+^I-1, n_-^I+1)\\ &- [\gamma_{+-}(n_+^I n_-^I) + \gamma_{-+}(n_+^I, n_-^I)]P_I(n_+^I, n_-^I)\end{aligned} \tag{4.16}$$

The probability $P_B(n_+^B, n_-^B, t)$ of the number of believers in group B with positive opinion n_+^B or negative opinion n_-^B is driven by the following equation:

$$\begin{aligned}\frac{\partial P_B}{\partial t} &= \gamma_{+-}^B(n_+^B+1, n_-^B-1)P_B(n_+^B+1, n_-^B-1)\\ &+ \gamma_{-+}^B(n_+^B-1, n_-^B+1)P_B(n_+^B-1, n_-^B+1)\\ &- [\gamma_{+-}^B(n_+^B n_-^B) + \gamma_{-+}^B(n_+^B, n_-^B)]P_B(n_+^B, n_-^B)\end{aligned} \tag{4.17}$$

Let's introduce the excess number of people with a positive opinion relative to those with a negative opinion in each group:

$$\begin{aligned}x_D &= (n_+^D - n_-^D)/n_D\\ x_I &= (n_+^I - n_-^I)/n_I\\ x_B &= (n_+^B - n_-^B)/n_B\end{aligned} \tag{4.18}$$

Here

$$n_D = n_+^D + n_-^D$$
$$n_I = n_+^I + n_-^I$$
$$n_B = n_+^B + n_-^B \tag{4.19}$$

Here n_D, n_I and n_B represent the total number of people in each group (assumed constant for simplicity), with different beliefs in the common information and different degrees of conformity and social interaction.

Using the same approach as in Equation (4.9) for a single group, we expand Equations (4.15)–(4.17) in terms of the excess variables x_D, x_I, x_B. In the limit of small excess x, Equations (4.15)–(4.17) can be approximated to second order by the diffusion Equation (4.10) for the three groups.

The stationary solution to Equation (4.10) for each individual group D, I, and B, when the positive and negative populations reach their equilibrium, is determined by Equations (4.11) and (4.12), with the group-specific parameters:

$$P_k(x) = \left(\frac{n_k C_k}{\gamma_{+-}^k + \gamma_{-+}^k}\right) exp\left[-n_k \int_{-1}^{x} \left(\frac{\gamma_{+-}^k - \gamma_{-+}^k}{\gamma_{+-}^k + \gamma_{-+}^k}\right) ds\right] \quad k = D, I, B \tag{4.20}$$

Here C_k is a normalization constant, and

$$z_k = \alpha_k x_k + \beta_k, \alpha_k = n_k \theta_k / \sigma_k, \beta_k = \delta_k / \sigma_k \quad k = D, I, B \tag{4.21}$$

Equations (4.20) and (4.21) also have three distinct regimes for the interaction between groups, reflecting convergence and divergence of opinions: (a) a bell-shape distribution centered around zero, reflecting a balance between positive and negative views; (b) a transition regime where there is significant dispersion of views; and (c) a polarized regime with distinct populations for positive and negative views. Note, however, that each population ($k = D, I, B$) will show a different shape driven by the credibility of the systemic opinion preference δ_k, which partially captures common information, and θ_k, which measures the strength of adaptation to other people's opinions in the group, that is, the degree of group conformity. These drivers are reflected in the normalized parameters α_k and β_k.

Figure 4.2 illustrates the issues described above for the positive and negative populations when each group assigns a different credibility to the systemic opinion β_k, where group D adopts a doubter view ($\beta_D = -0.1$), group I remains

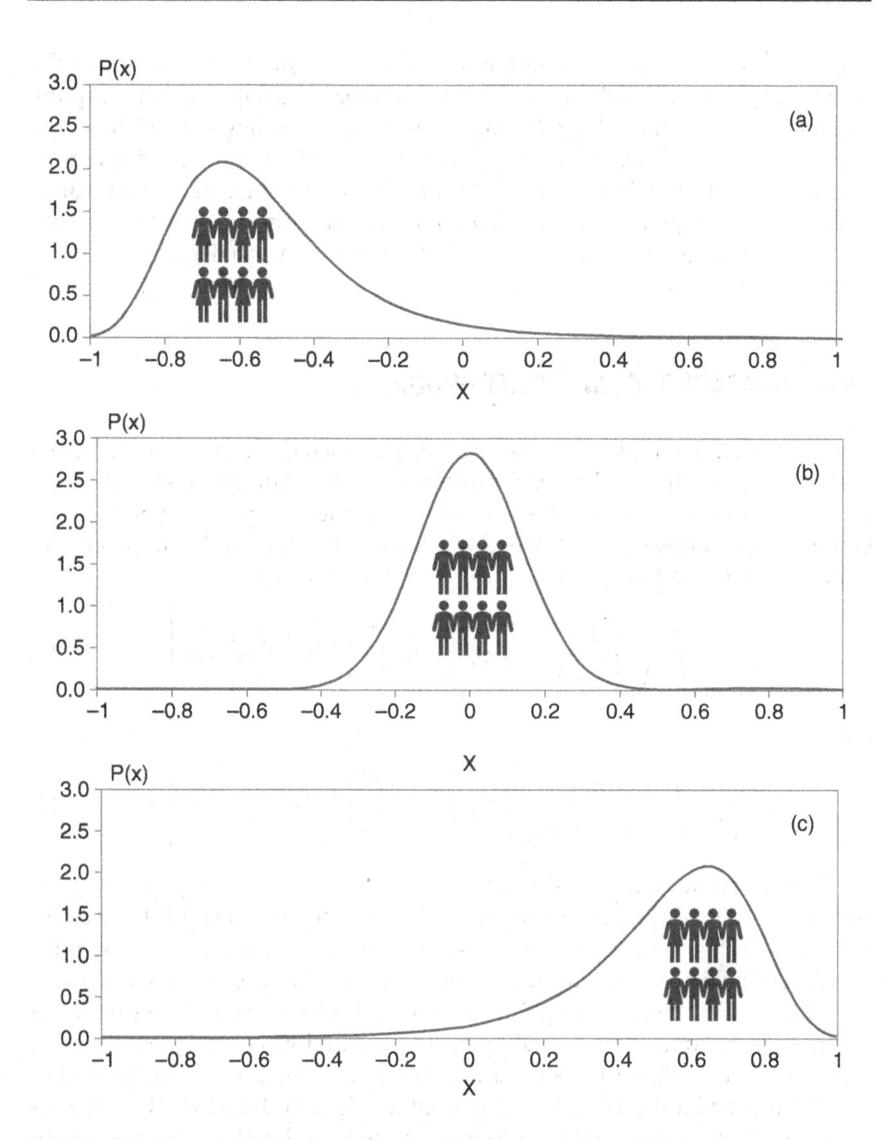

FIGURE 4.2 (a) group D (doubters) – polarized view with negative bias (α_D=**1** and β_D= – **0.1**); (b) group I (independents) – neutral view with no systemic bias (α_I = **0** and β_I=**0**); and (c) group B (believers) – polarized views with positive bias (α_B = **1** and β_B = **0.1**).

neutral ($\beta_I = 0$), and group B adopts a believer view ($\beta_B = 0.1$). Assuming the uncertainty of opinion changes σ is the same for all groups but the coupling strength between individuals (group conformity) for groups D and B is high and the same ($\alpha_D = \alpha_B = 1$) while group I is mainly composed of independent individuals with low social conformity ($\alpha_I = 0$). Notice that a wide range of regimes can be generated using different combination of parameters α_k and β_k describing different beliefs in common information and degrees of conformity in each group.

CHOICE IMPACT FOR DIFFERENT GROUPS

Equations (4.20) and (4.21) can be used to understand the impact of information and conformity on choices for different groups. More precisely, the total number of people N_+ with a positive or neutral view is the cumulative population across groups D, I, and B for $x \geq 0$, while the total number of people N_- with a negative view is the cumulative population for $x < 0$.

$$N_+ = \sum_{k=D,I,B} \int_0^1 \left(\frac{n_k C_k}{\gamma_{+-}^k + \gamma_{-+}^k} \right) exp \left[-n_k \int_{-1}^x \left(\frac{\gamma_{+-}^k - \gamma_{-+}^k}{\gamma_{+-}^k + \gamma_{-+}^k} \right) ds \right] dx \tag{4.22}$$

and

$$N_- = \sum_{k=D,I,B} \int_{-1}^0 \left(\frac{n_k C_k}{\gamma_{+-}^k + \gamma_{-+}^k} \right) exp \left[-n_k \int_{-1}^x \left(\frac{\gamma_{+-}^k - \gamma_{-+}^k}{\gamma_{+-}^k + \gamma_{-+}^k} \right) ds \right] dx \tag{4.23}$$

The difference between the positive population N_+ and negative population N_- is driven by different degrees of credibility and social influence for different groups, which can determine the ultimate outcome of economic, social, political, or reputational choices. Table 4.1 illustrates changes in the positive and negative populations for positive values of the strength of conformity α_k and different values of the credibility β_k of statement A in Equation (4.1). Table 4.1 assumes that the total population is composed of 50% of independents (group I), 25% of doubters (group D) and 25% of believers (group B). Table 4.1 includes the following cases: (a) baseline case for groups D, I and B, (b) groups D, I and B agree on the credibility of information supporting statement A, (c) groups I and B agree on the credibility of the information but group D assigns low credibility to it, (d) group D assigns no credibility to the information, and (e) group D assigns negative credibility to

TABLE 4.1 Changes in the positive and negative populations for positive values of the strength of conformity α_k and different credibility β_k of information supporting statement A in Equation (4.1)

	Alpha			Beta			Choices	
	L	C	R	C	L	R	N_- (%)	N_+ (%)
A	0	0	0	0	0	0	50	50
B	0	0	0	0.05	0.05	0.05	37	63
	0.5	0	0.5	0.05	0.05	0.05	35	65
	0.75	0	0.75	0.05	0.05	0.05	32	68
	1	0	1	0.05	0.05	0.05	28	72
	1.25	0	1.25	0.05	0.05	0.05	23	77
	1.5	0	1.5	0.05	0.05	0.05	22	78
C	0	0	0	0.025	0.05	0.05	38	62
	0.5	0	0.5	0.025	0.05	0.05	37	63
	0.75	0	0.75	0.025	0.05	0.05	35	65
	1	0	1	0.025	0.05	0.05	31	69
	1.25	0	1.25	0.025	0.05	0.05	26	74
	1.5	0	1.5	0.025	0.05	0.05	24	76
D	0	0	0	0	0.05	0.05	40	60
	0.5	0	0.5	0	0.05	0.05	39	61
	0.75	0	0.75	0	0.05	0.05	38	62
	1	0	1	0	0.05	0.05	36	64
	1.25	0	1.25	0	0.05	0.05	34	66
	1.5	0	1.5	0	0.05	0.05	33	67
E	0	0	0	−0.05	0.05	0.05	43	57
	0.5	0	0.5	−0.05	0.05	0.05	44	56
	0.75	0	0.75	−0.05	0.05	0.05	44	56
	1	0	1	−0.05	0.05	0.05	44	56
	1.25	0	1.25	−0.05	0.05	0.05	44	56
	1.5	**0**	**1.5**	**−0.05**	**0.05**	**0.05**	**44**	**56**

the information under the presumption it is deceptive. Notice the impact of the strength of conformity to adopt other people's view, which changes the size of the population with positive and negative opinions. Table 4.2 shows similar results for the changes in the positive and negative populations for negative values of the strength of conformity α_k and different credibility β_k of information supporting statement A in Equation (4.1).

TABLE 4.2 Changes in the positive and negative populations for negative values of the strength of conformity α_k and different credibility β_k of information supporting statement A in Equation (4.1)

	Alpha			Beta			Choices	
	L	C	R	C	L	R	N_- (%)	N_+ (%)
A	0	0	0	0	0	0	50	50
B	0	0	0	0.05	0.05	0.05	37	63
	−0.5	0	−0.5	0.05	0.05	0.05	37	63
	−0.75	0	−0.75	0.05	0.05	0.05	38	62
	−1	0	−1	0.05	0.05	0.05	38	62
	−1.25	0	−1.25	0.05	0.05	0.05	38	62
	−1.5	0	−1.5	0.05	0.05	0.05	38	62
C	0	0	0	0.025	0.05	0.05	38	62
	−0.5	0	−0.5	0.025	0.05	0.05	39	61
	−0.75	0	−0.75	0.025	0.05	0.05	39	61
	−1	0	−1	0.025	0.05	0.05	39	61
	−1.25	0	−1.25	0.025	0.05	0.05	39	61
	−1.5	0	−1.5	0.025	0.05	0.05	39	61
D	0	0	0	0	0.05	0.05	40	60
	−0.5	0	−0.5	0	0.05	0.05	40	60
	−0.75	0	−0.75	0	0.05	0.05	40	60
	−1	0	−1	0	0.05	0.05	40	60
	−1.25	0	−1.25	0	0.05	0.05	40	60
	−1.5	0	−1.5	0	0.05	0.05	40	60
E	0	0	0	−0.05	0.05	0.05	43	57
	−0.5	0	−0.5	−0.05	0.05	0.05	43	57
	−0.75	0	−0.75	−0.05	0.05	0.05	43	57
	−1	0	−1	−0.05	0.05	0.05	43	57
	−1.25	0	−1.25	−0.05	0.05	0.05	43	57
	−1.5	**0**	**−1.5**	**−0.05**	**0.05**	**0.05**	**42**	**58**

WHAT'S NEXT? TACKLING THE CLIMATE CHANGE CHALLENGE

The world population is estimated to have exceeded eight billion people. It took roughly 300,000 years for the human population to reach one billion people but only took about 220 years more to reach its current 8 billion level. The world's population grows at exponential pace and requires more food, water, energy, and other resources. This growth has an impact on the environment as societies struggle to grow and increase the standard of living of their citizens. Both the population growth and economic growth to provide prosperity are driven by an ever-increasing need for energy. Although there are multiple alternative sources of energy with different degrees of reliability and sustainability, the major sources of energy to support the current growth are fossil fuels (oil, gas, and coal), whose byproducts of fuel burning can accelerate natural climate change processes as indicated by the persistent rise in atmospheric CO_2 and other greenhouse gases over the last 170 years. The increase in the emission of greenhouse gases could contribute to climate changes beyond the natural climate variability, which in turn can affect society and its prosperity globally.

In response to this dilemma, politicians, statesmen, and policymakers around the world have been debating and introducing policies, incentives, and penalties to reduce carbon emissions, to develop alternative sources of energy, and to limit the use of fossil fuels, primarily in developed economies. The shift away from fossil fuels in pursuit of renewable and alternative energy sources will have a significant impact on the energy sector, primarily on coal mining and oil and gas extraction, and oil-, gas-, and coal-fired power generation. The proposed reduction of emissions will require a complete transformation of how energy is produced, transported, and consumed, affecting all sectors of the economy. Furthermore, these policies can impact a wide range of activities by restricting the use of internal-combustion engine vehicles while promoting the use of electric vehicles (EV) for public transportation and private use; by restricting the use of fossil fuels as sources of energy for transportation, heating, or cooking; or by promoting low-carbon emission construction for housing projects. The use of these policies, incentives, and penalties to drive a large-scale transition to a low-carbon economy can result in unintended detrimental consequences for society due to limited understanding of climate change and its drivers, or due to politically or economically motivated agendas disguised as environmental concerns. This can be exacerbated due to lack of competence to address these complex issues, disrupting economic activity and society's wellbeing without appropriately mitigating the effects of climate change. The public debate on these issues has resulted in highly polarized views on climate change and its effects, which are strongly influenced by information, misinformation, and social interaction.

The discussion in the previous sections focused on social pressure, misinformation, conformity, and credibility using a behavioral approach that quantifies people's trust. People have a variety of degrees of trust in the information they receive and how they react to social influence, which affects the probability they assign to uncertain events and how they perceive credibility and reputation. This is important in the context of climate risk and ESG since people are frequently presented with edited, censored, or distorted versions of the facts, and even fabricated stories motivated by self-interest, ideology, or economics, and rarely have access to sound evidence that can corroborate these facts.

The social pressure and misinformation effects that impact credibility and trust affect consumers, investors, financial regulators, policymakers and the public. Companies, regulatory and policy-making institutions should be aware of the business and reputation risks due to social pressure and misinformation for which strategic proactive action can be taken to avoid costly crises. For most institutions it takes years to build a solid business and acquire a good reputation and recognition, but it could take a single adverse event to destroy an institution's reputation and franchise value completely and put the institution out of business.

REFERENCES

Ariely, D. (2008). *Predictably Irrational.* HarperCollins, New York.

Ariely, D. (2013). *The (Honest) Truth about Dishonesty.* Harper Perennial, New York.

Bernstein, P.L. (1992). *Capital Ideas: The Improbable Origins of Modern Wall Street.* The Free Press, New York, pp. 17–38, 149–180.

Bernstein, P.L. (1998). *Against the Gods: The Remarkable Story of Risk.* John Wiley & Sons, New York, pp. 173–214.

Buonomano, D. (2011). *Brain Bugs: How the Brain's Flaws Shape Our Lives.* Norton Company, New York.

Calvin, W.H. (2011). *How Brains Work: Evolving Intelligence, Then and Now.* Basic Books, New York.

Coates, J.M., Gurnell, M., and Sarnyai, Z. (2010). From Molecule to Market: Steroid Hormones and Financial Risk-Taking. *Philosophical Transactions of the Royal Society of London, B* 365: 331–343.

Coates, J.M., and Herbert, J. (2008). Endogenous Steroids and Financial Risk Taking on a London Trading Floor. *Proceedings of the National Academy of Science, U.S.A.* 105(16): 6167–6172.

Czerner, T.B. (2010). *What Makes You Tick? The Brain in Plain English,* John Wiley and Sons, New York.

Damasio, A.R. (1994). *Descartes' Error: Emotion, Reason, and the Human Brain,* Grosset-Putnam Books, New York.

Dorner, D. (1996). *The Logic of Failure: Recognizing and Avoiding Error in Complex Situations.* Perseus Books, Cambridge, MA, pp. 37–102.

Elan, S.L. (2010). Behavioral Patterns and Pitfalls of US Investors. *Investor Behavior,* Federal Research Division, Library of Congress, Washington, DC.

Evans, J.B.T. and Feeney, A. (2004). The Role of Prior Belief in Reasoning. In: *The Nature of Reasoning* (Eds.) J.P. Leighton and R.J. Sternberg Cambridge University Press, Cambridge, pp. 78–102.

Fama, E.F. (1965). The Behavior of Stock-Market Prices. *Journal of Business* 38(1): 34–105.

Fama, E.F. (1970). Efficient Capital Markets: A Review of Theory and Empirical Work. *Journal of Financial Economics* 25(2): 383–417.

Fama, E.F. (1998). Market Efficiency, Long-Term Returns and Behavioral Finance. *Journal of Financial Economics* 49: 283–304.
Gardiner, C.W. (1985). *Handbook of Stochastic Methods*. Springer-Verlag, Berlin.
Gigerenzer, G. (2000). *Adaptive Thinking*. Oxford University Press, Oxford, pp. 93–266.
Gray, P. (2007). *Psychology*. Worth Publishers, New York, pp. 501–514.
Haken, H. (1975). Cooperative Phenomena. *Review of Modern Physics* 47(1): 94.
Hirshleifer, D. (2001). Investor Psychology and Asset Pricing. Ohio State University, Working Paper.
Jaynes, E.T. (1958). Probability Theory in Science and Engineering. *Colloquium Lectures in Pure and Applied Science*, No. 4, Field Research Laboratory, Socony Mobil Oil Company Inc, Dallas, Texas.
Jaynes, E.T. (2003). *Probability Theory: The Logic of Science*. Cambridge University Press, Cambridge, pp. 122–133.
Kahneman, D. (2011). *Thinking Fast and Slow*. Farrar, Strauss and Giroux, New York, pp. 19–146.
Lerner, J. (2009). *How We Decide*. Houghton Mifflin Harcourt, Boston.
Malkiel, B.G. (2003). The Efficient Market Hypothesis and Its Critics. Princeton University, CEPS Working Paper No. 91.
Mlodinow, L. (2009). *Subliminal: How Your Unconscious Minds Rules Your Behaviour*. Pantheon Books, New York.
O'Hara, M. (1998). *Market Microstructure Theory*. Blackwell Publishers, Oxford.
Pinker, S. (1997). *How the Mind Works*. W.W. Norton & Company, New York, pp. 3–148
Ruddiman, W. (2005). *Plows, Plagues and Petroleum*. Princeton University Press, Princeton, NJ.
Shiller, R.J. (2000). *Irrational Exuberance*. Princeton University Press, Princeton, NJ, pp. 135–168.
Shleifer, A. (2000). *Inefficient Markets: An Introduction to Behavioral Finance*. Oxford University Press, Oxford, pp. 1–52, 112–174.
Sobehart, J.R. (2003). A Mathematical Model of Irrational Exuberance and Market Gloom. *GARP Risk Review* (July/August): 22–24.
Sobehart, J.R. (2012). Market Reaction to Price Changes and Fat Tailed Returns. *Risk* (June): 76–81.
Sobehart, J.R. (2014). Rumor Has It: Modeling Credibility, Reputation and Franchise Risk. Special issue on Behavioral Finance. *Journal of Risk Management in Financial Institutions* 7(2): 161–173.
Sobehart, J.R. and Farengo, R. (2003). A Dynamical Model of Market Under- and Overreaction. *Journal of Risk* 4(5): 91–114.
Sobehart, J.R., and Keenan, S.C. (2005). Capital Structure Arbitrage and Market Timing Under Uncertainty and Trading Noise. *Journal of Credit Risk* 1(4): 1–29.
Thaler, R.H. (1999). The End of Behavioral Finance. *Financial Analyst Journal* 55(6): 11–17.
Tversky, A., and Kahneman, D. (1974). Judgment Under Uncertainty: Heuristics and Biases. *Science* 185(4157): 124–131.

CHAPTER 5

Global Shift in Response to Climate Change

THE SHIFT IN THE GLOBAL ECONOMY IN RESPONSE TO CLIMATE CHANGE

Although multiple sources of energy contribute to the generation of electricity globally, most of the generated power comes from burning *fossil fuels* (coal, crude oil, and petroleum products). Fossil fuels are composed of hydrocarbons formed from the remains of animals and plants (diatoms) that lived millions of years ago. The remains of these animals and plants were covered by layers of silt, sand, and rock. Heat and pressure from these layers turned the biological remains into crude oil and petroleum. The energy created from fossil fuel combustion can be used to produce steam, which drives the steam turbines of electrical generators. The electricity generated by the turbines can be transferred across vast distances to satisfy energy demand. Burning fossil fuels releases carbon dioxide (CO_2) and other greenhouse gases into the atmosphere, including ozone, sulfur dioxide (SO_2), nitrogen dioxide (NO_2), and aerosols (particulate matter). The estimated CO_2 emissions from the world's electrical power industry is about 10 billion tons annually, which contributes to greenhouse effects.

Figure 5.1(a) shows the contributions of different sources of energy to the world's electric power generation in 2000, 2010, and 2022, covering coal, gas, oil, nuclear, hydropower, wind, solar, geothermal, biomass, and other sources. Figure 5.1(b) show the same contributions in the United States. The world's utility-scale electricity production is about 29,000 terawatt-hour (TWh), of which about 4,200 TWh are produced in the United States. While coal and fossil fuels are still dominant for energy generation in developed and developing countries, in the United States natural gas was the largest source (about 40%) for electricity generation in 2022.[1] Natural gas is used in steam turbines and gas turbines to generate electricity. Renewable sources were the second-largest energy source (over 21%; 6% from hydropower, 10% from wind power, 3.4% from solar power, 1.3% from biomass, and the rest from geothermal and other sources). Renewables, including solar, wind,

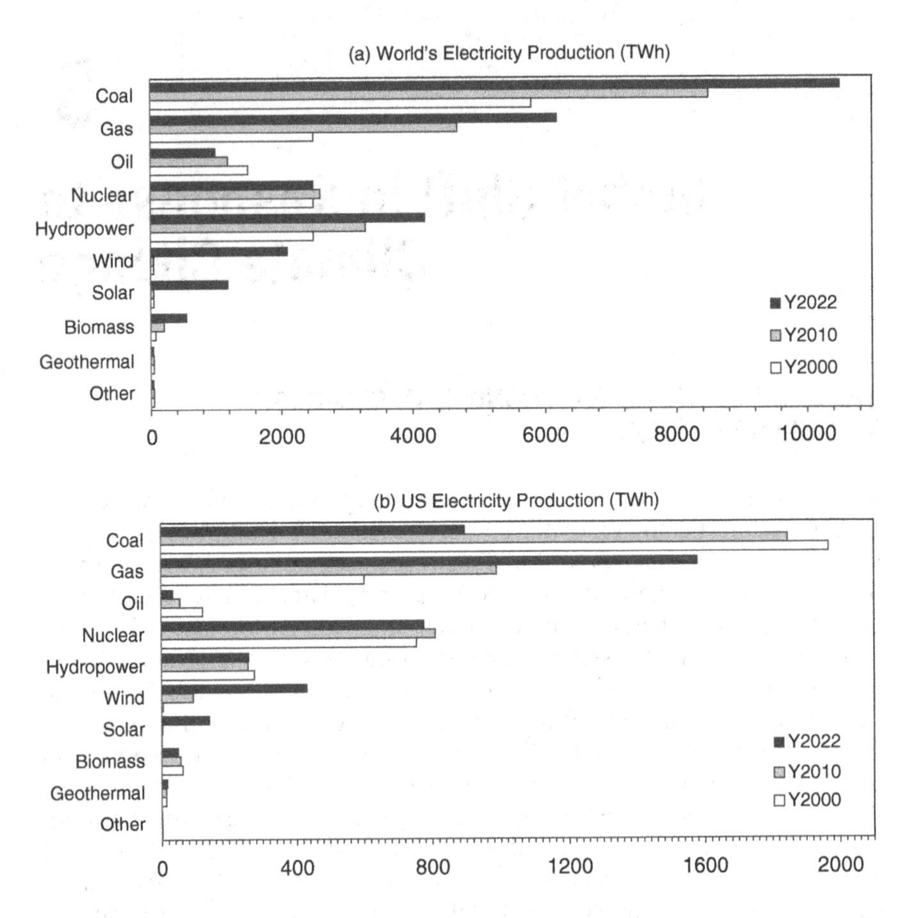

FIGURE 5.1 (a) Contributions of different sources of energy to the world's electric power generation: coal, gas, oil, nuclear, hydropower, wind, solar, biomass, geothermal, and other sources; (b) same contributions in the United States.

hydropower, biofuels, and others, are at the center of the transition to less carbon-intensive energy sources. Coal was the third-largest energy source (19%), followed closely by nuclear (18%). Nearly all coal-fired power plants in the US use steam turbines, with a few coal-fired power plants that convert coal to a gas for use in gas turbines to generate electricity.

The increased attention to climate change and the search for alternative sources of energy that can replace the use of fossil fuels have shifted the focus to the development of renewable sources of energy and the revival of nuclear energy, which includes enhancing conventional nuclear plants and developing next-generation reactors.

Energy can be stored in different forms and can be converted to other forms of energy (gravitational, mechanical, heat) or released through different types of reactions (chemical, electrochemical, nuclear). For example, the typical energy density for electrochemical reactions in batteries is about 0.2–1.5 megajoules/kg (MJ/kg), while the typical energy density for chemical reactions from burning fossil fuels, gas, and biofuels is about 20–50 MJ/kg. In contrast, the energy density that can be released in nuclear reactions is about 30,000–500,000 MJ/kg. Because nuclear reactions produce large amounts of energy for relatively small amounts of fuel, nuclear power plants have low land-footprint relative to other low carbon emission energy sources. A one gigawatt (1 GW) nuclear power plant may require about 1–3 square miles of land, which is significantly smaller than a typical wind farm of the same power, which may require about 7–15 square miles. Nuclear energy is a viable and cost-effective low-carbon emission source of energy, despite some setbacks due to rare but catastrophic accidents over the years. Public sentiment regarding nuclear energy is improving, in part due to the unreliable nature of renewable sources of energy for generating and storing vast amounts of energy, and due to the uncertain future of fossil fuels in response to policies and restrictions that drive investments in the energy sector.

To understand nuclear energy, we need to describe the atomic structure of matter. Atoms are composed of basic building blocks: electrons, protons, and neutrons glued together by different fundamental forces. Protons and neutrons form the atomic nucleus, which is surrounded by a cloud of electrons. Electromagnetic forces keep atomic pieces together, binding electrons and the protons in the atomic nucleus. Furthermore, there are two fundamental forces of nature that are active within the confines of atomic nuclei: the strong and weak nuclear forces. The strong nuclear force is responsible for linking protons and neutrons to form the atomic nucleus, overcoming the natural electric repulsion of positively charged protons. The weak nuclear force is responsible for the decay of neutrons into other particles.

The binding energy of the atomic nucleus shows an inverted U-shape curve moving from the lightest of the atomic elements to the heaviest, with the highest energy point observed around the element iron (Fe), which has a stable atomic nucleus. Nuclear energy can be harnessed through two different processes: nuclear fission and nuclear fusion. Nuclear fission occurs when the atomic nucleus of a heavy element splits into two or more lighter nuclear fragments, releasing energy. Nuclear fusion occurs when two nuclei of light elements coalesce into a single heavier atomic nucleus, also releasing energy. Both reactions create more stable nuclei, releasing excess energy. These two reactions differ in their basic fuel: ordinarily fission uses heavy elements, such as uranium, plutonium, or thorium, while fusion relies on lighter elements, such as deuterium and tritium (forms of hydrogen), or helium and boron.

Nuclear fission drives energy generation in existing nuclear power plants. The conversion of nuclear energy into electricity takes place indirectly, as in conventional thermal power plants. Fission reactions heat the reactor coolant, which could be water or gas, or even liquid metal, depending on the reactor type. The reactor coolant transfers heat to a steam generator and heats water to produce steam. The heat contained in steam is eventually converted into mechanical energy that feeds electrical generators which produce electricity. On the other hand, nuclear fusion has several advantages over nuclear fission, including higher energy yield and potentially lower radioactive waste. However, mastering nuclear fusion has been very elusive and is still in the development stage despite decades of research and heavy investment. Fusion requires large amounts of energy to overcome the repulsive electric forces between the protons in the two light atomic nuclei being combined. Once protons are brought into proximity, the nuclear force outweighs their mutual electrical repulsion. Inside the core of the Sun and other stars, where the temperature is extremely high and gravity keeps the star together, light nuclei collide with each other at very high energy, overcoming the electrical repulsion and generating fusion reactions. In a nuclear fusion reactor, the fuel needs to be contained for the fusion reactions to occur at high energies (Sobehart et al., 1989; Sobehart and Farengo, 1990). There are two basic approaches for the confinement of fusion fuel:

1. *inertial confinement*, where the fuel is compressed to incredible pressure and temperature using laser beams until nuclear ignition occurs
2. *magnetic confinement*, where fuel (in the form of hot plasma) is trapped in a strong magnetic field and heated until nuclear reactions start (Sobehart, 1989a, 1989b, 1990; Farengo and Sobehart, 1994a, 1994b, 1995).

Mastering nuclear fusion for energy generation will have a profound impact on society and on the transition to a low carbon emission economy.

There have been numerous developments in nuclear energy in the last decades, primarily for nuclear fission. Advanced fission power plant concepts include a range of high-power advanced reactors and small modular reactors (SMR), typically around 100–300 megawatts of power and even smaller microreactors that can help manage the increasing demand for energy.

Another source of energy is based on renewable power technologies (solar, wind, hydroelectric, geothermal, biomass, or biofuels), which can lead to significant carbon-emission reductions but can also produce environmental impact. Unlike fossil fuels (coal, oil, and natural gas), these alternative energy sources can generate power without releasing significant amounts of CO_2 and other GHGs that contribute to climate change. However, the GHG savings from biofuels seems much less than originally anticipated.

Let's discuss the most common alternative sources of energy: hydropower, wind, solar, biomass, geothermal, and other alternative sources. Hydropower taps into energy stored in large volumes of water raised against gravity. Energy is released (converted into motion) as water flows downhill. The water motion is used to move electrical generators that generate electricity. Hydroelectric dams with water reservoirs are a significant source of energy that can be stored for later electrical production. The combination of large-scale energy storage, production on demand and low operating costs makes hydropower a good source of renewable energy. Some dams can also operate as pumped-storage plants balancing the supply and demand of energy in the generation system. Hydropower dams have some disadvantages, including the disruption of aquatic ecosystems and wildlife, and their impact on the river environment. In rare situations, they can result in catastrophic failure of the dam.

Wind power, captured primarily by wind turbines, provides another source of renewable energy. As a result of the global patterns of atmospheric circulation, regions in the higher northern and southern latitudes have the highest potential for generating wind power. Wind supplies over 2000 TWh of electricity, which is about 7% of world electricity output and about 2% of world energy. With wind farms constructed mostly in China, the United States, and Europe, global installed wind power capacity exceeded 800 GW.

A wind farm is a group of wind turbines distributed over an extended area. Ordinarily, large wind turbines have a similar design, composed of a horizontal-axis wind turbine with an upwind rotor driven by three long blades. The turbine is attached to a nacelle on top of a tall tower mounted on land for onshore power generation, or on the sea floor for offshore power generation. Individual turbines are interconnected with a power collection system and communications network. Large wind farms may have tens or hundreds of individual wind turbines. Wind farms can be located onshore and offshore. For onshore wind farms, the land between the wind turbines may be used for a variety of agricultural and other purposes. Wind power generation can be produced during nighttime and in winter, when solar power output is low, which makes the combination of wind and solar power farms complementary sources of energy.

Solar power is cleaner than the power generated from burning fossil fuels and does not lead to harmful GHG emissions during operation. However, solar power generation carries a large upfront cost to the environment. The production of solar panels contributes significant amounts of pollution, including both GHG emissions and chemical pollution. These levels of pollution are expected to decline as manufacturers use cleaner energy sources and recycle materials for their processes. However, inconsistent standards across jurisdictions and countries for controlling pollution from manufacturing can provide an unfair competitive advantage for some manufacturers that could result in an economic benefit at the expense of environment degradation.

Biomass energy refers to electrical power generated by burning crops that are grown specifically for this purpose. Ordinarily biomass fuel (or biofuel) is obtained by fermenting plant matter to produce ethanol, which is then burned. Alternatively, biofuel can be produced by allowing organic matter to decay, generating biogas, which is then burned. Burning wood is also a form of low-level biomass fuel. Burning biomass generates many of the same emission by-products produced by burning fossil fuels. However, some of the emitted carbon dioxide can be captured by the remaining growing biomass, so that the contribution to GHG emissions is expected to be smaller. In practice, the GHG savings from biomass fuels seems lower than originally anticipated, and the process of growing biomass for fuel raises similar environmental concerns as for other types of agricultural activities: use of land, fertilizers, and pesticides.

Geothermal energy is an additional energy source that can be tapped into for power generation. The source of geothermal energy is the heat generated deep in the earth, which creates underground warm water and steam. Underground steam can be used to run a steam turbine for as long as the underground water is not depleted. The recirculation of surface water through rock formations to produce hot water or steam can make this energy source renewable.

Although geothermal power plants do not burn fuels to generate power, they do produce emissions from the gases and substances that come up from the geothermal wells.

Clean energy from sources such as hydropower, wind, solar, biomass, geothermal, or nuclear power is mainly generated and distributed in the form of electricity, which can be transmitted across long distances. Furthermore, switching from fossil fuels to clean energy requires that end uses, such as transport and heating, be electrified. Electrification, which is the process of powering systems and processes with electricity, has been a critical driver of industrial and economic development, and is one of the most important strategies for reducing CO_2 emissions. Furthermore, electrification allows replacing internal combustion engines used for transportation with electrically powered equivalents, such as electric vehicles. More than 26 million electric cars[2] were on the road in 2022, which represents more than five times the stock in 2018. Electrification is not only a driver of economic development, but also a process that helped transform and shape society (Mokyr, 1990; Nye, 1990). Electrification made possible modern life, impacting activities such as private and mass transportation, street and home electric lightning, cooling, and heating, cooking, assembly-line factories, automated production, communication, and entertainment.

TECHNOLOGY CHANGE: FIRST MOVERS, COMPETITIVE LANDSCAPE, AND ECONOMIC ENVIRONMENTS

As the impact of climate change and CO_2 emissions gain more visibility globally, there is increased interest in understanding how the industry landscape will be reshaped in response to the changing physical, economic, and political environment. Energy-related CO_2 emissions account for a significant fraction of global greenhouse gas emissions, where CO_2 emissions from fuel consumption, transportation, and for production of industrial materials are key contributors. The reduction of these sources of high-carbon emissions will depend on the replacement of fossil fuels as primary sources of energy with renewable energy sources and alternative sources of energy, and the introduction of more energy-efficient technologies for power generation, transportation, and production of industrial materials. Heavy investments and large subsidies were required to support the development of hydropower, onshore and offshore wind farms, and solar farms, driving costs down as the technologies are scaled up.

The adoption of renewable and alternative energy sources and new technologies to transition to a low carbon economy depends on industrial and political decisions but, most importantly, people's behavior. People's decision to adopt new sources of energy, reduce their consumption of fossil fuels, purchase electric vehicles (EVs) for private transportation, improve home heating and cooling efficiency, adopt high-efficiency appliances, reduce their carbon footprint, and minimize environmental pollution are often driven by government policies, regulations, incentives, and restrictions on energy sources and business activities, are influenced by the costs and benefits of these activities, and guided by markets and social influence. However, adoption decisions are ultimately a matter of personal choice for individual consumers. For example, manufacturing of EVs (cars and trucks) for private transportation has grown at an accelerated pace since they became available on the market. EVs include a wide range of road and rail vehicles, boats and underwater vessels, and aircraft. EVs use one or more electric motors for propulsion and can be powered by a collector system, with electricity provided by an external source (e.g. electric trains, trams, and trolley buses), can be powered by internal batteries or by converting fuel to electricity using a generator or fuel cells.

Early electric cars for private transportation were introduced in the late nineteenth century, following advances in science and technology driving the Second Industrial Revolution that brought forth mass electrification (Nye, 1990). The use of electric power for transportation provided an ease of operation and overall level of quietness that could not be matched by the internal combustion engine vehicles of the time. However, limitations in energy storage offered by early battery technologies impacted the mass adoption of EVs for private transportation. Because of the higher energy density offered by fossil fuels for long-range transportation and the reliability of operation in all weather conditions, internal combustion engines became the dominant technology for private transportation throughout the twentieth century. Electricity-powered locomotion focused on mass transit vehicles such as electric trains, trams, and trolley buses in urban areas or dedicated railroads for long-distance transportation.

As awareness on CO_2 emissions and climate change increased, governments across the world provided different incentives to reduce environmental impact and promote the adoption of electric vehicles. In the late 1990s, hybrid electric vehicles (HEV), where electric motors provide supplementary propulsion to internal combustion engines (ICE), became more widespread. Plug-in hybrid electric vehicles (PHEV), where electric motors provide the primary propulsion, became available in the late 2000s. Advances in battery technology, electric motors, power electronics and control systems have made modern battery EVs a practical option since the 2010s. Figure 5.2 presents the sales of EVs since 2015.

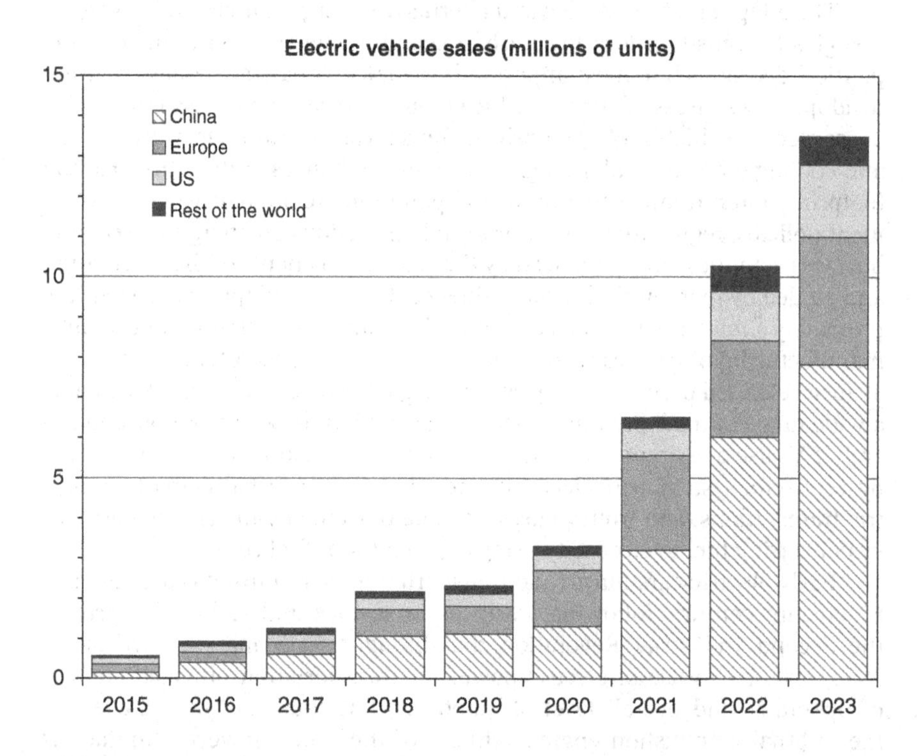

FIGURE 5.2 Sales of electric vehicles for private transportation.

The initial growth in EV use has been limited primarily by supply, not by demand. Soaring requirements for critical minerals (e.g. lithium, nickel, copper, cobalt) and rare earths needed in batteries and related EV technologies have not been fully met by rising supply. High purchase and maintenance costs of EVs, their limitations on reliable performance in all weather conditions and the general lack of adequate electrification and charging-station infrastructure, even in developed areas, impact the adoption of this technology. Furthermore, the transition to EVs will impact electricity consumption bills and presents a real challenge in highly urbanized areas, where high-rise apartments house people who will be unable to charge their EVs without new electric power infrastructure. Although energy costs of renewable sources have been declining as these technologies scale up, the cost of electricity may not include all external costs borne indirectly by society as a consequence of using those energy sources. These indirect costs may include enabling costs for new technologies, environmental impact, energy storage, recycling costs, infrastructure costs, and other costs. Current limitations of the existing electric power infrastructure and battery technology also create EV *social anxiety* due to the uncertainty about reaching one's destination for long trips or in severe cold weather conditions, or the time demand associated with long battery charging times relative to the simplicity of fueling traditional ICEs. The issues above, combined with the devastating social and environmental impact of mining and processing EV-related elements, and the potential monopolistic concentration and manipulation of the global supply chain for EVs and battery technology by a few manufacturers or nations, have started to affect public opinion and their preferences for these vehicles.

Solar farms and onshore and offshore wind farms face similar credibility and preference issues over their true environmental impact, ranging from pollution resulting from manufacturing solar panels or wind turbines to high noise levels generated by wind farms and their impact on wildlife. Furthermore, public concern about environmental issues tends to weaken in times of economic distress as people's priorities change.

To better understand the impact of people's behavior on transition to a low-carbon economy, it is important to understand basic features of the adoption of products and services, and their impact on both demand and the competitive environment of the companies offering these products and services. Many of the models used to estimate changes in supply and demand of services and products and assess the impact of transition risk are based on common quasi-equilibrium assumptions on economic and population growth that do not reflect the complex non-linear effects of competitive business landscapes (IAE, 2021). For example, energy consumption in high-carbon-emission or energy-intensive industry sectors is driven by the demand for products such as iron and steel, aluminum, car manufacturing, chemicals and petrochemicals, cement, and pulp and paper. Changes in the demand for products and services in these sectors can impact the ability of companies to compete, and their effects can cascade to other sectors of the economy through the supply chain or due to an impact on the overall economy (Twiss, 1974; Porter, 1985; Foster, 1986; Mokyr, 1990; Zangwill, 1993; Rogers, 1995; Day and Schoemaker, 2000; Case, 2007; Dicken, 2007). Furthermore, the incremental demand of energy and products could reflect changes in demand patterns in developing economies as they evolve, or due to the retirement of old units or replacement of obsolete technologies, as is the case for advanced economies. Here we discuss an approach for quantifying the risks and uncertainty associated with the release and adoption of competitive products and services for different climate risk scenarios. The approach can be leveraged for assessing the risks of business strategies whose revenues are used for the repayment of obligations in industries affected by climate risk or CO_2 emission costs, and for estimating the probability of default on those obligations under different scenarios.

CLIMATE RISK UNCERTAINTY IN COMPETITIVE BUSINESS ENVIRONMENTS

Climate change is one of the most recognized and polarizing issues facing society today, with scores of environmental, economic, and social studies performed by international organizations, the scientific community, regulatory and industry groups for many years (Mayeski and White, 2002; Ruddiman, 2005; ICC, 2018; NGFS, 2020; FSB, 2021; IEA, 2021). Despite some debatable assumptions found in many of these studies, they generally suggest that unless actions are taken now to address climate risk, the potential harmful effects are only expected to increase.

The shift away from fossil fuels toward renewables and alternative energy sources will have a significant impact on the global economy, directly affecting oil and gas extraction, coal mining, and oil- and coal-fired power production (Smith, 2020; IEA, 2021; USTD, 2021). Some banking institutions have already implemented restrictions on financing construction or expansion of coal-fired power plants. This includes transactions supporting financing of equipment, materials, and services directly required for the construction of such plants. In countries with serious energy deficiency, where the national electrification rate is below critical levels, some exceptions may be considered for plants with unabated CO_2 emissions. In general, there will be a significant impact on the energy sector and industries with high CO_2 and other GHG emissions. These changes will reshape the global economy and the competitive business landscape (Sobehart, 2022).

The proposed global reduction of GHG emissions to net zero by 2050 to limit the increase in average global temperature to about 1.5 °C will require a complete transformation of how energy is produced, transported, and consumed (ICC, 2018; NGFS, 2020; FSB, 2021; IEA, 2021; USTD, 2021). Notice the distinction between (absolute) zero carbon emissions and *net zero carbon emissions*, which implies that companies could still produce significant GHG emissions as long as they are offset by reforestation, carbon-capture or similar activities, or by the purchase of carbon offsets.

Net zero targets have become key drivers for climate action. Many companies, organizations, and countries are setting net zero targets to reduce carbon emissions, removing, or offsetting their remaining residual emissions. However, without committed investments or funding, clear transitions plans, or enforcement mechanisms requiring a transition to net zero, the credibility of many of the net zero commitments remains low or, at least, questionable. Since many net zero commitments defer current emission reductions by relying on future low-carbon emission technologies, claims on business strategies are somewhat dubious.

Net zero targets cover a wide range of activities for companies divided into three basic scopes:

1. *Scope 1* refers to all direct emissions from activities performed by a company or under its control.
2. *Scope 2* refers to indirect emissions from power for electricity, heating, cooling, or steam purchased and used by the company.
3. *Scope 3* refers to all indirect emissions resulting from activities performed by a company and from sources that are not under its control. Transportation of goods and products, and other indirect emissions are also part of this scope. These emissions could be the most significant contributions to a company's carbon footprint, and the hardest to estimate because they cover suppliers, distributors, and customers.

Creating accurate accounts of GHG emissions presents several challenges. For example, scope 3 emissions can be difficult to estimate, and scope 2 or scope 3 activities for a company could be scope 1 activities for the company's suppliers. Without a detailed analysis of activities, these challenges could lead to overestimating the effects of emissions.

The effects of emission reductions to achieve the proposed GHG reductions by 2050 can be analyzed under a wide range of orderly and disorderly transition scenarios:

1. *Orderly:* consistent regulations, policies, and incentives reflect early and ambitious actions toward a net zero emissions economy.

2. *Disorderly:* inconsistent regulations, policies, and incentives reflect actions that are late, disruptive, sudden, unanticipated and/or uncoordinated across countries.
3. *Hothouse world:* inconsistent regulations, policies, and incentives result in limited actions that lead to a hothouse world with significant global warming and increased exposure to physical risks.
4. *Too little, too late:* late, limited, or non-existing regulations, policies, and incentives lead to actions that are not enough to meet the emission reduction goals, and increasing physical risks spur a disorderly transition.

While an orderly transition may follow a more gradual and predictable path to a low-carbon economy, a disorderly transition might lead to significant uncertainty in outcomes and increasing risks. Sudden, unexpected, or uncoordinated increases in the price of carbon emissions or other measures to limit emissions would hit carbon-intensive sectors directly and would cascade into other sectors of the economy indirectly through production, distribution, and consumption channels. Disorderly transitions could also affect the financial sector through their credit and market risk exposures to companies in carbon-sensitive sectors. Disorderly transition scenarios allow uncooperative countries to take an unfair competitive advantage of emission restrictions that other countries may impose on their own industries and economies to meet commitments on GHG emission reductions.

To illustrate this point, in the disorderly transition scenario of the Network for Greening the Financial System (NGFS, 2020), global policy measures to reduce carbon emissions are delayed until 2030. For nations to still achieve the Paris Agreement targets under this scenario, an unexpected and sudden increase in the price of carbon would be needed. In the NGFS's disorderly scenario, this increase takes place around 2030.

The potential reshaping of the competitive landscape across industry sectors and its impact on the economy will drive financial institutions to reassess their long-term investment strategies. This will require a re-evaluation of their portfolios and risk management practices to identify vulnerabilities to GHG emissions under different scenarios. This risk assessment will require detailed information from borrowing companies on scopes 1, 2, and 3 emissions related to their activities, power purchases, and business strategy to a net zero or low-carbon energy future.

Assessing a financial institution's net zero strategy and its overall impact requires understanding three basic issues:

1. The business environment and competitive landscape for different sectors and climate change scenarios.

2. The ability of individual borrowers to adapt to their competitive landscape, which impacts their demand for credit.
3. The institution's own net zero commitments and strategy, which impact the supply of credit.

In the following we discuss the uncertainty introduced by the unpredictable nature of highly competitive business environments, which can be affected by climate risk and GHG emission costs and can precipitate a firm into financial distress or default on debt obligations. We follow closely previous work on uncertainty in highly competitive environments for disruptive technologies and climate change (Sobehart, 2016, 2022). Since revenues generated from the sale of products and services are the main source of income for most businesses, it is important to understand how competition for market share affects a company's value.

The analysis below focuses on two areas: (1) the interaction between competitors and their impact on product adoption and market share, and (2) the impact on the company's market share of the time to market new products and services. The combination of these two issues with the non-linear aspects of competitive environments and the breadth of climate change scenarios can lead to a substantial impact on outcomes. Notice that our approach differs from traditional credit risk approaches (Merton, 1974; Kim et al., 1993; Cathcart and El-Jahel, 1998; Leland, 1999) that do not include changes to the industry-competitive landscape, which are critical for understanding the long-term impact of climate change.

Products are introduced to the market at different times, reflecting their own technology development timelines, marketing strategies, and different carbon-emission costs. Products may also reflect competitive advantages when manufactured in countries that do not follow the same emission reduction standards as other countries or have lax regulations with poor enforceability. Customers adopt products based on their utility, quality, price, and availability in the market. This approach allows us to quantify the uncertainty and non-linear effects introduced by the unpredictable nature of highly competitive business environments (Twiss, 1974; Porter, 1985; Foster, 1986; Zangwill, 1993; Rogers, 1995; Day and Schoemaker, 2000; Case, 2007; Dicken, 2007; Sobehart, 2016, 2022).

INNOVATION AND PRODUCT DYNAMICS

Bringing new technologies to market can often take years or even decades of research, development, and implementation. The commitment to achieve net zero emissions globally by 2050 would require that innovation on alternative energy sources and low-carbon emission technologies must evolve at a faster and more aggressive pace than the deployment of any other technology implemented at a global scale. In some of the proposed transition scenarios (IEA, 2021; Sobehart, 2022), the time from first prototype to market introduction is assumed to be about 20% faster than the fastest energy technology developments in the past, and around 40% faster than was the case for solar photovoltaic (PV) technology.

Technologies that are still at the demonstration stage, such as carbon capture, utilization, and storage (CCUS) in cement production or low-emissions ammonia-fueled ships, are assumed to enter the market in less than five years. Hydrogen-based steel production, direct air capture (DAC), and other technologies at the large prototype stage are assumed to reach the market in about six years, while other technologies at the "proof of concept" stage, such as (refrigerant-free) solid state cooling, or solid-state batteries, are assumed to be achievable within 10 years. An innovation acceleration of this breadth and magnitude is clearly extremely ambitious (if not unrealistic) and will require technologies to be implemented quickly at scale across multiple applications and geographies. This contrasts with typical technology development, where learning is usually transferred across consecutive projects before widespread deployment at scale. The accelerated innovation would also require a large increase in investment to bring these technologies to market across sectors and geographies.

Technologies available today may help provide a temporary bridge to a net zero emissions economy by 2050. However, reaching the global net zero commitments by 2050 would require significant innovation in clean energy and other technologies. For example, in IEA's Net Zero Emission (NZE) scenario for 2050, almost half of GHG emissions reductions come from new technologies or technologies currently at their proof-of-concept or prototype stage. The need for these technologies is even more significant in high-carbon sectors such as heavy industries and transportation.

The role of governments and international organizations and their cooperation are also crucial in shortening the time needed to bring new technologies to market and to diffuse them widely. Governments can provide incentives and opportunities to private sectors, eliminate obstacles and unfair manufacturing and trade practices, and protect intellectual property, enabling the infrastructure for new energy production and transfer, and setting adequate regulatory frameworks for markets and finance to support investments.

In highly competitive environments, characterized by innovation and a limited ability to transfer costs to consumers, product development strategies that reduce risk over time and protect the firm's market share must consider the competitive dynamics of revenue generation. Understanding business competition is important in the early phases of product adoption when new technologies are typically not competitive with incumbent or mature technologies, and uncertainty about outcomes is relatively high. Understanding the impact of highly competitive business environments on product adoption can help companies determine strategies covering:

1. product research and development
2. strategies for releasing products to market
3. project financing and its related credit risk.

The approach described below focuses primarily on items (1) and (2) above, driven by the following production factors:

1. time to develop new products and services
2. quality and price of released products
3. time to market new products.

By modifying these production factors, companies affect the business environment in which their competitors evolve, introducing synergies between the competitors can have a significant impact on the adoption of products and market share. When markets are new or are growing slowly, the finite size of these markets can introduce saturation effects in the adoption of products, which can impact a company's capacity to generate revenues. Companies can lose their market share partially or can be exposed to high levels of revenue uncertainty, forcing them to withdraw from these markets. This situation is not contemplated in most standard credit models based on the valuation of the firm and could lead to a higher probability of default on financial obligations.

PRODUCT ADOPTION

The typical product cycle is a several-year period in which a variety of business, production, and marketing decisions need to be made before and after the early version of a product has been released to the market (Twiss, 1974; Porter, 1985; Foster, 1986; Mokyr, 1990; Zangwill, 1993; Rogers, 1995; Day and Schoemaker, 2000; Case, 2007; Dicken, 2007; Day, 2013; Thomke and Reinertsen, 2013). Ordinarily, after a new product is released, initial growth in market share is usually slow since the product is not widely known and prices are usually high, reflecting the costs of development, marketing launch, and low production. After a period of initial customer growth, market share grows, production costs often drop and profits rise sharply, attracting competitors. Increasing levels of competition can cause an erosion of the profits of the

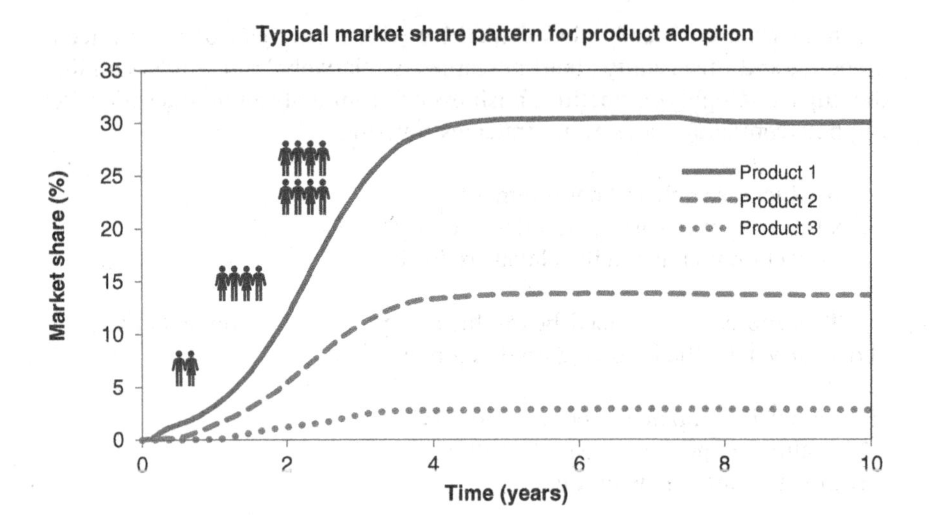

FIGURE 5.3 Typical S-shape growth pattern of market share.

company by reducing its market share and margins. Eventually a stage of market saturation is reached when companies with high production costs may see their profit margins disappear, forcing them to withdraw from the market (D'Aveni, 1994; Rogers, 1995; Sobehart, 2016, 2022). Even without competition, saturation of a company's product adoption can occur as the company eventually takes all the available market share in a new or slowly growing market with finite size. At this stage the company's market share exhibits a logistic, S-shape, pattern as shown in Figure 5.3.

In some situations, after a period of initial growth, demand increases sharply due to inflated expectations (creating a transitory excess demand), and market growth attracts competitors. At this stage, customers may start switching to other products or reducing their demand for the product because it is no longer appealing or needed, causing rapid erosion of profits and market share to a lower level as the hype cycle comes to an end, as shown in Figure 5.4.

How clearly customers can articulate their preferences on what new products they will adopt seems related to their knowledge of the product or related products and their market presence. This behavior can be analyzed using models for product adoption (Sobehart, 2016, 2022).

To gain insight into product adoption, let's start our analysis focusing on the release of a single new product in a market of limited size and no competition. Let's assume that, for a given product, product adoption is a function of both the number of customers $Z(t)$ who have already adopted the product and

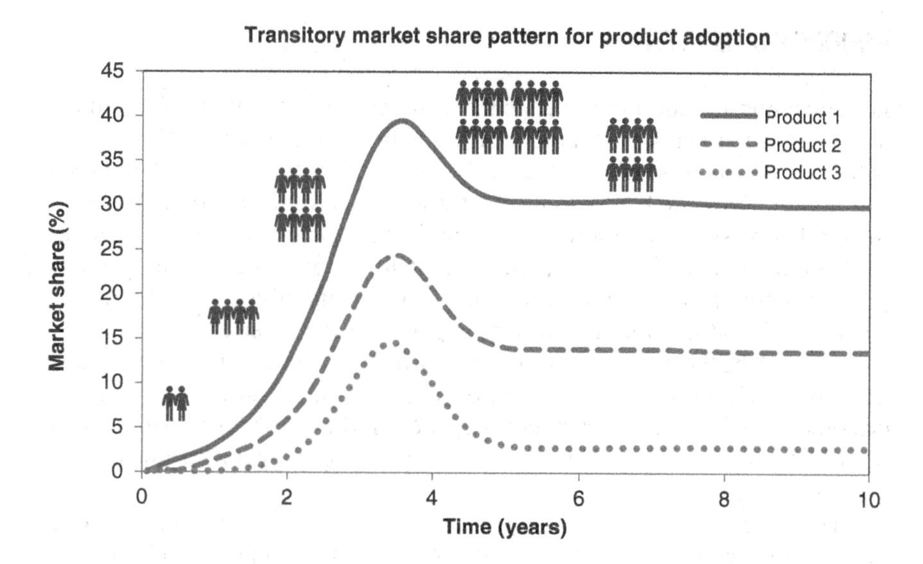

FIGURE 5.4 Growth pattern of market share with transitory excess demand.

the number of potential new adopters $N(t) - Z(t)$ remaining in the market of size $N(t)$. The rate of new adopters per unit time is: $\lambda Z(N - Z)$, with product adoption rate λ, which depends on the company's reputation, product quality, price, and other key characteristics.

Also let's assume that there is no cost for customers who switch between products, and that some customers simply stop using the product because it is no longer useful to them (product rejection). The fraction of customers per unit time that rejects the product (product rejection rate) is γ.

The customer growth rate is the net effect of customer adoption and rejections:

$$\frac{dZ}{dt} = \lambda Z(N - Z) - \gamma Z \tag{5.1}$$

The term $N - Z$ in Equation (5.1) describes the saturation of the customer population to a final level:

$$Z_F = N\left(1 - \frac{\gamma}{\lambda N}\right) \tag{5.2}$$

In this situation, all the potential customers accept the new product except for a fraction $\gamma/\lambda N$ of the population. Because here there is no competition, the acceptance of the new product is determined only by the adoption rate λ. The effects of competition are introduced next.

PRODUCT COMPETITION

Let's turn our attention to the effects of competition when there is no uncertainty in the market. We start our analysis with the simplest case of two competitive products or services, P_1 and P_2, which are introduced to the market at different times and have adoption rates λ_1 and λ_2, which are determined by the marketing strategy, product quality and price. In our example, product P_2 could be of greater or inferior perceived quality than product P_1 in terms of price or actual product quality. This difference could affect both the rate of adoption of products and the number of potential customers.

Let $Z_1(t)$ be the number of adopters of product P_1, and $Z_2(t)$ be the number of adopters of product P_2. As a result of competition, the number of potential customers who can adopt product P_1 depends on the number of customers for product P_2 and vice versa. Let's assume both products target the same market segment of size N.

The simplest interaction between products P_1 and P_2 representing competition for the same market segment is given by a customer-switching term proportional to both number of customers Z_1 and Z_2. At any time, the rate of adoption of P_1 is reduced by a higher level of adoption of P_2 and vice versa. Thus, the equations for the rate of change of the number of customers of products P_2 are:

$$
\begin{aligned}
\frac{dZ_1}{dt} &= \lambda_1 Z_1(N - Z_1 - C_{12}Z_2) - \gamma_1 Z_1 \\
\frac{dZ_2}{dt} &= \lambda_2 Z_2(N - Z_2 - C_{21}Z_1) - \gamma_2 Z_2
\end{aligned}
\tag{5.3}
$$

Here $C_{12}, C_{21} \leq 1$ are the fractions of adopters not willing to switch to the competitor's product. When products have the same quality and price and there are switching costs or other limitations, customers who adopted one product may not be willing or able to switch to other products. In this case, these customers must be removed from the pool of potential customers available to adopt the product, leading to $C_{12} = C_{21} = 1$. The linear terms in γ_1 and γ_2 represent customers who stop using either product because they are no longer useful or become obsolete.

An increase (or decrease) of the adoption rates λ_1 and λ_2 impacts the number of customers adopting products P_1 and P_2 in a non-linear way as described in Equation (5.3). Although real competitive environments do not follow this simplistic model exactly, they do fluctuate around such curves. The impact of deviations and uncertainty in business environments is described in later sections.

Finding the (equilibrium) fixed points for Equation (5.3) is an important first step in understanding the competitive landscape. The non-trivial equilibrium points of Equation (5.3) are:

a. Product P_1 emerges as the market leader

$$\begin{cases} Z_1 = N\left(1 - \frac{\gamma_1}{\lambda_1 N}\right) \\ Z_2 = 0 \end{cases} \tag{5.4}$$

b. Product P_2 emerges as the market leader

$$\begin{cases} Z_1 = 0 \\ Z_2 = N\left(1 - \frac{\gamma_2}{\lambda_2 N}\right) \end{cases} \tag{5.5}$$

c. Both products coexist $C_{12}, C_{21} \neq 1$

$$\begin{cases} Z_1 = N\left(1 - \frac{\gamma_1}{\lambda_1 N} - C_{12}\left(1 - \frac{\gamma_2}{\lambda_2 N}\right)\right) \\ Z_2 = N\left(1 - \frac{\gamma_2}{\lambda_2 N} - C_{21}\left(1 - \frac{\gamma_1}{\lambda_1 N}\right)\right) \end{cases} \tag{5.6}$$

Whether products P_1 and P_2 coexist, or one forces the other to withdraw from the market depends upon the size and correlations C_{12} and C_{21}. For complex situations with multiple competitors, fixed points may be difficult to identify analytically (Nicolis and Prigogine, 1977; Kreuzer, 1981; Hoefbauer and Sigmund, 1998).

The stability of the fixed points above can be analyzed introducing small perturbations $\delta_k = \varepsilon_k e^{\omega t}$ $(k = 1, 2)$ around the solutions to Equations (5.4)–(5.6):

$$\frac{d\delta_k}{dt} = \sum_{j-1}^{2} \frac{\partial G_k}{\partial Z_j} \delta_j k = 1{,}2 \tag{5.7}$$

Here $G(Z_1, Z_2)$ represents the non-linear terms in Equation (5.3). The stability condition for evolving away from a selected solution (Equations 5.4, 5.5, or 5.6) is that the determinant of Equation (5.7)

$$\det\left|\frac{\partial G_k}{\partial Z_j} - \omega\right| = 0 \tag{5.8}$$

has at least one root ω with a positive real part.

To illustrate this issue, suppose that product P_2 is introduced to the market when the producer of P_1 is already the market leader as described in Equation (5.4). The condition needed for the number of customers Z_2 to grow exponentially from a small initial population is:

$$1 - \frac{\gamma_2}{\lambda_2 N} > C_{21}\left(1 - \frac{\gamma_1}{\lambda_1 N}\right) \tag{5.9}$$

If in addition to Equation (5.9), the following condition is satisfied:

$$1 - \frac{\gamma_1}{\lambda_1 N} > C_{12}\left(1 - \frac{\gamma_2}{\lambda_2 N}\right) \tag{5.10}$$

the populations of customers Z_1 and Z_2 are both positive and the system evolves to a fixed point where product P_1 and P_2 coexist. In contrast, if Equation (5.10) is not fulfilled, then the system can evolve to the fixed point (Equation 5.5) in which product P_2 forces product P_1 to withdraw from the market. In this case, product P_2has a competitive advantage over product P_1. This example highlights that the interaction between competitors can have an adverse impact on market share and revenue generation.

Notice that the time at which products are released into the market is a key source of risk, which can impact product adoption and market share. If the two products compete for the same market segment, their price and quality are similar, and no customer is willing or able to switch to the other product ($C_{12} = C_{21} = 1$), then market timing becomes crucial. In this case, the only non-trivial fixed points are Equations (5.4) and (5.5) that induce that one product will force the other to withdraw from the market.

When products are similar (in terms of quality, price, or appeal), once one of the products becomes the "market leader," customers have no incentive to switch to the other product. In this situation, the later a product is introduced to the market, the lower its chances of capturing any significant market share. The situation may be exacerbated in the presence of switching costs, entry barriers or unfair competitive advantages due, for example, to uneven restrictions on carbon emissions or environmental policies that apply to some competitors but not to others.

MULTIPLE COMPETITOR ENVIRONMENTS

Achieving global net zero emissions by 2050 will require an unprecedented level of international cooperation as well as technical and financial support to ensure the implementation of GHG emission reduction technologies and infrastructure across high emission economies. This will also require adequate enforcement mechanisms for limiting political manipulation of rogue governments that can take an unfair advantage of a reshaped economic landscape while other countries are forced to absorb the burden of high emission costs by limiting their own economic activity. These issues create uncertainty for climate risk modeling, ranging from sensitivity to the non-linearities of complex competitive environments to ambiguity in scenario specification. The economic and social costs of carbon emissions could be more pronounced when all sources of uncertainty are acknowledged.

Here we limit our analysis to the uncertainty in the competitive landscape from variations in product prices, quality, and their release time to market. Our general model includes several companies, each with many competitive products targeting different local or global market segments. Each market segment, m has size $N_m(t)$ and is assumed to change at rate $\mu_m(t)$ with random fluctuations η_m

$$dN_m = N_m(\mu_m dt + d\eta_m) \tag{5.11}$$

The growth rate $\mu_m(t)$ and the properties of the random fluctuations $d\eta_m(t)$ depend on the nature of the industry and expansion of the customer base.

The general equation of product adoption driven by market presence is

$$\frac{dZ_{k[j,m]}}{dt} = \lambda_{k[j,m]} Z_{k[j,m]} \left(N_m - \sum_{h[i,n]} C_{h[i,n]} Z_{h[i,n]} \right) - \gamma_{k[j,m]} Z_{k[j,m]} \tag{5.12}$$

Here $Z_{k[j,m]}$ is the number of customers of product $P_{k[j,m]}$ of firm j in market segment m, $\gamma_{k[j,m]}$ is the rejection rate of product $P_{k[j,m]}$ and $\lambda_{k[j,m]}$ is its adoption rate, which can depend on the product quality $q_{k[j,m]}(t)$ and price $p_{k[j,m]}(Z_{k[jm]}, t)$.

The relationship between the product price $p_{k[j,m]}$ and the number of customers $Z_{k[j,m]}$ adopting the product depends on the product characteristics, production, and distribution costs, GHG emission costs, and other costs. Product price $p_{k[j,m]}$ is expected to be a decreasing function of the number of customers $Z_{k[j,m]}$, reflecting that increasing production volume generally reduces variable production costs.

The product quality $q_{k[j,m]}(t)$ is determined by the type of technology involved in developing the products and production controls, and customer service when required. In the following, we assume that all the products have similar quality and price with some variations assigned at random from normal distributions of quality and price $N(q, \sigma_q)$ and $N(p, \sigma_p)$ with means q and p, and deviations σ_q and σ_p, except for product P_1, which is selected as the reference product for our analysis. The quality of product P_1 can be changed independently from the other products to study the relationship between market timing, product quality, price, and competitor characteristics under different scenarios.

The adoption rate $\lambda_{k[j,m]}$ for generic product P_k is approximated by an S-shape (logistic) function of quality and price that describes the saturation of customers' preference for extreme values. Intuitively, if product P_k has very low quality ($q_k \ll q_0$, where q_0 is the typical quality for similar products), customers have no incentive to select it. In contrast, when the product quality is very high ($q_k \gg q_0$), customers' perceived marginal gain for getting a product with quality exceeding q_0 may also saturate. In the middle of the quality range, where $q_k \sim q_0$, the adoption rate changes significantly in response to changes in perceived quality above or below the typical value q_0. Similar arguments apply to the product's price p_k.

When changes in quality and price have similar impact on customer preferences but with different sensitivities, a simple functional form for the adoption rate is:

$$\lambda_k = \lambda_k^0 + \delta_k \tanh(x)$$
$$x = \left(a\frac{(q_k - q_0)}{\Delta q_{k0}} - b\frac{(p_k - p_0)}{\Delta p_{k0}}\right) \tag{5.13}$$

Here we dropped indices $k[j,m]$ for simplicity. Notice that λ_k^0, δ_k, q_0 and p_0 are reference parameters that can depend on all the competing products in the market segment, and a and b are the relative sensitivities to quality and price. In the following, we assume $a = b = 1$ for simplicity. Parameter Δq_{k0} is the customers' sensitivity to differences in quality and Δp_{k0} is the sensitivity to differences in price. Note that when the difference in quality $| q - q_0 |$ is smaller or of the order of Δq_{k0}, the adoption rate changes significantly. In contrast, when the difference in quality $| q - q_0 |$ is much larger than the value Δq_{k0}, the adoption rate saturates. A similar effect occurs for differences in price. Although the S-shape functions for adoption rates are intuitive, other functions can also be used to accommodate product adoption for specific situations.

The term $C_{k[j,m]h[i,n]}$ in Equation (5.12) represents the coupling of product $P_{k[j,m]}$ of firm j in market m and product $P_{h[i,n]}$ of the firm i in the market segment n. The coupling coefficients C_{kh} reflect switching costs and customers' loyalty or limitations to switch products, and are approximated by logistic functions of quality and price to describe saturation effects in customers' preference:

$$C_{kh} = \alpha_{kh} + \varepsilon_{kh} \tanh(y)$$
$$y = \left(c\frac{(q_k - q_h)}{\Delta q_{kh}} - d\frac{(p_k - p_h)}{\Delta p_{kh}}\right) \tag{5.14}$$

Here $\alpha_{kh}$is a preference parameter, ε_{kh} represents the strength of the coupling, Δq_{kh} is the customers' sensitivity to differences in quality and Δp_{kh} is the sensitivity to differences in price. Parameters c and d are the relative sensitivities to quality and price. For simplicity, we assume $c = d = 1$. Notice that any pair of products with the same change in price-quality utility y would have the same coupling coefficients C_{kh}. When $C_{kh} = 1$, customers of product P_h are not willing or able to switch to product P_k, reducing the population of potential adopters for this product. Also notice that $C_{kk} = 1$ since customers already have adopted P_k.

Finally, when the difference in quality is of the order of Δq_{kh}, the product couple coefficients are very sensitive to changes in quality or price. In contrast, when the difference in quality is much larger or much smaller than Δq_{kh}, the product coupling coefficients saturate. A similar effect occurs for differences in price. For markets where customers are not biased toward any specific product or brand, $\alpha_{kh} = \alpha$, $\varepsilon_{kh} = \varepsilon$, $\Delta q_{kh} = \Delta q_{hk}$, and $\Delta p_{kh} = \Delta p_{hk}$ for each pair of the products $\{k, h\}$ in any given market segment.

UNCERTAINTY IN COMPETITIVE ENVIRONMENTS

Understanding how the business competitive landscape will be reshaped in response to economic and social changes driven by the transition to a low-carbon economy requires having a forward-looking vision of the way technologies will evolve under different scenarios. Here we discuss the use of Monte Carlo simulations to solve the system of Equations (5.11)–(5.14) for stylized scenarios of business competition, roughly aligned to three transition risk scenarios:

1. *orderly transition* – represented by a level playing field for competitors
2. *mildly disorderly transition* – represented by increasing levels of uneven competition where some companies have a competitive advantage
3. *strongly disorderly transition* – represented by significant levels of uneven or unfair competition.

The stylized scenarios are constructed by modifying product quality and price, which affect product adoption rates and the coupling coefficient between products. These scenarios also include variability in the release of the products to the market, which occurs at random times $T_{k[i,j]}$ for product k of company $1 \leq i \leq n$ in market $1 \leq j \leq m$, and can provide insight on how new products may threaten established products.

The Monte Carlo simulations of the competitive environment can be described as follows. First, the number of competitors and products are selected for each market segment. For simplicity, we assume that there is only one product P_k for each company k (among n companies), and only one market segment ($m = 1$) of size N.

Second, we select product P_1, which is exposed to different levels of competition from asymmetric costs or unfair competitive advantages as reflected in the quality or price of the other products. Product P_1 is used to gain insight into the market share uncertainty associated with market timing under different scenarios. The first competitor's product introduced to the market is labeled P_2 regardless of which competitor releases it. The release of the first competitor's product defines the reference time for our analysis $T_2 = 0$. The properties of product P_1 (quality, price) are assigned and a release time T_1 is selected. The release time of product P_1 can be changed to study the impact of market timing on market share. Thus, the condition $T_1 < 0$ indicates that product P_1 is released to the market before product P_2 (that is, before any other product in the market), while $T_1 > 0$ indicates that the product is released only after product P_2 is released.

Third, the quality, price, and release time of all the other products ($2 \leq k \leq n$) are assigned according to the distributions of product characteristics discussed above. Except for product P_1 and P_2, all the other products are released at random times $T_k \geq 0$ following a Poisson distribution with mean release time τ, which reflects the nature of the products, market segment, maturity of the technology, and level of competition. Because each product P_k is released at a random time T_k, it is possible for product P_1 to be released before an arbitrary product P_k ($T_1 \geq T_k$) for some of the random Monte Carlo scenarios, while product P_1 could be released after product P_k ($T_k \geq T_1$) for other random scenarios. The likelihood of these events depends on the relation between the release time T_1 and the mean value τ of the distribution of product release times. This is a more realistic situation than scenarios with fixed product releases from competitors since ordinarily companies do not have control over their competitors' product releases, quality, or price and need to determine the level of uncertainty in the market.

The evolution of the market share for all competitors is evaluated using Equations (5.11)–(5.14) through the product lifecycle. For simplicity, we assume that once products are released onto the market, their quality and price remain constant throughout each stochastic simulation. At the end of each simulation period new values are assigned to the competitors' products keeping the properties and release time of product P_1 fixed. The process is repeated for thousands of possible scenarios (Figure 5.5).

Now let us turn our attention to the impact of the stylized orderly, mildly disorderly, and strongly disorderly transition scenarios on the market share of

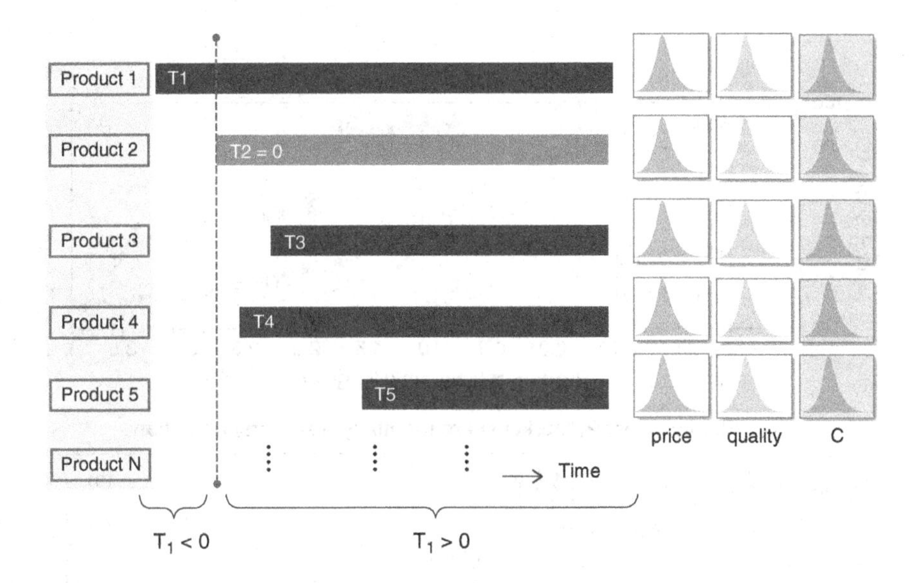

FIGURE 5.5 Distribution of competitive products for different quality, price, and release times T_k. Here $T_1 < 0$ indicates that product P_1 is released to the market before any other product, while $T_1 > 0$ indicates that product P_1 is released only after the first competitor product P_2. Other products are released at random times.

product P_1. In the stylized orderly transition scenario, all competitors are subject to similar carbon-tax costs, GHG emission limitations or business restrictions from environmental policies. A weak competitive regime is observed when product P_1 has a superior quality, that is, when the quality of product P_1 is at two or more standard deviations σ_q above the average quality of other products. Once P_1 is introduced to market, customers are eager to switch to product P_1 until a saturation of the market is reached. Note that when product P_1 has a superior quality and all competitors are affected by similar carbon-tax costs, emission limitations or business restrictions, competition remains weak and adequate market timing becomes important only for significant delays in the release of new products. Figure 5.6(a) shows the upper and lower values of the market share of product P_1 at the end of the product lifecycle (assumed to be a 10-year period) for different release times T_1(in years). The level of business uncertainty is given by the spread between the upper and lower values of the market share (which impacts earnings). For illustration, the market includes only $n = 5$ competitors to provide a dynamic business environment. The adoption rate for product P_1 is $\lambda_1^0 = 0.14$, $\delta_1 = 0.03$ (representing roughly a potential 50% market share change over a 1.5yr period), $\tau = 2$yrs, and the customer coupling parameters are $\alpha = 2$ and $\varepsilon = \pm 0.25$. Product P_1 can be

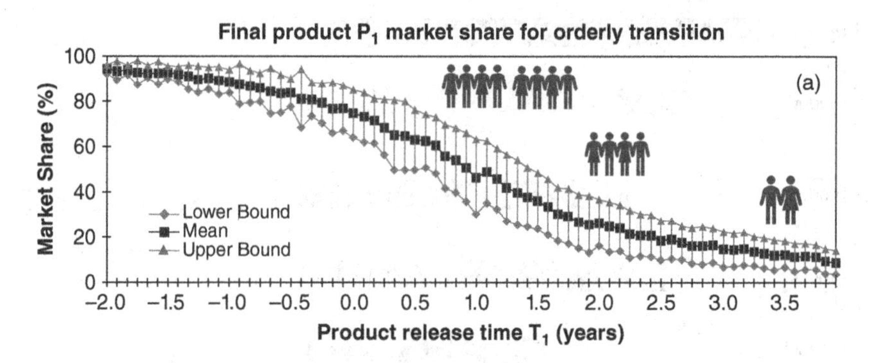

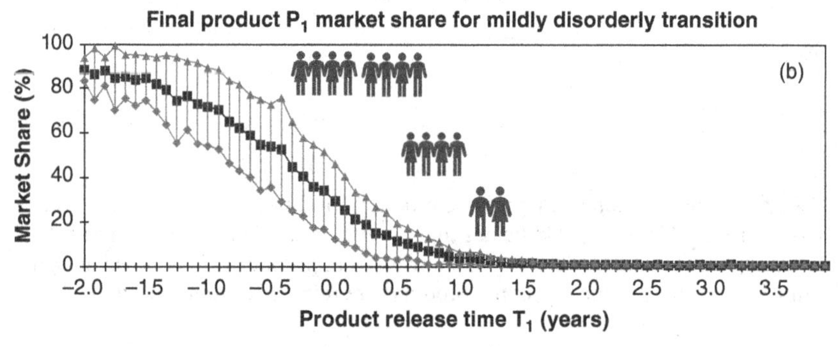

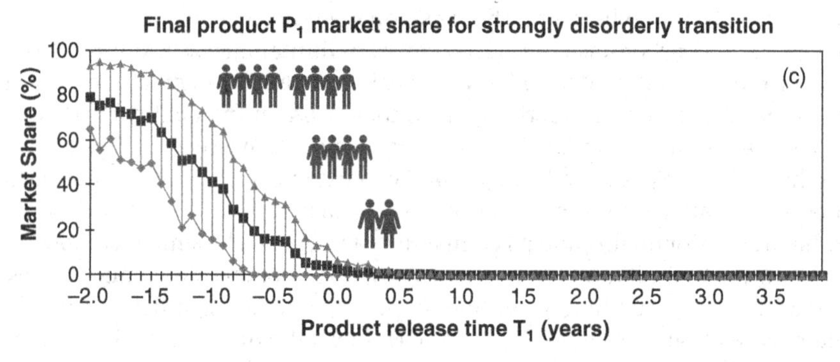

FIGURE 5.6 Market share for product P_1 at the end of the product lifecycle (10 years) for release time T_1 (in years). The transition scenarios are (a) orderly transition – product P_1 remains the dominant product, (b) mildly disorderly transition – product P_1 is about the same or slightly better than other products that have mild competitive advantages, and (c) strongly disorderly transition – product P_1 is worse than other products that have unfair competitive advantages.

released to market as early as two years ahead of the next competitor's product P_2 ($T_2 = 0$), and as late as 4 years after P_2 and other products are released.

In the stylized mildly disorderly scenario, carbon-tax costs, emission limitations and business restrictions are not coordinated completely across sectors and governments, creating competitive advantages for some companies. A moderate competitive regime is observed when product P_1 has a similar or only slightly better quality/price than other products. That is, when the quality/price of product P_1 is roughly no more than one standard deviation above the average quality/price of other products, some of which have a competitive advantage from uncoordinated business restrictions, emission limitations and costs. After P_1 is introduced to the market, some customers switch to better products as they are released to the market. Because product P_1 has roughly average quality/price relative to its competitors, poor market timing for releasing products could result in a loss of market share and an increase in credit risk as shown in Figure 5.6(b).

Finally, in the stylized strongly disorderly scenario, there is a wide range of emission limitations, costs, business restrictions, and incentives across economies or countries. Some countries can take advantage of the economic benefits of a reshaped global economic landscape by delaying or avoiding emission limitations while other countries take on the burden of high emission costs and impact on their economic activity. A strong competitive regime is observed when product P_1 has a quality/price worse or not much different from the average within one standard deviation due to competitive disadvantages. After P_1 is introduced to market, customers are not eager or willing to switch to product P_1, and product P_1 is unable to capture a large market share over time. Figure 5.6(c) shows the market share of product P_1 for this scenario.

Despite the simplicity of the model, the results above provide insight into the broader issue of uncertainty in climate risk modeling, sensitivity to non-linearities, and ambiguity in scenario specification that can lead to more pronounced economic and social costs of CO_2 emissions.

NOTES

1. US Energy Information Administration. https://www.eia.gov/energyexplained
2. International Energy Agency. https://www.iea.org/energy-system/renewables

REFERENCES

Case, J. (2007). *Competition: The Birth of a New Science.* Hill and Wang, New York, pp. 115–144.

Cathcart, L., and EI-Jahel, L. (1998). Valuation of Defaultable Bonds. *Journal of Fixed Income* 8(1): 65–78.

D'Aveni, R.A. (1994). *Hyper-Competition: Managing the Dynamics of Strategic Maneuvering*. The Free Press, New York, pp. 39–70.

Day, G.S. (2013). Is It Real? Can We Win? Is It Worth Doing? *Innovation*. HBR's 10 Must Read Series. pp. 59–81.

Day, G.S., and Schoemaker, P.J.H. (2000). *Wharton on Managing Emerging Technologies*. John Wiley & Sons, New York, pp. 125–149.

Dicken, J. (2007). *Global Shift: Mapping the Changing Contours of the World Economy*. The Guilford Press, New York, pp. 73–136.

Farengo, R., and Sobehart, J.R. (1994a). Minimum Ohmic Dissipation and DC Helicity Injection in Tokamak-like Plasmas. *Plasma Physics and Controlled Fusion* 36: 1691–1700.

Farengo, R., and Sobehart, J.R. (1994b). Minimum Dissipation States in Tokamak Plasmas: Inductively Sustained Tokamaks. *Plasma Physics and Controlled Fusion* 36: 465–471.

Farengo, R., and Sobehart, J.R. (1995). Determination of Minimum Dissipation States with Self-Consistent Resistivity in Magnetized Plasmas. *Physical Review E* 52(2): 2102–2105.

Foster, R. (1986). *Innovation: The Attacker's Advantage*. Summit Books, New York, pp. 87–111.

FSB (Financial Stability Board). (2021). FSB Roadmap for Addressing Climate-Related Financial Risks. Financial Stability Board, pp. 1–29. www.fsb.org/2023/07/fsb-roadmap-for-addressing

Hoefbauer, J., and Sigmund, K. (1998). *Evolutionary Games and Population Dynamics*. Cambridge University Press, Cambridge, pp. 42–66.

IEA (International Energy Agency). (2021). Net Zero by 2050: A Roadmap for the Global Energy Sector. IEA Special Report, pp. 1–222. www.iea.org/events/net-zero-by-2050-a-roadmap

IPCC (Intergovernmental Panel on Climate Change). (2018). Global Warming of 1.5 °C. IPCC Special Report. www.ipcc.ch/srl5/

Kim, J., Ramaswamy, K., and Sunderasan, S. (1993). Does Default Risk in Coupons Affect the Valuation of Corporate Bonds? A Contingent Claims Model. *Financial Management* 22(3): 117–131.

Kreuzer, H.J. (1981). *Nonequilibrium Thermodynamics and Its Statistical Foundations*. Clarendon Press, Oxford, pp. 98–107.

Leland, H.E. (1999). The Structural Approach to Credit Risk. In *Frontiers in Credit-Risk Analysis (AIM R Conference Proceedings)*, pp. 36–46.

Mayeski, A., and White, F. (2002). *The Ice Chronicles*. University Press of New England, Hanover, NH, pp. 80–199.

Merton, R.C. (1974). On the Pricing of Corporate Debt: The Risk Structure of Interest Rates. *Journal of Finance* 29: 449–470.

Mokyr, J. (1990). *The Lever of Riches: Technological Creativity and Economic Progress*. Oxford University Press, New York, pp. 81–191.

NGFS (Network for Greening the Financial System). (2020). NGFS Climate Scenarios for Central Banks and Supervisors. (June): 1–39. https://www.ngfs.net/sites/defau It/fi Ies/meelias/documents/8 20 I 84 ngfs scenarios final version v6.pdf

Nicolis, G., and Prigogine, I. (1977). *Self-Organization in Non-Equilibrium Systems.* Wiley-Interscience Publication, New York.

Nye, D.E. (1990). *Electrifying America: Social Meanings of a New Technology.* MIT Press, Cambridge, MA, pp. 138–286.

Porter, M.E. (1985). *Competitive Advantage.* The Free Press, New York, pp. 164–314.

Rogers, E.M. (1995). *Diffusion of Innovations.* The Free Press, New York, pp. 252–280.

Ruddiman, W. (2005). *Plows, Plagues and Petroleum.* Princeton University Press, Princeton, New Jersey, pp. 177–189.

Smith, P. (2020). The Climate Risk Landscape: A Comprehensive Overview of Climate Risk Assessment Methodologies. United Nations Environmental Programme Finance Initiative, pp. 1–48. www.unepfi.org/wordpress/wp-content/uploads/2021/

Sobehart, J.R. (1989a). Vacuum Magnetic Field Structure of Compact Torus Equilibria. *Physics of Fluids B (Plasma Physics)* 2(1): 222–224.

Sobehart, J.R. (1989b). Analytic, Two Fluid, Field Reversed Configuration Equilibrium with Sheared Rotation. *Physics of Fluids B (Plasma Physics)* 1(2): 470–473.

Sobehart J.R. (1990). Flux Trapping Efficiency During the Early Formation Phase of a Field Reversed Configuration. *Physics of Fluids B (Plasma Physics)* 2(9): 2268–2270.

Sobehart, J.R. (2016). The FinTech Revolution: Quantifying Earnings Uncertainty and Credit Risk in Competitive Business Environments with Disruptive Technologies. *Journal of Risk Management in Financial Institutions* 9(2): 163–174.

Sobehart, J.R. (2022). Quantifying Climate Risk Uncertainty in Competitive Business Environments. *Journal of Risk Management in Financial Institutions* 15(1): 26–37.

Sobehart, J.R., and Farengo, R. (1990). Low Frequency Drift Dissipative Modes in Field Reversed Configurations. *Physics of Fluids B (Plasma Physics)* 2(12): 3206–3208.

Sobehart, J.R., Maqueda, R., and Rodrigo, A.B. (1989). Early Formation Phase of Field Reversed Configurations in a Fast Theta Pinch, *Plasma Physics and Controlled Fusion Research,* IAEA, vol. 2, pp. 697–703.

Thomke, S., and Reinertsen, D. (2013). Six Myths of Product Development. *Innovation.* HRB's 10 Must Read Series, pp. 59–81, 83–100.

Twiss, B. (1974). *Managing Technical Innovation.* Longman, London, pp. 26–94.

USDT (U.S. Department of the Treasury). (2021). Fossil Fuel Energy Guidance for Multilateral Development Banks. https://home.treasury.gov/system/files/136/Fossil-Fuel-Energy-Guidance-for-the-MultilateralDevelopment-Banks.pdf

Zangwill, W.I. (1993). *Lightning Strategies for Innovation: How the World's Best Firms Create New Products.* Lexington Books, New York, pp. 135–209.

CHAPTER 6

Risk Management for Climate Risk and ESG

Risk reflects any event or action that may adversely affect an institution's ability to achieve its goals, execute its strategies, or result in adverse variations of profitability or in losses. Risk taking is a fundamental activity for financial institutions. To achieve their economic and strategic goals, financial institutions must carefully measure, monitor, and manage all their risks, including the impact of climate risk and ESG activities.

OVERVIEW, PURPOSE, SCOPE OF RISK MANAGEMENT, AND GOVERNANCE

Risk management in financial institutions covers processes, approaches, tools, and systems that allow the institution to implement risk-based policies and practices for measuring, monitoring, and controlling risks. The spectrum of risk types includes credit risk, market risk, interest rate risk, liquidity risk, operational risk, reputation risk, and conduct risk, to mention only a few major areas. The main aim of a risk-based approach is to provide safe and sound practices and to enhance the risk-return profile of the institution's portfolio. The financial industry has a wide range of risk management practices across banks, broker-dealers, asset management, insurance firms, and other types of financial institutions.

Tracking risks for risk management purposes requires models, practices, and processes for better measuring and monitoring risks and relating them to compensating controls. Without links to compensating controls, the models, risk measures, and monitoring are of limited use because they do not help in taking forward-looking corrective actions for risk remediation.

The move toward risk-based processes and practices covering climate risk and ESG now extends to the entire financial industry. While banking institutions and their approach to investing and managing capital have evolved rapidly over the last few decades, the pace of change has accelerated in recent years due to significant changes in government policies and the regulatory environment around the world, massive investments in alternative sources of energy, transportation, and infrastructure, and competitive pressure in a highly dynamic and uneven playing field for financial and non-financial sectors. From a competitive perspective, having the ability to differentiate short- and long-term risk characteristics of clients and counterparties and determining their contributions to the institution's strategies based on risk-return metrics are key issues. Without a balanced view on risks and returns, the institutions could have a limited myopic view of the consequences of climate risk and ESG impact on their own strategies, future income, and reputation.

The accelerated pace of change in the industry reflects the fact that financial institutions have shifted from identifying and developing improved methods for risk measurement and capital allocation to creating the necessary infrastructure for implementing these methods in their risk management practices and systems. While these approaches were developed primarily in the industrialized countries, this transformation is now migrating also into institutions around the world. As the field of climate risk and ESG risk management continues to evolve, companies adopt the innovations and lessons learned in the industry into their business strategies and risk management practices to enhance their expertise and create a sound framework to manage and mitigate risks.

Models for assessing various types of risks at the product, borrower, and portfolio level have a long track record of research in the industry and have been available for decades. However, advanced analytical methods for climate risk and ESG are still in the early stages of development as the industry moves forward to determine best practices. Many financial institutions now collect and process sufficiently rich and timely flows of data to make model implementation feasible to support their quantitative capital management and allocation strategies. They can identify portfolio characteristics that match their strategic goals, measure the degree to which their portfolio and business strategies matches these goals, and identify specific transactions and internal allocations to steer their investments in the desired direction.

Capital markets too have evolved to provide the investment channels needed to implement these strategies. Financial institutions can now manage their risks by selling loans or issue debt securities in an increasingly liquid debt market. They can also issue collateralized debt obligations (CDOs) to transfer risks off their balance sheet or sell risk exposure while retaining asset ownership, can tailor transactions to meet investor needs and issue credit default swaps (CDS) and other structured instruments to efficiently go long or short specific risks. This provides greater flexibility for financial institutions in hedging their portfolios and counterparty risks.

While overall strategic goals and management of common risks may be easily communicated in statements from management, the technical underpinnings and implications of risks related to climate risk and ESG are more difficult to define and communicate. Concepts such as reputation or conduct risk, driven by climate risk, or ESG issues, such as physical risk due to more extreme weather events, and risk due to the transition to a low-carbon economy, require sophisticated approaches for predicting social behavior and for risk measurement and monitoring. Management may not be able to articulate or reveal the details of how such metrics are determined. However, these concepts are now as important to investors, regulators, clients, and the general public as more traditional strategic goals. Therefore, institutions need to find effective ways to communicate to investors what they are doing, why, and how these techniques are likely to benefit their shareholders.

RISK IDENTIFICATION, MEASUREMENT, AND MANAGEMENT

Risk reflects the uncertainty in outcomes determined by the likelihood and severity of events that can lead to loss of capital and/or less-than-expected returns. In simple terms, the possibility of bad outcomes. How an institution interprets this definition helps to determine the strategic goals of risk management. The first step is to identify and classify the specific risks faced by the institution. The second step is to determine appropriate metrics so that risks can be measured and monitored. The third step is to apply portfolio analytics to determine the appropriate levels of capital to hold against these risks and manage down risks that seem excessive or are not supported by appropriate returns.

An explicit identification and classification scheme is needed in order to do the following:

- Determine who (which department) will be responsible for managing the risks.
- Develop methodologies for measuring specific risk types.
- Develop aggregation approaches to provide an integrated risk profile for the institution.
- Deploy capital within the institution on a risk-return basis.

The identification and classification of existing and emerging risks also lead innovation by:

- Recognizing the need for quantification of risks for measuring and monitoring.

- Providing risk managers with a proactive view on risk to enhance controls.
- Providing banking supervisors with a reference to develop guidelines, policies, and disclosure requirements.
- Allowing the development of new techniques or products for managing risk.
- Providing a reference for new processes or organizations for integrating and controlling risks.

Risks can be segmented into idiosyncratic and systemic components. *Idiosyncratic risk* is the risk due to unique circumstances as opposed to due to the overall market, industry, product, or group of borrowers or assets. In principle, idiosyncratic risk can be mitigated through diversification (*diversifiable risk*). In contrast, *systemic risk* arises from common factors or correlations affecting an entire financial market, industry, product, or group, and not just specific borrowers or assets. In practice, financial institutions may be unable to completely eliminate systemic risk through diversification or hedging (*undiversifiable risk*). Moreover, the situation could even worsen as the portfolio grows, since undiversifiable risks with residual correlations may compound their effects as business activities expand in scope.

Risks can be classified into broad types, which include market risk, credit risk, liquidity risk, asset-liability management risk, revenue or profit and loss (P&L) risk, operational risk, reputation risk, conduct risk, and compliance risk, to name a few (Bessis, 2005; Sobehart and Keenan, 2005). Institutions usually define their own multi-level hierarchy of risk types, where the most fundamental risks affecting the institution are listed in the first level of the hierarchy. Subsequent levels of the hierarchy may include more detailed risk sub-types to cover all the business activities. Figure 6.1 illustrates typical risk types.

Market Risk

Market risk results from the possibility that the price of an asset may decline simply because of specific issues affecting the asset or economic events that impact large portions of the market. Because assets are acquired with a specific purpose in mind, fluctuations in asset prices affect primarily those assets that must be liquidated or are marked at fair value. Buy-and-hold investors may ignore short-term price movements except in situations where they may be forced to sell the assets on short notice.

Climate-related events can result in changes to the risk preferences of market participants, leading to changes in market value of assets, the value of investments, higher volatility, credit spreads, or funding costs. Acute events driven by climate change can affect the revenue of individual companies (e.g. due to a severe hurricane) or the entire industry (e.g. reduced production for

Risk type	Description
Market risk	Market risk results from the possibility that the price of an asset may decline simply because of specific issues affecting the asset or broader economic events.
Credit risk	Credit risk covers both the possibility that a borrower will default, and the possibility that the borrower's credit quality will deteriorate, leading to an economic loss.
Liquidity risk	Liquidity risk can manifest itself in three areas: (1) the inability to raise funds at reasonable costs, (2) liquidity limitations in the markets, and (3) asset liquidity.
Asset-liability management risk	Asset-liability management risk reflects the mismatch between assets and liabilities that can cause a liquidity shortfall or result in losses due to balance sheet adjustments.
Profit & loss risk	Profit and loss risk reflects unexpected revenue volatility and may precipitate a tenor mismatch between expected cash flows from assets and liabilities.
Operational risk	Operational risk reflects losses resulting from inadequate or failed internal processes, people, and systems or from external events.
Reputation risk	Reputation risk is the potential that negative publicity will cause a decline in the customer base, costly litigation, or a reduction of revenue or market capitalization.
Conduct risk	Conduct risk reflects actions of a financial institution, or its individuals that lead to customer detriment, poor outcomes for customers, or negatively impact market stability.
Compliance risk	Compliance risk is the inability to fulfill regulatory requirements, rules, and laws, resulting in economic penalties, fines, or incremental capital requirements.

FIGURE 6.1 Risk types and their impact.

agriculture or fishing) in different geographies. This could lead to a reduction in market prices, increased volatility of commodities, equities, and debt, limitations of credit supply or availability of insurance and risk mitigation products. Climate-related events that are more chronic in nature could lead to a sustained impact on supply chains, import or export activities, regional or global economic conditions affecting capital markets, equities, debt issuance and borrowing, and interest rates. Changes in technologies and migration to a low-carbon emission economy can also create market risk by impacting the valuation of stranded assets and revenue generation (e.g. thermal coal mining, oil and gas, coal-fired power plants) and, consequently, the market value of the affected companies.

Credit Risk

Credit risk covers both the possibility that a borrower will default by failing to repay the principal and interest in a timely manner, and the possibility that the credit quality of the borrower will deteriorate, leading to an economic loss. Borrowers could go bankrupt or be forced to restructure debts under duress, which could result in substantial losses. In other situations, the risk faced by an institution is not related to the financial instrument itself but to a counterparty responsible for some aspect of the transaction. The risk that this party will prevent the settlement of the obligation for its full value, either when due or at an earlier time, is called *counterparty risk*.

Credit risk can affect the banking portfolio (or banking book) as borrowers delay loan payments, restructure their debt due to credit deterioration, or default on their obligations. Even a limited number of defaults for important corporate clients or systemic losses for retail customers can generate significant losses affecting an institution's solvency. Credit losses can also affect the trading portfolio (or trading book) as capital markets value the credit risk of debt issuers and borrowers, requiring a higher yield on their debt obligations. For over-the-counter instruments, such as options, swaps, or credit derivatives, sale may not be readily feasible, and institutions may face the risk of losing the value of the instruments and having to incur additional hedging or replacement costs.

Credit risk in a portfolio of assets arises from two basic sources: (1) idiosyncratic risk, and (2) systemic risk. Idiosyncratic risk reflects risks that are peculiar to individual companies, such as uncertain marketing strategies, new products, or managerial changes. Systemic risk reflects unexpected changes in macroeconomic and financial market conditions on the performance of borrowers. Borrowers may have different sensitivities to idiosyncratic and systemic risk, but few firms are completely indifferent to the wider economic conditions in which they operate. Therefore, the systemic component of portfolio risk is unavoidable and undiversifiable. The separation of risk into systemic and idiosyncratic risk components is useful because of the large-portfolio properties of idiosyncratic risk. As a portfolio becomes more and more fine-grained (i.e. the largest individual exposures account for a smaller and smaller share of total portfolio exposure), idiosyncratic risk is diversified away at the portfolio level. In the limit, when a portfolio becomes "infinitely fine-grained," idiosyncratic risk vanishes at the portfolio level, and only systemic risk remains.

No real-world portfolio is infinitely fine-grained. There is always some residual idiosyncratic risk that cannot be fully diversified away. If this residual risk is ignored, then an institution just satisfying minimum capital requirements based on their expected losses only may be undercapitalized when large loss deviations due to large concentration materialize. To avoid such an undercapitalization scenario, expected loss estimates may need a size/concentration adjustment based on the borrowers' exposure size effects and the discrete nature of default and credit events.

Climate-related events can lead to credit risk through multiple channels: (1) impacting the client's ability to pay interest, repay debt, or fulfill contractual obligations, (2) impacting the institution's ability to recover the value of outstanding debt in full or close out transactions with no credit losses, and (3) affecting the institution's reputation and ability to raise funds from capital markets and investors, which are used to fund investments and business activities. We will discuss these issues in more detail in later sections.

Liquidity Risk

Liquidity risk can manifest itself in three distinct areas: (1) the inability to raise funds at reasonable costs, (2) liquidity limitations in the markets, and (3) asset liquidity. First, funding liquidity depends on how capital markets perceive the creditworthiness and reputation of the borrowing institutions, which may limit the market participants' willingness to lend funds and may result in higher funding costs. This is closely related to the asset-liability management described below. Second, market liquidity depends on the willingness of counterparties to trade among themselves, which could lead to lack of traded volume, volatile market prices, and credit crunches. This can materialize as an impaired ability to execute transactions at reasonable prices and costs. Third, asset liquidity can result from the nature of the assets or as the result of a high level of asset customization as opposed to overall market liquidity. This can result in depressed prices, which could also affect the institution's funding liquidity. Despite extensive work done in this area over the years, modeling liquidity risk for market liquidity and asset liquidity remains a major conceptual challenge.

Climate-related events can result in unexpected demand for funds from corporate and consumer clients, counterparties, governments, and related entities for emergency loans to repair damage to assets or infrastructure due to physical risk. They can also result in increased demand for credit due to transition risk, including investment in new technologies or infrastructure, alignment to regulatory requirements, implementing net-zero strategies, or transitioning products and services to a low-carbon emission economy. The unexpected demand for funds could affect the ability of financial institutions to support these activities and raise funds from depositors, institutional investors, and capital markets.

Asset-Liability Management Risk

Asset-liability management (ALM) risk is related to funding liquidity as it reflects the risk that a mismatch between assets and liabilities will cause a liquidity shortfall or result in loss-generating balance sheet adjustments to avoid a liquidity shortfall. ALM risk is dominated by the risk that changes in the level of interest rates or their term structure will negatively affect the balance of assets and liabilities, leading to a reduction in portfolio returns and/or to a potential loss. However, ALM can also result from tenor mismatch or differences in the cash flow relationships between assets and liabilities due to inattention, operational and reputation risks. Banks and other lending institutions are particularly sensitive to ALM risk due to the nature of their business, often borrowing in the short term from depositors or capital markets and lending long-term loans or investing in riskier assets. To support these activities, they rely heavily on their reputation to ensure stability of the liabilities side.

Related to liquidity risk, climate-related events can result in unexpected demand for funds from corporate and consumer clients, counterparties, governments, and related entities for short- and long-term funds and investments. The unexpected demand for funds could affect the ability of financial institutions to balance the maturity and risk profile of their assets and liabilities.

Profit and Loss Risk

Revenue or profit and loss (P&L) risk is not the risk that the tenor or expected cash flows from assets will not match with those from liabilities, but that unexpected revenue volatility may precipitate such a mismatch, especially in businesses with significant fixed costs. For financial institutions, most shocks to revenue are the result of changes in market conditions or in their client base. For example, reduction in revenue due to falling interest rates combined with sluggish economic growth, or due to increases in loan defaults during economic downturns. Political, economic, or reputation crises can also lead to P&L volatility as markets react to these events. There is considerable overlap between P&L risk and other types of risk, as described above.

Climate-based events could impact the institution, its clients, counterparties, suppliers, institutional investors, and other critical stakeholders. Climate risk resulting from more extreme weather conditions, disruptive technologies, and innovation, or from transitioning to a low-carbon economy can result in deteriorating economic conditions for individual companies or for the industry as a whole and can trigger changes in regulatory requirements, policies, market reaction, and social behavior. This could affect the institution's execution of business and risk mitigation strategies, its ability to generate revenue, impact assets and investments, operations, reputation and, ultimately, the overall franchise.

Operational Risk

Operational risk is the risk of loss resulting from inadequate or failed internal processes, people, and systems or from external events. Examples include massive losses due to unauthorized rogue trading, internal and external fraud, criminal mismanagement, and corporate theft. This also includes failures of electronic trading, clearing or wire transfer systems, trading execution, or legal losses. Failures of an institution's plant and equipment or damage to its physical assets also reflect operational risk. Because of the rare and heterogeneous nature of these events, operational risk modeling presents multiple challenges.

Climate-related events, ranging from extreme weather events to activity restrictions on high-carbon-emitting sectors, can disrupt operations for the institution, counterparties, clients, and suppliers of products and services. These events can also impact the health, safety, productivity, and mobility of the workforce and critical infrastructure.

Reputation Risk

Reputation risk is the potential that negative publicity regarding a company's practices and actions will cause a decline in the customer base, costly litigation, revenue reduction, liquidity constraints, or significant depreciation in market capitalization. Reputation is one of the most valuable assets a company has, one of the most difficult to protect and the easiest to lose. If not managed appropriately, reputation risk can result in significant losses, a decline in franchise value, or even can force the company out of business. Reputation risk can be affected significantly by the impact of climate risk and ESG issues and social behavior. The avoidance of events that may damage an institution's reputation lies properly in the sphere of operational risk. The crucial components required to support an active reputation risk management regime include the identification of key constituents and the understanding of social influence and behavioral aspects that can affect the risk perception of clients, regulators, markets, and the general public.

Climate-related events can impact an institution's reputation from existing activities in high-carbon-emitting sectors or from providing financial support for those activities, from limited or insufficient commitments to transitioning to a low-carbon economy, or simply from associating with clients or counterparties that promote detrimental activities or contribute to high-carbon emissions. Reputation risk can also arise from providing limited support to clients who are trying to manage their own transition risk and require extensive financing.

Conduct Risk

Conduct risk is the risk that the actions of a financial institution or its individuals lead to customer detriment, poor outcomes for customers, or negatively impact market stability. Conduct risk is closely related to the institution's culture on safety and soundness and compliance practices.

Since financial institutions act as intermediaries of funding and capital for the economy, climate-related events can lead to increased risk for clients when financial institutions do not meet their commitments or are perceived not be meeting their climate-based commitments, or do not provide transparency to their clients on investments, risks, and lending practices.

Compliance Risk

Compliance risk is the inability to fulfill regulatory requirements, rules, and laws, resulting in economic penalties, fines, or incremental capital requirements. This can also result in reputation impact, which could also affect market perception and liquidity funding.

Climate-related events can lead to changes in policy and regulatory requirements, limitations of activities for high-carbon-emitting sectors, and increased scrutiny of industries perceived not to be meeting low-carbon emission commitments.

We now turn to climate risk and ESG, which are rapidly becoming top emerging risks for most institutions, growing beyond legal and social risks into financial, franchise, and reputation risks.

Climate Risk and ESG

Climate risk is the risk resulting from climate change or from efforts to mitigate climate change, its economic and financial consequences, and impact on the institution.

Climate risk is closely related to the *environmental, social, and governance* (ESG) aspects driving an institution's ESG policies and practices. As the impact of climate change and carbon dioxide (CO_2) emissions gain more visibility globally, there is increased interest in understanding how the industry landscape will be reshaped in response to the changing physical, economic, and political environment.

Climate risk can be divided into two key types: (1) *physical risk* (which refers to the impact of changes in climate and weather and their economic consequences), and (2) *transition risk* (which arises from actions to mitigate the economic impact of transitioning to a low-carbon economy). Table 6.1 describes the physical and transition risks.

Below we describe physical and transition risk in more detail.

TABLE 6.1 Climate risk contributions: physical and transition risks

Risk Type	Description
Physical risk	▪ Physical risk refers to financial losses that an institution may incur, directly or indirectly, as result of incremental climate changes (e.g. temperature increase) or due to extreme weather events (severe hurricanes or floods) that can result in damage or destruction of fixed assets, negative impact on supply of commodities, and decline in productivity or revenues
Transition risk	▪ Transition risk refers to financial losses that an institution may incur, directly or indirectly, as result of the transition to a sustainable lower-carbon economy ▪ Transitioning to a lower-carbon economy to counter the effects of climate change may result in extensive policy, legal, social, technological, and market changes ▪ Changes may occur at varying speeds, impacting how business operate or costs are incurred, rendering certain product offerings or capital equipment obsolete

Physical risk refers to the financial impact of a changing climate, including gradual changes in climate and increased frequency of extreme weather events, such as catastrophic floods, severe storms, and prolonged droughts or water scarcity. It also includes environmental degradation and pollution of water, air, and land, loss of biodiversity, and deforestation. Physical risk can result in damage to property, reduced productivity, or disruption of supply chains. The effects of physical risks could be both short- and long-term. Furthermore, physical risk is *acute*, if it arises from extreme events, such as floods, storms, and droughts; and *chronic*, if it is the result of progressive shifts with detrimental outcomes, such as increases in global temperature or sea level, loss of biodiversity, land and habitat destruction, or water and resource scarcity.

Acute physical risk includes:

1. **Droughts**, which are periods of extreme dry conditions that can last over long periods of time (from days to months to years).
2. **Wildfires**, which are fires fueled by brushes, trees, grasses, and other types of vegetation, usually when the weather is hot, dry, and windy.
3. **Floods**, which are events where severe rainfall, storms, or unusually high river flows can result in severe property inundation in urban or rural areas, damage to infrastructure, such as roads, schools, or hospitals.

4. **Severe storms**, which form primarily over tropical ocean areas and get their strength from heat driven by changes in ocean surface temperature. Severe storms (hurricanes, tropical cyclones, and typhoons) can have wind speeds of 40–150mph, with devastating consequences when passing over land, or creating storm surges and strong winds capable of leveling houses and damaging infrastructure.
5. **Heat waves**, which are periods of extreme high temperatures that can result in infrastructure failure, power shortages, reduced worker productivity, and loss of life.
6. **Cold waves**, which are periods of severe cold temperatures that can result in agriculture failure, disruptions of businesses and infrastructure or power shortages, also leading to loss of life.

Acute physical risk can result in sudden disruptions to operations, supply chain issues, logistics, and distribution, leading to reduced revenues or increased costs for companies (e.g. due to stronger hurricanes, more severe floods, or prolonged droughts). Higher frequency or more severe occurrences of events may impact critical infrastructure or assets (e.g. energy and power, buildings, roads, water and sewage, health facilities) or the quality of life for individuals (e.g. due to damaged infrastructure, limited access to alternative support, higher costs of living or insurance costs, decrease in the quality of transportation, reduced compensation, or income for affected businesses). It could also lead to health issues, injuries, or loss of life for severe events (e.g. stronger hurricanes or storms).

Chronic physical risk includes:

1. Changes in atmospheric, ocean, and land temperature, leading to social impact, changes in businesses, loss of work productivity, and energy and power shortages.
2. Changes in rainfall patterns, resulting in increases in precipitation in some regions while other regions become much drier. This could impact agriculture, forest and land management and other business areas.
3. Changes in sea levels, which may impact coastal areas, cities and their infrastructure, real estate properties, and the relocation of affected populations.

Chronic physical risk can lead to disruptions to operations, logistics or supply chain issues, which could lead to reduced revenues or increased costs. It could also lead to damage of assets and infrastructure or limit the ability to access or maintain property (e.g. due to frequent droughts or floods, rising sea levels, or higher temperatures). In addition, it could result in job loss or reduced worker income and productivity due to deterioration of working conditions (e.g. due to increase in humidity and temperature above health safety standards) and could lead to increased operating costs or reduced income.

Transition risk refers to direct or indirect financial losses that a company may incur as result of changes leading to a low-carbon emission economy. This could be triggered by different events, ranging from the abrupt adoption of climate and environmental policies and regulations to advances in carbon-related technologies and alternative sources of energy, to changes in market preferences, or in social influence and social acceptance of environmental issues. Policymakers, financial institutions, investors, and consumers are already making changes in response to climate risk concerns, a wide variety of environmental impact assessments and information driven by social and political factors. For example, policymakers and statesmen from many countries signed the Paris Agreement as a commitment to limit greenhouse gas (GHG) emissions and restrict future temperature changes to a range of 1.5°C to 2°C, as an effort to mitigate the impact of physical risks. This commitment will drive transition risk across different areas:

1. Regulations and supervisory activities:
   a. Policy changes focused on decarbonization, reduction of greenhouse emissions, impact of carbon taxes, energy restrictions, or business incentives.
   b. Supervisory activities focused on environmental compliance, emission restrictions, or other policy related issues.
2. Legal, reputation, and operational risks;
   a. Risks related to legal actions against the institutions, impact on reputation or operational risk resulting from business strategies and operations that are not fully aligned to climate policies or supervisory requirements.
3. Stakeholders, clients, and consumers:
   a. The effectiveness of low-carbon and other climate risk mitigation strategies is driven by their adoption by policymakers and regulatory entities, businesses, financial institutions providing capital, and consumers of the affected products and services.
   b. The adoption of the mitigation strategies reflects changes in the stakeholders' risk preferences, risk perception, and social influence, and could result in the following risks:
      i. **Financing and funding risk** due to financial institutions limiting their support to provide loans, debt or equity underwriting or other financial services to companies in high-carbon-emitting sectors.

ii. **Supply chain risk** due to key company partners' restrictions to provide or receive products or services from high-carbon-emitting companies.
iii. **Talent retention and employee risk** due to existing employees and potential candidates seeking to avoid working for companies perceived to be in high-carbon-emitting sectors, which also impacts company reputation.
iv. **Client and consumer risk** due to clients of financial services or consumers of products and services moving away from companies in high-carbon-emitting sectors and shifting to companies in low-carbon-emitting sectors. This can result in the loss of revenue or higher unexpected costs and in the acceleration of changes to the competitive landscape for the affected sectors.

Transition risk can result in changes to government policies, regulatory mandates, and emission restrictions, which could lead to the reduction or termination of company operations in specific sectors or industries, reduction in the demand for workers, or significant job loss. It could also result in social pressure, reputation issues, loss of franchise value, and legal action against institutions due to the impact of its operations and strategies on climate change. Transition risk could also affect the preferences and behavior of investors, customers, suppliers, employees, regulatory supervisors, and the general public toward the institution's business strategy and low-carbon emissions and net-zero commitments. The demand for low-carbon emission products and services relative to existing technologies could be disrupted by gradual or sudden changes in technologies and the infrastructure to support these new technologies (e.g. reducing the production and maintenance costs of electric vehicles (EV) and increasing the number and power efficiency of charging stations for the EVs).

Understanding the consequences of the issues described above for physical and transition risks is critical, as there is growing social and political pressure to divest from greenhouse emission activities, such as fossil fuel production and thermal coal mining, where stranded assets create significant credit risk in addition to climate risk.

The accumulated scientific evidence over the last few decades indicates that climate change results in measurable effects (Ruddiman, 2005; McCarthy, 2009; Plimer, 2009; IPCC, 2018; Smith, 2020). However, there is still significant uncertainty about the relative contributions of man-made and natural processes driven by solar activity, a tectonically active planet, periodic orbital changes of the Earth and the Sun, and other external factors that lead to cycles of global warming and cooling, which impact human activities and have resulted in the rise and demise of many civilizations over the ages.

Climate risk could be far-reaching in terms of the breadth and magnitude of its impact and may lead to significant short-term and long-term consequences (NGFS, 2020; FBS, 2021; IEA, 2021; USTD, 2021).

In this book we introduce an approach for quantifying the risks associated with the uncertainty generated by climate change and ESG issues and present a behavioral model of the impact of social influence on perception and decision making, which could provide insight into the social response and adoption of emission-reduction efforts. However, we recognize the highly uncertain nature of long-term forecasts as new technologies and rapid divestment of existing technologies and practices in high-carbon-emitting sectors could result in a disorderly and abrupt transition to a low-carbon emission economy with dramatic consequences for these sectors and their business competitive environment, the global economy, and society (Porter, 1985; Mokyr, 1990; Rogers, 1995; Day and Schoemaker, 2000; Case, 2007; Dicken, 2007; Day, 2013). In contrast, an orderly transition will require a balanced risk mitigation strategy with the appropriate political and social support to move to a low-carbon emission economy.

The management and mitigation of the risks described above require accurate measurement and monitoring, which are provided by the development of models and methodologies. Note, however, that models and their outcomes are subject to model risk. As analytics and models continue to evolve, supplemental considerations will be required for adequate model risk management and governance. This is particularly important for climate risk models, which usually cover both short-term and long-term effects for loss estimation and are sensitive to a wide range of assumptions that cannot be fully tested or can be tested with sparse or limited historical data.

Model Risk

Model risk covers all risk types above and may materialize as gaps between predicted model outcomes and realized values, leading to the model failure to reproduce observed values with reasonable accuracy. Models are abstractions of reality, subject to limited or incorrect assumptions, simplifications, misspecification, errors in statistical methods and hypothesis testing, limited or sparse data used for calibration and testing, and errors in model design, documentation, or implementation. Model risk is significant in the market and credit areas, which make relatively intensive use of models not only for measuring and monitoring risks but also for decisions on trading execution and extension of credit.

Model risk can result in significant adverse consequences from incorrect business decisions that rely on limited or deficient models of climate change for physical and transition risks and their impact on the institution. This book presents different advanced quantitative approaches covering the risk types described above in the context of climate risk and ESG, focusing on their benefits and limitations.

Critical to the identification, measuring, and monitoring of climate-based risks is the determination of transmission changes across different risk types. The transmission channels provide a means for climate risk drivers to materialize as a source of financial and non-financial risks to the institution, its assets and revenue generation, its counterparties, and clients. Table 6.2 illustrates the transmission channels and affected risk types.

REGULATORY ENVIRONMENT AND CLIMATE RISK

There is a wide range of global and local supervisory regulations and guidelines for financial institutions, covering banking, insurance, broker-dealers, securities issuance and trading, asset management, and other financial activities. The primary aim of these regulations and guidelines is:

1. to promote safe and sound business and risk management practices, including capital requirements aligned to the institution's risks
2. to protect customers from unfair or unlawful practices
3. to protect the integrity of the financial system by providing clear guidelines on trading and lending activities
4. to set common benchmarks, metrics, disclosures, and reporting standards to create a level competitive playing field across institutions, and measure and monitor their activities consistently
5. to create reproducible and sustainable practices, controls, and corporate governance
6. to drive institutions to develop and enhance their internal risk models and practices, and create incentives for their use in business activities and decision making.

The list above is by no means exhaustive but provides a sensible range of supervisory goals.

The combination of sound institutional practices and strong supervision in risk management, and improved disclosures, improves market discipline, the creditworthiness of financial institutions, and the overall stability of the financial system.

TABLE 6.2 Transmission channels for different risk types

Risk Type	Transmission Channel
Market risk	Climate risk drivers could impact a company's revenues or assets, affecting its creditworthiness and reputation, leading to market volatility or decline in equity prices, increase in credit spreads, or limits in the availability of funding
Credit risk	Climate risk drivers could impact a company's revenue or assets directly or indirectly, impacting the company's ability to service or refinance its debt and use of credit lines, therefore, affecting its creditworthiness. This may result in a higher probability of default on its obligations or the severity of losses.
Liquidity risk	Climate risk drivers can result in an unanticipated demand for incremental credit funding, a reduction on the value of assets posted as collateral or other issues that could limit the ability to gain funding. This may also result in higher probability of default on obligations and higher severity of losses.
Asset-liability management risk	Climate-related risks could lead to a mismatch between assets and liabilities that can cause a liquidity shortfall or an economic loss. This could result from increased credit risk from lending or providing services to companies in high-carbon-emitting sectors, or from reputation and liquidity risk to fund activities. Lending institutions often borrow in the short term from depositors and/or capital markets and lend long-term loans or invest in riskier assets. These activities rely heavily on their reputation to ensure stability of the liabilities side, which could be affected by their climate risk strategy and perceived risks.
Profit and loss risk	Climate risk drivers could have a significant impact on a company's ability to generate revenues, compete with other companies, adapt to changing economic conditions, retain customers, and comply with regulatory requirements
Operational risk	Climate risk drivers could affect a company's assets (e.g. physical damage to the company's buildings, commercial or residential real estate investments or other impact to assets), operations (e.g. disruption of services or products), or employees (e.g. productivity and ability to provide services)

(*Continued*)

TABLE 6.2 *(Continued)*

Risk Type	Transmission Channel
Reputation risk	Climate risk drivers can impact reputation if the company is perceived as unable or unwilling to implement low-carbon emission strategies, provide transparency on practices and alignment to low-carbon emission and net-zero commitments, or is perceived not to be able to meet regulatory requirements or industry practices
Conduct risk	Lax practices on safety, soundness, and compliance, and inadequate understanding of climate change drivers could lead to actions resulting in customer detriment, poor outcomes for customers, or negatively impact market stability
Compliance risk	Inadequate understanding of climate-related policies and guidelines may lead to the inability to fulfill regulatory requirements, rules, and laws, resulting in economic penalties, fines, or incremental capital requirements. This could also result in reputation impact, also affecting the institution's liquidity and funding ability.

OPERATIONAL READINESS AND RESILIENCE FOR CLIMATE-RELATED EFFECTS

The operational readiness and resilience for climate-related effects should be the result of a strong climate risk management framework. The general objective of a climate risk management framework is to promote a consistent approach to managing climate risk and providing governance, principles, and requirements for the adequate identification, measurement, monitoring, reporting, and controlling climate-related risks. The framework should describe the roles and responsibilities of employees across different lines of defense within the company, executive management, and the company board. The framework should also be able to accommodate evolving industry standards and practices in response to changes in regulatory requirements, competitive landscape, the institution's net-zero and related commitments, the institution's risk profile and organizational structure.

Key to the issues above is the role that the institution's board and senior management play in determining the institution's strategic view and oversight, setting the pace of change, and creating a culture that fosters responsible business and risk management practices.

1. **Board of Directors oversight:** A company's Board of Directors has ultimate oversight on the overall business strategy and the identification, assessment, and implementation of ESG and climate-related efforts and opportunities. The Board also determines the company's risk appetite tolerance, and limits based on materiality, franchise, reputation, and operational risks and on the company's commitments to transition to a lower carbon economy. The Board should receive regular reports regarding sustainability activities, identification, and monitoring of risks, including the institution's own plans on low-carbon emission and net-zero efforts and impact on businesses.
2. **Senior management oversight:** A sound governance framework for climate risk and ESG issues should also include the role of senior management at the highest level of the organization to oversee the governance and integration of climate risk efforts into the business strategies, and the review of risk management processes and controls, operations, and technology. The climate risk and ESG governance structure may continue to evolve for most institutions in response to regulatory, political, and social changes, which will require senior management to keep abreast of the latest industry developments to identify, assess, and manage climate-related risks. Senior management oversight could be supported by appropriate working groups, whose primary goals are to provide guidance and technical expertise on the integration of climate risk management and to assist in the coordination of resources across the institution. These specialized supporting groups and sub-groups could focus on risk identification, risk assessment, analytics and methodologies, risk measurement and monitoring, analysis under different regulatory and market scenarios, stress testing, risk controls, and governance. In addition, specialized groups dedicated to environmental, social, and climate issues could provide support on the institution's low-carbon emission and net-zero commitments and their impact on businesses and operations. This is an important issue as institutions could have different target dates for their long-term commitments on financing clients or supporting capital market activities in low-carbon emission sectors and their own net-zero commitments affecting their operations, which could be shorter-term.
3. **Business and risk management:** Business and risk management (Risk) play a critical role in supporting the integration of climate-related efforts into the institution's business strategy and risk management practices. These organizations work closely with clients to support their business plans as they transition to low-carbon emission or net-zero plans, which may require direct financing or access to capital markets. Corporate and investment bankers and risk managers with sound knowledge of

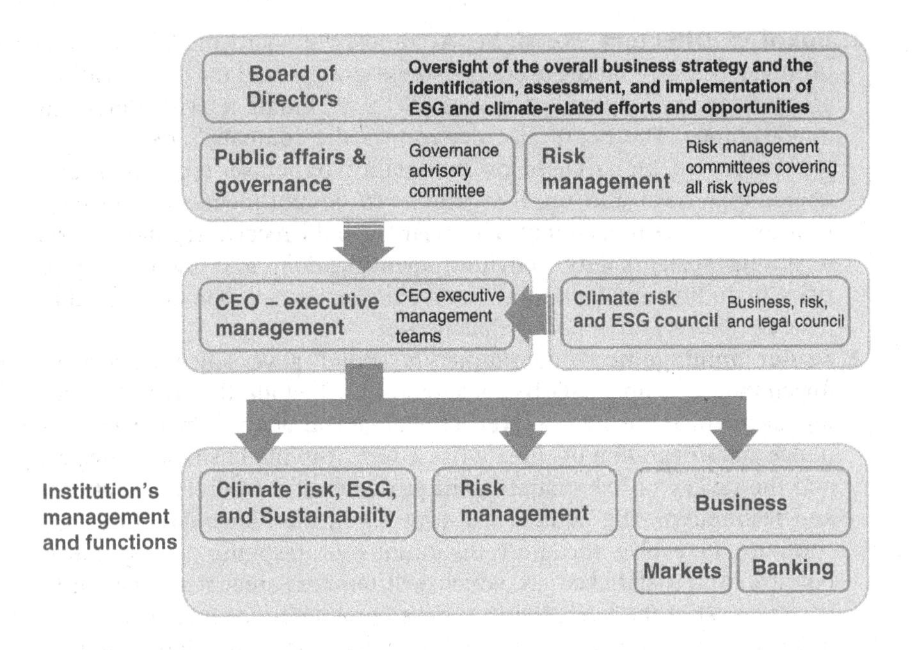

FIGURE 6.2 Idealized governance structure for climate risk.

high-carbon-emitting sectors, such as energy and power generation, oil and gas, metals and mining, car manufacturing, and transportation can support clients in their transition to a low-carbon emission economy and support their own institution's business strategy and commitments following closely their climate risk management framework.

Figure 6.2 shows an idealized governance structure for financial institutions with emphasis on the top-down organization hierarchy.

An institution's climate risk management framework should align to its internal policies and governance on risk management, taking into consideration risk management principles and practices, impact on firmwide programs, governance, and the overall institution's culture.

The framework should include clear guidelines on the following:

1. Development of policies, procedures, and risk appetite of climate risk-related activities aligned to supervisory guidelines and regulatory requirements for all legal entities of the company.
2. Defining key principles for assessing climate risk, including the identification of connections and *transmission channels* to other risk categories,

such as credit risk, market risk, operational risk, reputation risk, or conduct risk.

3. Providing guidelines for identifying risk concentrations and correlations across risk types.
4. Defining oversight functions on the impact of climate risk on credit risk, market risk, operational risk, reputation risk, conduct risk, and other risk types.
5. Defining stress-testing methodologies, scenarios, metrics, and thresholds for assessing risk.
6. Development of climate risk scorecards, risk assessment guidelines, training, and questionnaires for business activities.
7. Development of reporting, monitoring, and escalation guidelines, and recommendation of follow-up actions.
8. Establishing a strong culture of risk awareness and compliance and determining audit functions on climate-related activities.

The framework should also provide roles and responsibilities across the institution's units that create risk through their business activities (or first line of defense), those that assess and manage risk (second line of defense), and those units that provide independent assessments and assurance (third line of defense), such as internal audit.

The responsibilities for the first line of defense (e.g. corporate and consumer banking or loan underwriting) may include the following:

1. Identifying and managing climate risk and related risks in relation to their regular business activities.
2. Assessing and monitoring controls for the mitigation of climate risk.
3. Understanding relevant supervisory guidance on climate risk and ensuring compliance with relevant regulatory requirements.

The responsibilities for the second line of defense (e.g. risk management) may include the following:

1. Identifying relevant climate-related supervisory guidance and regulatory requirements.
2. Developing policies, processes, and practices for identifying, measuring, and monitoring activities related to climate risk.
3. Providing independent review and identification, measurement, reporting, monitoring, and escalation of relevant risks.
4. Advising and providing guidance to first line on risk management practices to measure, monitor, and control risks.

The responsibilities for the third line of defense (e.g. internal audit) may include the following:

1. Providing effective challenge to the institution's practices and their alignment to regulations.
2. Assessing the adequacy and timeliness of remediation plans for closing compliance gaps and monitoring progress against the implementation of these plans.

RISK MONITORING AND REPORTING

Financial institutions use a range of scenario analysis and stress-testing approaches to assess the impact of climate risk drivers on the company's franchise, reputation, and operations through different transmission channels. Scenario analysis provides a sound means to gain insight into possible climate-related outcomes and can uncover synergies across climate risk drivers, the competitive landscape of climate-impacted sectors, the demand for credit and financing for companies affected by physical or transition risk, low-carbon emission, and net-zero commitments, and their impact on the supply of credit.

The unprecedented consequences of physical and transition risks due to climate change strongly suggest that reliance on past events alone may not provide the breadth of scenarios required for detailed analysis of the likelihood and severity of climate-related risks. A more comprehensive and conceptually sound approach will require the combination of historical data, expert judgment, computer simulations, and industry benchmarks to create a broader range of plausible scenarios. These scenarios can be tailored to a company's specific risk profile to capture the effects of physical and transition risks, leveraging internal risk identification processes to prioritize areas of concern or vulnerability and their risk drivers. The granularity of the scenarios should consider the institution's portfolio or client base segments that share common climate risk drivers to identify geography, industry, product, or borrower concentrations.

Scenario analysis should be able to provide a forward-looking view of the economic environment and competitive landscape under different plausible assumptions for long-term horizons, which can drive the behavior and demand for credit of individual companies and the institution's strategy and supply of credit. The analysis of scenario assumptions and interpretation of scenario outcomes should lead to actionable results to help mitigate the impact of climate risk and help define target markets or opportunities for future business.

The design and implementation of climate-related scenarios should adhere to the institution's stress-testing framework and governance practices, including alignment to sound economic forecasting and the effective challenge of assumptions and outcomes by both business and risk management. This is relevant, given the significant uncertainty related to climate risk modeling (both for physical and transition risks) and their impact on the competitive landscape of different industries and the economy.

The design and implementation of reliable climate-based scenarios can be leveraged to evaluate changes to the economic environment and business opportunities, including the identification and assessment of material risks facing the institution. Physical and transition risks may impact the company's revenue, franchise, reputation, or operations; however, their effects may not materialize for years.

Climate risk-related analysis provides valuable information for the strategic planning process, allowing evaluation of existing business strategies in the context of a changing economic environment, material risks facing the institution, and potential business opportunities. Potential new business activities require a review of risks and the analysis of how these activities could impact the institution's risk profile and strategies. Business and risk management need to integrate climate considerations into the monitoring of these activities, including identification, escalation, measurement, and remediation of breaches of the institution's risk limits or risk appetite targets. Adequate controls and risk reduction strategies should be in place to mitigate material risks associated with climate risk drivers. This requires the implementation of processes and governance for ensuring that the controls are designed and are operating effectively, and that open issues are remediated on a timely basis. Table 6.3 shows key impact considerations for franchise, credit, and reputation risks.

RISK REPORTING AND ANALYSIS

Climate risk reporting and analysis should provide complete, comprehensive, and timely information on the institution's efforts to identify, measure, monitor, and manage climate risks and new business opportunities. Table 6.4 shows the focus areas for climate-related reporting activities.

Critical to managing climate risks effectively is gathering data and implementing processes, technology infrastructure, and governance that can support these efforts. There are significant data, analytics, and technology challenges for implementing such an approach. Data identification, data integrity, and data completeness can impact the ability to determine damage to physical assets (e.g. for geo-location of assets in flood zones) or to determine

TABLE 6.3 Impact considerations for key risk types

Risk Type	Impact
Strategic and franchise risk	■ Would climate risk-related activities impact the franchise or strategic plans? ■ Would the activities impact client relationships or the institution's ability to generate revenues? ■ Would the activities impact low-carbon emission or net-zero commitments?
Credit risk	■ Would the activities lead to increased exposure to physical or transition risks that can materialize as credit losses? ■ Would the activities impact the institution's credit quality of assets? ■ Would the activities result in increased funding costs to the firm?
Reputation risk	■ Are the activities aligned to the institution's sustainability or net-zero strategy? ■ Are the activities consistent with low-carbon emissions commitments? ■ Could the activities impact the firm's reputation and public perception?

TABLE 6.4 Focus areas for climate-related reporting activities

Focus Area	Activities
Low-carbon emissions and net-zero reporting	■ Progress on commitments and efforts toward a low-carbon emission economy or net-zero strategies ■ Reports on financed emission targets, metrics, and breaches ■ Progress against exposure reduction targets for high-carbon-emitting sectors (e.g. thermal coal mining, coal-fired power, oil extraction and refining, transportation)
Risk reporting	■ Progress against regulatory requirements or regulatory remediation ■ Alignment to risk appetite targets, limits, and breaches ■ Reports on the institution's risk profile, specific products, or individual business units ■ Reports on the institution's climate risk initiatives ■ Reports on credit, market, operational, franchise, or reputation risks resulting from climate risk drivers

the full scope of emissions for companies in high-carbon emission industries. There are many data providers in the industry today who offer a wide range of data services with different coverage, quality, and analytics capabilities. Notice, however, that, while there are multiple sources of very detailed information and forecasts of flood zones, severe heat and humidity, wind and other physical perils, the translation of their severity into credit loss estimates is still subject to significant uncertainty due to the lack of refined risk models. Until common standards and best practices on disclosure and data collection are firmly established in the industry, institutions need to work with multiple sources of data, including with clients and data vendors, to collect relevant information for data analysis, risk assessment, modeling, and reporting (Keenan, 2015). Also, as data analytics and models continue to evolve and mature, additional considerations about model complexity and sustainability will be needed for adequate risk management and governance of analytics and modeling.

Appropriate data governance standards and strategies should be in place to ensure adequate planning for implementation and support of data sources, data integrity and quality assurance, and data analytics (Keenan, 2015). The supporting data and technology infrastructure should focus on the key areas described in Table 6.5.

RESHAPING THE INDUSTRY LANDSCAPE: WINNERS, LOSERS, AND SYNERGIES

Transition to a low-carbon economy will have profound implications for the affected industries, which will need to adapt by effectively identifying their climate-related risks and opportunities.

The recent shift away from fossil fuels toward alternative energy sources has already impacted the energy sector landscape. Adapting to a changing industry landscape will require more detailed information for transactions, products, or services involving high-carbon emission sectors, such as oil and gas or thermal carbon mining and coal-fired power generation. For example, high-carbon emission power generation companies may be exposed to transition risks as their primary revenue stream can be impacted by new regulatory restrictions or changes in policies with respect to consumption of oil and gas. Similarly, low-carbon emission power companies that rely heavily on renewable energy sources with an unproven record of reliability may be unable to meet future customer demand or suffer reputational damage due to the inability to service customers.

These assessments should include both physical and transition risks since they can impact collateral management due to the repricing of stranded

TABLE 6.5 Key areas for data and technology infrastructure

Area	Description
Economic environment	▪ Fundamental information on the economic and market environment ▪ Description of competitive landscape for the institution and its clients
Risk profile	▪ Risk profile of the institution's businesses and portfolios, including exposure and risk concentrations to products, industries, or geographies ▪ Transaction-, borrower-, and counterparty-level risk assessments of climate-related impacts, emissions, and asset-specific analysis (e.g. risk assessment of real estate properties affected by physical risk) ▪ On-demand capability to quantify climate-related impacts for individual borrower, counterparties, products, asset classes, or business segments.
Scenario-based risk assessment	Scenario-based risk assessments used for portfolio reviews, strategic planning, internal and regulatory stress testing, focusing on: ▪ Competitive landscape and scenarios: Information on industries, technology development, and business demand across low- and high-carbon emission sectors ▪ Credit demand: Client needs for funding and use of credit lines and financial products, investment strategies, securities issuance, and access to capital markets ▪ Credit supply: Institution's business strategy, target markets, and their impact of low-carbon emission and net-zero commitments ▪ Franchise, strategic and reputation risks driven by climate risk drivers
Disclosures	▪ Regulatory requirements and disclosures, including commitments on transition to low-carbon emission and net-zero economies
Risks and opportunities	▪ Emerging business risks and opportunities across different sectors and geographies ▪ Regulatory and policy changes and their business impact ▪ Social and political changes and their impact on the franchise and reputation

assets for fossil fuel extraction, thermal carbon mining, or coal-fired power plant operations or the repricing of real estate assets in locations exposed to severe climate events. Accounting for the impact of stranded assets due to climate-related write-downs is an important aspect of the industry reshaping and the ability to fund investments.

The impact of transition to a low-carbon economy will not be limited to high-carbon emission sectors but will spill over to other industries through direct and indirect dependencies (e.g. through energy consumption, car manufacturing, or logistics and transportation). Companies seeking funding for their business activities will be exposed to a potential reduction in revenues, increased operating costs, supply-chain issues, greater regulatory burden, or reputation issues that can impact their franchise and ability to service their customers. Financial institutions may need to enhance their due diligence to identify industry and geography concentrations, including having a deeper understanding of energy use, emission data, low-carbon emission commitments, plans for production and operations and mitigation strategies for supply-chain exposure. Figure 6.3 describes a potential reshaping of the industry landscape as a result of a transition to a low-carbon economy.

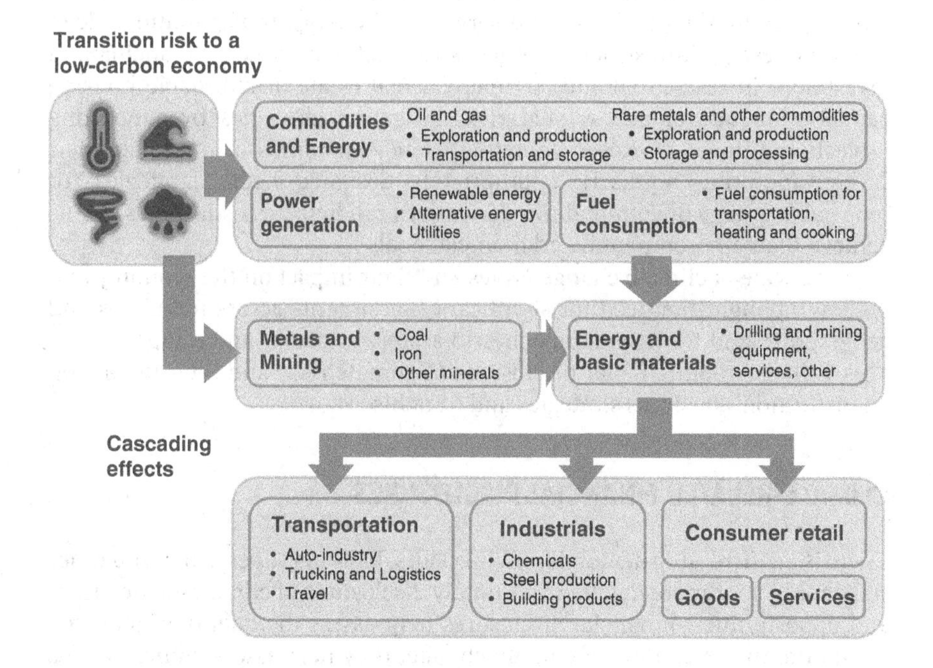

FIGURE 6.3 Potential reshaping of the industry landscape as a result of a transition to a low-carbon economy.

SUSTAINABILITY, COMPETITIVE ENVIRONMENT, AND A LEVEL PLAYING FIELD

Environmental finance and sustainability, which reflect financial activities that reduce or mitigate the impact of climate-related risks, are currently the focus of many financial institutions. The increased interest in environmental finance and sustainability has been driven primarily by the accelerated growth in renewable energy sources and the need for funding, changes in policies and regulatory requirements on high-carbon emission activities, the development of new technologies, changes in investor preferences and consumers, and new financial products oriented to a low-carbon economy (such as green bonds or ESG-linked bonds with incentives if certain sustainability targets are met or penalties if they are not). Another contributing factor has been the accumulated efforts by institutions and various organizations to promote and achieve their own sustainability targets. This requires helping their clients achieve their sustainability goals by providing funding and services while also supporting their low-carbon emission commitments toward a sustainable economy. The need also extends to national and subnational governments, states, municipalities, government-related entities, and utilities companies that require funding and financial services for energy, transportation, clean water, or sewage infrastructure projects to meet local needs. Financing and investment in residential and commercial real estate or affordable housing can also include climate-related risk mitigation strategies by supporting projects that meet or exceed energy-efficiency standards or minimize their carbon footprint. Similarly, appropriately designed hedge strategies for renewable energy or commodities derivatives can help an institution's clients achieve their environmental and financial goals.

The scale of climate change issues and their impact on the economy and society highlight the need for significant investments across industries and geographies and for new approaches to assess, measure, and mitigate risks. This will also require better disclosures of climate risks and their impact on an institution's business strategies and franchise value.

CLIMATE-RELATED FINANCIAL DISCLOSURES

There is an increasing demand from investors, banking supervisors, and other stakeholders for more detailed and timely disclosures of climate-related risks and opportunities. Financial disclosures help assess the quality of management and, in the context of climate change, they help assess vulnerabilities to physical and transition risk and the company's readiness for transition to a low-carbon economy. Unfortunately, without a consistent and comprehensive international regulatory framework to steer the content of financial disclosures for climate-related issues, the financial industry and markets continue to struggle with inconsistencies in the reported climate and ESG disclosures needed to accurately assess risks.

Although there is no common industry standard for climate risk disclosures, there are a few established organizations that provide disclosure guidelines and principles. Among these organizations is the Sustainability Accounting Standards Board (SASB) and the Global Reporting Initiative (GRI), which issue guidance on ESG topics. In Europe, the Sustainability Finance Disclosure Regulation focuses on sustainability. At a global level, the Task Force on Climate-Related Financial Disclosures (TCFD), created by the G20 Financial Stability Board (FSB) whose mandate is to monitor the stability of the global financial system, provides guidelines on measuring, managing, and reporting climate risk across four dimensions: (1) governance, (2) strategy, (3) risk management, and (4) metrics and targets. The TCFD group also calls on financial institutions to disclose emissions following the guidance issued by the Partnership for Carbon Accounting Financials (PCAF), which focuses on harmonizing accounting methodologies for disclosing greenhouse gas emission across institutions. Among these options, the financial industry and market participants have coalesced mainly around the SASB and TCFD approaches because of their industry focus, and the scope and usefulness of the disclosed information.

It is important to recognize that assessing the materiality of climate risk requires thoughtful consideration of metrics and the purpose of the assessment to provide clear communication to a wide range of stakeholders such as clients, investors, auditors, supervisory entities, and the general public. An institution's disclosures of environmental, social, and governance (ESG) and climate-related risks should include topics relevant to businesses, investments, and decisions relevant to investors and other stakeholders.

For communicating ESG and climate-related risks and investment opportunities, institutions could adopt a comprehensive framework aligned to disclosure rules and regulations from different jurisdictions and applicable exchange listing standards (e.g. US Securities and Exchange Commission or SEC) combined with the recommendations from entities such as the TCFD and similar organizations. The disclosure framework should cover two interconnected issues:

1. the impact of ESG and climate risk on the institution's decisions and activities
2. the impact of the institution's decisions and activities on ESG and climate risk.

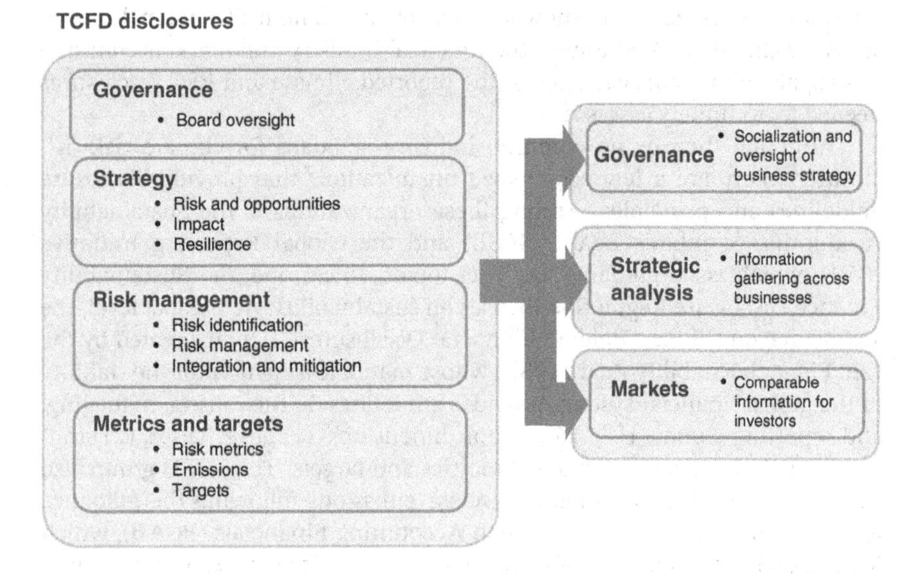

FIGURE 6.4 Key components of a climate risk and ESG disclosure framework.

For financial institutions, the former is mainly related to the regulatory environment, competitive landscape across industries, and demand for credit from clients, while the latter is primarily related to the institution's strategic decisions and supply of credit aligned to the institution's low-emission or net zero commitments. Figure 6.4 shows key components of a climate risk and ESG disclosure framework.

To communicate clearly and accurately to all stakeholders how the various climate risk drivers may impact an institution, the institution's disclosure framework should include a categorization of relevant physical and transition risks, their key drivers, metrics, and targets, and their short- and long-term impact, as shown in Table 6.6. The time horizons for short- and long-term periods can be defined in terms of the institution's business strategy. For illustration purposes, here we defined them roughly as less than three years for short to medium term, and more than three years for long term, which are aligned to the typical strategic planning exercises, and a variety of regulatory stress testing exercises.

The inherently uncertain nature of short- and long-term forecasts of climate-related issues limits the reliance on materiality assessments. Many assumptions and estimates are likely to change over time to reflect changes in the general understanding of climate impact and political and social changes. In addition, low-carbon emission and net-zero strategies will also evolve

TABLE 6.6 Short- and long-term impact of climate risk for different risks

Risk Type	Category	Description	Short- and Long-term Impact
Physical risk	Acute physical risk	Acute risk arises from extreme events such as floods, storms, and droughts	■ Disruptions to operations, supply chain issues, logistics and distribution ■ Impact on critical infrastructure or assets ■ Impact on quality of life for individuals ■ Health issues, injury, or loss of life for severe events
	Chronic physical risk	Chronic risk is the result of progressive shifts with detrimental outcomes, such as increases in global temperature or sea level, loss of biodiversity, land and habitat destruction, or water and resource scarcity	■ Disruptions to operations, logistics or supply chain issues. Damage of assets and infrastructure ■ Limited ability to access or maintain property ■ Job loss or reduced worker income and productivity due to deterioration of working conditions ■ Increased operating costs or reduced income
Transition risk	Regulations and supervisory activities	Policy changes focused on decarbonization, reduction of greenhouse gas emissions. Supervisory activities focused on environmental compliance, emission restrictions, or other policy-related issues	■ Impact of carbon taxes, energy restrictions, or business incentives ■ Environmental compliance ■ Emission restriction targets, limits, and breaches
	Legal, reputation and operational risks	Risks related to legal actions against the institutions, impact on reputation or operational risk resulting from business strategies and operations that are not fully aligned to climate policies or supervisory requirements	■ Legal actions ■ Impact on reputation or operations due to misalignment to climate policies or supervisory requirements or ESG activities

(Continued)

TABLE 6.6 *(Continued)*

Risk Type	Category	Description	Short- and Long-term Impact
	Stakeholders, clients, and consumers	The effectiveness of low-carbon and other climate risk mitigation strategies is driven by their adoption by policymakers and regulatory entities, businesses, financial institutions providing capital and consumers of the affected products and services. Key risks: **1.** Financing and funding activities in high-carbon-emitting sectors **2.** Supply chain risk to provide or receive products or services from high-carbon emission companies **3.** Talent retention risk due to climate-related reputation impact **4.** Clients and consumers of products and services moving away from companies in high-carbon-emitting sectors and shifting toward companies in low-carbon-emitting sectors	▪ Adoption of low-carbon and risk mitigation strategies by policymakers and regulatory entities, businesses, financial institutions providing capital and consumers of the affected products and services ▪ Financing and funding activities in high-carbon-emitting sectors ▪ Company's supplier restrictions to provide or receive products or services ▪ Talent retention risk due to reputation issues ▪ Clients and consumers of products and services moving away from companies in high-carbon-emitting sectors

as institutions adapt to changes and redefine investment targets. Related disclosures should be expected to be amended, restated, or updated to reflect changes in strategies, methodologies or improvements in the accuracy and completeness of data and estimates.

CLIMATE RISK FINANCIAL DISCLOSURES

Traditionally, climate change and financial markets have been seen as vaguely related. As market participants, governments, regulators, and the general public become more informed on climate change issues and their impact on different industry sectors, the development of alternative energy sources and technologies and new regulations, financial markets are increasingly considered to play a critical role in supporting the transition to a lower-carbon emission economy and the mitigation of climate risks. In response to this renewed interest in environmental issues, the volume and quality of information now available on climate change impact have increased significantly, providing both (1) additional visibility of this issue to market participants and investors, and (2) a positive feedback mechanism for markets to require increasing levels of information and disclosures from companies on their climate-related activities and risk mitigation strategies. Companies with better disclosures provide markets with adequate information on their carbon footprint, emission reduction, and net-zero strategies, which are used by market participant to select investment opportunities and can be leveraged by financial institutions and institutional investors to manage their own climate risk strategies and improve disclosures.

Several initiatives are currently underway in the industry to enhance information access and disclosures. Among these initiatives are the recommendations of the TCFD, which outline basic principles of financial disclosures regarding climate change. The TCFD recommendations are aligned to two basic goals:

1. Prompt companies to manage and mitigate their individual climate-related risks.
2. Allow markets to assess and price risk appropriately.

Note, however, that because disclosures are not presently comparable across companies or sectors and due to the uncertainty involved in long-term forecasting of climate change impact, the second basic goal is not completely fulfilled in today's markets. More precisely, financial markets are still far from being able to assess properly all aspects of climate-related risks and opportunities and price them correctly. This can create information arbitrage opportunities that can lead to an inadequate transfer of risks among market participants or even misrepresentation of risks and rewards. Below we outline some topics related to financial disclosure that need improvement.

Also notice that, as result of government actions and regulations, and the increasing media coverage of climate-related topics usually biased toward dramatic views that are often politically or economically motivated, for many market participants and the general public the ultimate goal of climate change risk mitigation seems a significant decarbonization of the economy. That is, the transition of the global and regional economies to lower carbon-intensive processes and operations. As an increasing portion of the financial system and many regulatory entities become more focused on this goal, market emphasis on decarbonization among companies and investors will keep growing.

The TCFD was created in 2015 by the Financial Stability Board (FSB) to address the risk of disorderly pricing events that could disrupt the financial markets as a climate change transition unfolds. In the worst-case scenario of poor, limited, or contradictory disclosures, markets could react to sudden news on physical risks or transition risks from decarbonization, which could cause a destabilizing run on trading activity for certain industry sectors or specific names. The TCFD is a 32-member global task force whose objective is primarily to recommend a comprehensive framework to guide companies on their financial disclosures of climate-related issues. In 2017, the TCFD released its final recommendations, which were followed by annual status reports on their industry adoption. The TCFD recommendations require companies to focus on four key areas for their climate-related disclosures:

1. Governance
2. Strategy
3. Risk management
4. Metrics and targets.

These four key recommendations, developed to address the concerns that the financial system could be underestimating the short-term and long-term economic impact of climate-related risks, could lead to disclosures that are consistent, reliable, and comparable across institutions and can help financial institutions, investors, and supervisory entities evaluate risks and opportunities. Under these recommendations, companies are encouraged to provide better climate-related disclosures that can help minimize the risks to the financial system and be used to support the transition toward a more resilient low-carbon emission economy. Table 6.7 illustrates key components of the proposed disclosures and supplemental information that can help gain insight into a company's climate risk vulnerability and mitigation strategies.

Climate risks are expected to impact the overall economy but there are some specific sectors that are more vulnerable due to their carbon intensity or

TABLE 6.7 Key components of the TCFD financial disclosure recommendations and supplemental information

Type	Vulnerabilities and Risk Mitigation
Governance	
Board oversight	■ Are climate risks presented to the Board and senior management regularly and with enough depth? ■ Are these risks evaluated as part of the regular decision-making process aligned to the company's risk appetite?
Strategy	
Risk and opportunities	■ What are the key climate-related risks and opportunities? ■ What are the company's scope 1, 2, and 3 emissions and transmission channels? ■ What are the company's sources of emissions and alternatives to mitigate risks? ■ What company assets, operations, or logistics are exposed to short- and long-term physical risks? ■ What company assets, operations, or logistics are exposed to short- and long-term transition risks? ■ How do climate-related risks affect other risk types, such as credit, operational, or reputational risks?
Impact	■ What are the likely impacts of physical and transition risks? ■ What analytical approach is used to identify vulnerabilities and their impact? ■ Which scenarios are used to assess the impact? ■ How reliable are the assumptions of scenarios and analytical approaches? ■ What are the magnitude and severity of physical risks and their channels? ■ What are the magnitude and severity of transition risks and their channels? ■ What are the magnitude and severity of other climate-related risks?
Resilience	■ Is the company resilient to its physical risks? ■ Is the company resilient to its transition risks and the regulatory environment? ■ What are the remaining vulnerabilities to physical and transition risks and other climate-related risks? ■ What is the plan for mitigating remaining vulnerabilities?

(*Continued*)

TABLE 6.7 *(Continued)*

Type	Vulnerabilities and Risk Mitigation
Risk management	
Risk identification	■ What is the process for identifying and monitoring physical and transition risks? ■ What is the process for identifying and monitoring other climate-related risks?
Risk management	■ How are physical and transition risks measured, monitored, mitigated, and managed? ■ How are other climate-related risks (such as credit, market, operational, and reputation risks) measured, monitored, mitigated, and managed?
Integration and mitigation	■ How are the risk identification and risk management processes connected for adequate governance, monitoring, and risk mitigation? ■ What are the gaps and vulnerabilities?
Metrics and targets	
Risk metrics	■ What are the critical dimensions for measuring risks? ■ What specific metrics are used to measure, monitor, and mitigate these risks? ■ What are the assumptions, benefits, and limitations of the selected metrics?
Emissions	■ What are the scope 1, 2, and 3 emissions for different business, operations and processes? ■ How are the emission metrics defined? ■ What are the observed changes over time?
Targets	■ What are the company's climate risk, decarbonization, and net-zero targets? ■ What are the observed emission trends and their alignment to the company's risk mitigation strategy?

exposure to physical or transition risks. Recall that physical risk refers to the physical impact of climate change, leading to incremental changes in precipitation or temperature, rising of sea levels or more frequent and severe weather events that can result in disruptions to operations, damage to physical assets, or the destruction of critical infrastructure. In contrast, transition risk reflects risks from the transition to a low-carbon economy resulting from changes in technology and innovation, policy changes and carbon pricing and taxing that can lead to shifts in market supply and demand.

For physical risk, for example, financial disclosures could reflect specific analysis on how companies may react under different scenarios with more frequent and severe weather events that can affect the productivity of assets in different locations, the company's ability to generate revenues, and the impact on credit quality. For transition risk, financial disclosures could reflect analysis on how companies in different sectors of the economy may react to climate transition scenarios (such as changes in carbon taxes or new regulations) and how their operations, revenues, and costs are affected. This type of impact is reflected ultimately in the company's credit quality, its viability as a firm, and ability to honor its obligations. To illustrate, a scenario with a transition to a low-carbon emission economy leading to a 1.5°C increase in global temperature could impose a disproportionate increase in carbon taxes paid by companies that emit greenhouse gases. This could not only affect the oil and gas sector (indirectly through a reduction in the demand for fossil fuels) but also the power generation and utilities sectors. An unregulated utilities company with a high-emission footprint due to heavy use of fossil fuel or thermal coal could be impacted directly by carbon taxes without the ability to transfer the costs to consumers. In contrast, a regulated utilities company could potentially pass part of the costs to consumers as permissible, partially mitigating the costs of carbon taxes or capital investment supporting the transition to low-carbon emissions. This situation could create an uneven playing field for competition that could reshape the industry.

Better financial disclosures across companies benefit all stakeholders. For lenders, having more reliable information about climate risks enables them to better assess the credit standing of their clients and to improve the lenders' own climate-related analysis and disclosures. For investors and financial supervisors, better disclosures enable them to compare across industry and portfolio segments and assess systemic issues to be able to price risk accurately. While there are aspects of the TCFD recommendations that companies have been disclosing consistently for many years, these recommendations provide a common framework to incorporate climate-related scenarios into a company's strategy and financial disclosures. The TCFD recommendations are now a major framework used by companies and market participants to disclose and review climate risk and the path to decarbonization of the economy. Table 6.8 illustrates relevant information categories leveraged by the markets to assess and price risk.

Among different types of information, the assessment and pricing of risk relies on the concept of *scope 1, 2, and 3 emissions*, which are used regularly in the climate change to describe the hierarchy of emissions starting from the company itself and then moving outwards to suppliers and customers.

TABLE 6.8 Relevant information categories used for assessing and pricing risk

Activity	Current Condition	Trend or Direction
Carbon-emission footprint	■ Share of scope 1 and 2 emissions. Assessment of scope 3 emissions	■ Changes in share of scope 1 and 2 emissions
Emission costs risk	■ Share of scope 1 and 2 emissions and emission costs.	■ Trend in changes for scope 1 and 2 emissions ■ Expected abatement costs
Revenue risk	■ Diversification of revenues, sources, and market share	■ Revenue generation forecasts
Stranded assets risk	■ Life cycles and break-even prices for production assets ■ Regulatory environment and policies	■ Change in contract structures and asset renewals ■ Changes in regulations and policies
Opportunities	■ Sources of revenues and competitive environment	■ Capital investments, R&D, and alternative sources of revenue
Vulnerability, resilience, and adaptability	■ Diversification of revenues ■ Changes in high-carbon emission production ■ Alternative low-carbon emission production approaches ■ Vulnerability to regulatory environment and polices	■ Changes in revenue diversification ■ Changes in high-carbon and low-carbon emission approaches ■ Adaptability to changes in regulations and policies

Scope 1 emissions refer to the company's operational emissions such as emissions from generating its own power, heating buildings, or from its vehicle fleet. Scope 2 emissions refer to emissions created by third parties to support the company's operations, for example, generating power for the company's buildings. Scope 3 emissions refer to all emissions in the company's value chain not covered in scopes 1 and 2. This category is very broad, covering emissions in the supply chain, indirect emissions from capital investments or the construction of a factory. Scope 3 emissions also refer to the emissions from the company's customers, the products they sell, and other downstream emissions. In recent years the focus on scope 3 emissions has grown as they are the most material for some sectors and the most difficult to assess.

Another relevant piece of information is the effectiveness of the company's actions to manage risks and opportunities. Having thoughtful and proactive sustainability programs with adequate risk identification and risk management policies that are not executed effectively highlights a disconnect with the company's strategy and decision making.

Adoption of the TCFD recommendations since their initial release has been quick but variable across sectors and jurisdictions, with numerous companies pledging to report their financial disclosures in accordance with the framework. Furthermore, the Network for Greening the Financial System (NGFS), which includes 36 major central banks as members, also encouraged companies to disclose financial information following the TCFD recommendations. The United Nations Principles for Responsible Investing (UN PRI), which covers roughly $80 trillion in assets under management, also includes key aspects of the TCFD disclosure approach.

Although the original purpose of the financial disclosure recommendations is to prompt companies to manage and mitigate their individual climate-related risks, and allow markets to assess and price these risks, in practice, markets and financial institutions leverage this information to ensure that individual companies not only disclose their risks and mitigation strategies but also manage them effectively. Thus, the alignment to common financial disclosures for measuring, monitoring, mitigating, and managing risks for individual companies can facilitate the tracking of the overall decarbonization of the economy.

This dual purpose represents a challenge for disclosures as they may require different levels of information. Also, risk pricing requires market participants to have the ability to compare the risks faced by different companies on a common scale and then assess how much they are willing to pay, based on their risk preferences and expected returns. Information relevant to pricing covers issues ranging from direct climate risk impact on operations, revenue, or assets, to sensitivity to different scenarios and break-even price of assets exposed to physical and transition risks, to reputation, legal, and regulatory vulnerabilities. Table 6.9 illustrates key items in the recommended TCFD financial disclosures that provide both lenders and markets with information on a sound risk management approach to climate risk.

Key to these financial disclosure recommendations is the ability to effectively identify relevant climate-related risks and opportunities and actively respond to them. This also requires a clear communication link that balances

TABLE 6.9 Information on TCFD disclosures that reflect a sound risk management approach to climate risk

Category	Assessment Adequacy and Relevance	Actions and Resilience
Governance	The company's senior management: ■ understands physical and transition risks ■ takes accountability for outcomes	The company's management: ■ communicates climate-related risks and the overall strategy proactively and effectively
Strategy	The company's management: ■ leverages sound methodologies and information to assess climate risk as a critical strategic issue affecting the company's clients, revenues, and reputation	The company's management: ■ plans and integrates climate risk management with the overall strategy
Risk management	The company's management: ■ has adequate controls to measure, monitor, and manage physical and transition risks ■ mitigates other climate-related risks affecting revenues, operations, and reputation	The company's management: ■ integrates controls to measure, monitor, and manage climate-related risks, affecting revenues, operations, and reputation into the business decision-making process
Metrics and targets	The company's management: ■ defines clear metrics and measures for identifying, monitoring, and managing risks ■ defines carbon-emission and other business targets for steering activities toward the company's low-carbon emission and/or net-zero strategies	The company's management: ■ uses relevant metrics to measure business performance and for decisions to mitigate climate-related risks ■ uses climate-related performance measures to steer activities toward low-carbon commitments

the benefits of business activities with risks and opportunities, supported by sound technical analysis, relevant scientific evidence, and subject matter expertise from competent professionals. For example, a high-carbon emission power generation company that acknowledges superficially the threat of transition risks to its primary revenue stream due to regulatory restrictions or changes in policies is unlikely to be managing its risks well. Similarly, a low-carbon emission power company that acknowledges superficially that it relies heavily on renewable energy sources with an unproven record of reliability is also unlikely to be managing its risks well or is misleading investors. Indicators that the company has avoided these pitfalls are: (1) the ability to identify and communicate relevant climate-related risks and opportunities, and (2) to be able to provide a clear vision on how the company manages risks toward its emission goals. Table 6.10 provides suggested responses for the key components of the financial disclosures described in Table 6.7.

In order for the framework for climate risk disclosures to be practical and useful for all stakeholders, there needs to be a common and comprehensive set of scenarios that could be leveraged by all companies, including by the financial institutions' own risk management. These scenarios, combined with a description of the competitive landscape for different industries, can be used for assessing the competitive business and regulatory environment across sectors and geographies for financial institutions and their clients. The scenarios need to be translated into a financial impact for the affected companies and, ultimately, into economic losses. This requires an approach for modeling the dynamics of companies and other financial institution's clients under different scenarios, their *demand for credit* and their loss likelihood and loss severity. To understand the impact of these scenarios and outcomes to the financial institution, the framework also needs an approach for modeling the lender's behavior and business strategy, that is, the *supply of credit*.

Furthermore, the impact analysis based on these scenarios needs to be provided over similar periods of time using consistent measures and metrics for companies across geographies and sectors. Notice, however, that metrics and measures should also reflect fundamental differences across industries. For example, metrics on the net present value of investments, asset break-even costs, or asset decommissioning costs are relevant for oil and gas, thermal carbon extraction and power producers but are less relevant for real estate companies. Similarly, assessments for physical risk are relevant for real estate companies and companies in the building and infrastructure sector but transition risk assessments are more relevant for energy extraction and power generation companies. The financial disclosures discussed above could be supplemented with stress testing and sensitivity analysis to different assumptions to provide insight on the robustness and reliability of the approach, as shown in Table 6.11.

TABLE 6.10 TCFD financial disclosure recommendations and suggested responses across four categories: (1) governance, (2) risk and opportunities, (3) risk management, and (4) metrics and targets

Type	Vulnerabilities and Risk Mitigation	Suggested Responses and Rationale
Governance		
Board oversight	■ Are climate risks presented to the Board and senior management regularly and with enough depth? ■ Are these risks evaluated as part of the regular decision-making process aligned to the company's risk appetite?	■ Description of the Board's responsibilities for overseen risk management of climate-related risks ■ Description of management's responsibilities for managing climate-related risks and incentives
Risk and opportunities		
Strategy	■ What are the key climate-related risks and opportunities? ■ What are the company's scope 1, 2, and 3 emissions and transmission channels? ■ What are the company's sources of emissions and alternatives to mitigate risks? ■ What company assets, operations, or logistics are exposed to short- and long-term physical risks? ■ What company assets, operations, or logistics are exposed to short- and long-term transition risks? ■ How do climate-related risks affect other risk types such as credit, operational, or reputational risks?	■ Description of the company's and downstream relevant emissions, including major sources of emission from operations, supply chain, and products sold. For example, power purchases from high-carbon emission sources or using raw material or products from overseas providers with lax or no standards on emissions ■ Description of identified risks and their impact on credit lending activity, operational issues, or negative reputation that can affect clients, investors, or regulators

TABLE 6.10 (*Continued*)

Type	Vulnerabilities and Risk Mitigation	Suggested Responses and Rationale
Impact	■ What are the likely impacts of physical and transition risks? ■ What analytical approach is used to identify vulnerabilities and their impact? ■ Which scenarios are used to assess the impact? ■ How reliable are the assumptions of scenarios and analytical approaches? ■ What are the magnitude and severity of physical risks and their channels? ■ What are the magnitude and severity of transition risks and their channels? ■ What are the magnitude and severity of other climate-related risks?	■ Description of the key impacts from physical and transition risks and other climate-related risks and opportunities ■ Conceptual and technical description of major approaches and scenarios used for assessing risks ■ Business and economic impact of individual scenarios on portfolios or activities
Resilience	■ Is the company resilient to its physical risks? ■ Is the company resilient to its transition risks and the regulatory environment? ■ What are the remaining vulnerabilities to physical and transition risks and other climate-related risks? ■ What is the plan for mitigating remaining vulnerabilities?	■ Description of the company's sensitivity to physical and transition risks for different scenarios ■ Description of vulnerabilities, mitigation strategies, and rationale for resilience to risks

(*Continued*)

TABLE 6.10 (*Continued*)

Type	Vulnerabilities and Risk Mitigation	Suggested Responses and Rationale
Risk management		
Risk identification	■ What is the process for identifying and monitoring physical and transition risks? ■ What is the process for identifying and monitoring other climate-related risks?	■ Description of key economic and risk drivers and carbon emission costs ■ Description of scenarios and drivers ■ Description of stress-testing approach
Risk management	■ How are physical and transition risks measured, monitored, mitigated, and managed? ■ How are other climate-related risks (such as credit, market, operational, and reputation risks) measured, monitored, mitigated, and managed?	■ Conceptual description of metrics, monitoring, and mitigation processes ■ Description of emission targets, risk transmission channels, and climate-related risks
Integration and mitigation	■ How are the risk identification and risk management processes connected for adequate governance, monitoring, and risk mitigation? ■ What are the gaps and vulnerabilities?	■ Description of key interaction points between risk management processes and the company's strategy
Metrics and targets		
Risk metrics	■ What are the critical dimensions for measuring risks? ■ What specific metrics are used to measure, monitor, and mitigate these risks? ■ What are the assumptions, benefits, and limitations of the selected metrics?	■ Conceptual and technical description of metrics to monitor risks, including critical assumptions, benefits, and limitations for different sectors or business

TABLE 6.10 (*Continued*)

Type	Vulnerabilities and Risk Mitigation	Suggested Responses and Rationale
Emissions	▪ What are the scope 1, 2, and 3 emissions for different business, operations, and processes? ▪ How are the emission metrics defined? ▪ What are the observed changes over time?	▪ Description of scope 1, 2, and 3 emissions, estimation assumptions, and uncertainty ▪ Description of trend analysis
Targets	▪ What are the company's climate risk, decarbonization, and net-zero targets? ▪ What are the observed emission trends and their alignment to the company's risk mitigation strategy?	▪ Description of all relevant targets ▪ Carbon emission targets ▪ Business and revenue targets ▪ Operational and activity targets ▪ Description of future target plans

BUILDING A STRESS-TESTING FRAMEWORK FOR CLIMATE RISK USING THREE CORE PILLARS

Financial institutions play a critical role in addressing climate risk challenges manifesting from physical and transition risks for their clients and their own low-carbon emission commitments. Integrating climate risk metrics into a risk management framework could be a significant undertaking for most institutions, but it is a necessary step toward a comprehensive and effective risk management approach. The exposure to climate risk for banks and other financial institutions can impact multiple businesses and products across their retail and wholesale portfolios.

As a starting point, institutions should be able to assess their business strategies to address climate-related risks, including an analysis of the markets, their clients and needs, the products they offer, the regulatory and competitive environment, and the innovations they will bring to the market to address client needs or mitigate risks. Their assessment should also include their own sustainability commitments to a low-carbon economy, and their impact on products, credit processes and practices, and risk appetite. For

TABLE 6.11 Basic components of financial disclosures for climate risk

Component	Description	Requirements
Competitive landscape and scenarios	Competitive landscape and climate risk scenarios for stress testing	Comprehensive and consistent climate and economic scenarios for stress testing Detailed description of the regulatory and competitive landscape affecting the financial institution (lender) and its clients (borrowers)
Demand for credit	Modeling borrower's behavior and demand for credit	Consistent models for borrower's behavior, use of credit lines and loss likelihood and severity based on risk characteristics and economic environment
Supply of credit	Modeling the lender's behavior, business strategies, and supply of credit	Lender's business and environmental commitments and their impact on funding and investment strategies
Financial impact analysis	Economic impact of physical and transition scenarios on the financial institution and its clients	Consistent time periods for financial outcomes, for example, 5 years, 10 years, 30 years Consistent measures and metrics across scenarios, geographies, and industries. For example, use of consistent carbon-emission ratios for scope 1, 2, and 3; or use of standardized loss provisions and capital ratios. Insightful measures to reflect sector differences

example, assessing the impact of offering better financing terms for building homes using sustainable materials, clean energy sources, or more efficient low-carbon emission heating and cooling systems.

The risks faced by companies seeking funding for their activities and institutions providing credit can manifest in multiple ways, including the potential reduction in revenues, increased operating costs, greater regulatory burden or reputation issues that can impact the franchise. For large companies, financial information, carbon emissions and ESG metrics are often available from data vendors. For smaller companies and clients, financial information may be available but climate-related metrics may require additional due diligence based on models and analysis of industry and geography concentrations. This may require banks to request new types of data from clients and borrowers such as energy use, emission data, low-carbon emission or net-zero commitments, future production plans, and availability of renewable energy sources to support their operations and supply-chain exposure. Enhanced due diligence may also require more detailed information for transactions, products, or services involving high-carbon emission sectors, such as oil and gas or thermal carbon mining and coal-fired power generation. The shift away from fossil fuels toward alternative energy sources in recent years has already impacted the energy sector significantly; in particular, coal-fired power production and coal mining among other areas. Additional market and economic data can be leveraged to understand the impact of climate events, from extreme weather events to the longer-term consequences of changes in policies and regulatory requirements due to transition to a low-carbon economy. Determining the appropriate level of climate-related information is the result of a delicate balance between the needs for detailed due diligence, the reliability of the models and analytics for leveraging this information accurately, and the costs to clients and the institution for collecting the information.

To integrate these critical considerations into a risk management framework, institutions should develop a practical taxonomy of climate risks and plausible transmission channels such as changes in customer preferences, capital and asset depreciation, business disruptions or indirect effects through changes in macroeconomic activity and regulatory requirements, as described in Tables 6.2 and 6.6. These assessments should take physical and transition risks into account as well as the resilience of borrowers and lending institutions to climate-related events, the threats to their business strategies and risk mitigation strategies that can affect both the supply and demand for credit. Climate-related risks can also impact collateral management and hedging activities, for example, due to the repricing of stranded assets for fossil fuel extraction, thermal carbon mining or coal-fired power plant operations or repricing of real estate assets in locations exposed to severe climate events. Accounting for stranded assets impacted by climate-related write-downs is an important aspect of stress testing and risk mitigation. Limited availability of data or limited reliability of models and analytics creates challenges for implementing appropriate risk mitigation and hedging strategies. An additional challenge for financial institutions assessing the impact of transition risk is aligning the long-term timelines of most transition scenarios (measured in decades) to the shorter duration of typical loan book commitments and market transactions. Transition risk will remain elevated in sectors that rely heavily on high-carbon emission activities or carbon-intensive technologies, need significant investments to become more energy-efficient, or have limited availability of insurance against climate-related events.

The vulnerability of companies to the economic impact from climate risk drivers could be assessed using loss likelihood and loss severity estimates differentiated by business, industry, geography, product, or portfolio (e.g. banking book or trading book) based on three basic analytical pillars:

1. **Competitive landscape and scenarios** for climate risk and ESG stress testing:
   a. Assessing the competitive business and regulatory environment across sectors and geographies for financial institutions and their clients
   b. Creating comprehensive climate and economic scenarios for stress testing
2. **Demand for credit** – modeling borrower's behavior:
   a. Borrower's behavior, demand for credit and use of credit lines
   b. Borrower's risk characteristics, loss likelihood, and loss severity
   c. Regulatory environment and public policies affecting borrowers
3. **Supply of credit** – modeling lender's behavior and business strategies:
   a. Business and environmental commitments, funding and investment strategies, and reputational risk
   b. Regulatory landscape and public policies affecting lenders and financial institutions

Table 6.12 illustrates the three key pillars of the stress-testing framework and components.

The development of these analytical components requires a combination of statistical techniques and expert judgment to make inferences on plausible climate-related scenarios and losses even if there is limited or sparse historical evidence to draw from. Notice that the primary purpose of analyzing past data is to obtain a reasonable judgment that can allow us to estimate the likelihood of future events with similar characteristics. For example, given a random sequence of events (such as tossing a coin), the likelihood of the next event is not necessarily derived directly from past observations but *assumed* to be somewhat determined by the frequency of the events already observed. This is

TABLE 6.12 The stress-testing approach, the three key pillars, and their components

Component	Description	Key Items
Competitive landscape and scenarios	Competitive landscape and climate risk scenarios for stress testing	■ Assess the competitive business and regulatory environment across sectors and geographies for financial institutions and their clients ■ Create comprehensive climate and economic scenarios for stress testing
Demand for credit	Modeling borrower's behavior and demand for credit	■ Borrower's behavior, demand for credit, and use of credit lines ■ Borrower's risk characteristics, loss likelihood, and loss severity ■ Regulatory environment and public policies affecting borrowers
Supply of credit	Modeling the lender's behavior, business strategies, and supply of credit	■ Business and environmental commitments, funding and investment strategies, and reputational risk ■ Regulatory landscape and public policies affecting lenders and financial institutions

a *sensible but subjective judgment* inferred from the available data and the *belief* that nothing else will change unexpectedly in the future. This is our mental model of the relationship between observed events and future events. Every time we infer that some property of a sequence of past events will hold also in the future, we are making an educated – but completely subjective – judgment of this type. This is exactly the problem currently faced by financial institutions for modeling climate risk and ESG issues. Institutions always have limited information on their clients, the economy, and the competitive environment, and often need to leverage historical data to obtain meaningful estimates of plausible future events. In the following we discuss these issues in more detail.

The framework described here can be used to support two distinct approaches to modeling the impact of climate risk: (1) a top-down approach, by assessing the impact of climate-related events on product or portfolio characteristics and the aggregated likelihood of default and loss severity, and (2) a bottom-up approach, by modeling the impact of climate-related events on energy demand and costs, carbon taxes, revenue generation, and capacity to service debt for individual borrowers and their transactions.

The bottom-up approach reflects climate scenario-adjusted financial statements at the borrower and transaction level, which affect the probability of default (PD) and loss severity (more precisely, the loss given default (LGD) and exposure at default (EAD)) through different transmission channels. The assessment of transmission channels should also consider the primary, secondary, and tertiary effects of climate-related events and their consequences for different risk types. For example, for consumer mortgage lending, lenders should assess the immediate impact of climate events such as severe floods in terms of direct damage to the property (primary effects). Secondary effects could include the economic impact of higher insurance premiums, declining property value due to remapping flood zones or damage to other properties in the neighborhood, schools, or hospitals, or changes in insurance-related government policies. Tertiary effects could include medium- and long-term damage to basic infrastructure such as roads and highways, water and sewage systems, power or communication lines that can impact indirectly the value of the property and can also result in the correlation of losses across multiple clients in the same location. Even if primary effects could be dominant for estimating idiosyncratic losses for individual properties, the secondary and tertiary effects could result in significant systemic losses due to correlation across a larger population.

Given the differential impact on each sector of the economy and geo-political environment, financial institutions should adopt a comprehensive and consistent sectoral approach to understanding climate risks across industries and geographies. For example, real estate may be more impacted by increased severity of extreme storms, hurricanes, or prolonged droughts resulting from physical risk, while the energy production sector may be at risk from transition to a low-carbon emission economy due to changes in regulations, increased carbon taxes and costs, the use of new technologies, or migration to alternative sources of energy.

While responses to climate-related physical and transition risks may vary for different institutions, they could follow common basic principles that can lead to a better transition to a low-carbon economy. The conceptual framework for modeling climate risk and ESG issues is based on the three pillars discussed above.

Pillar 1: Competitive Landscape and Climate Risk Scenarios for Stress Testing

- Assessing the competitive business and regulatory environment across sectors
- Creating comprehensive climate and economic scenarios

Pillar 2: Demand for Credit: Modeling Borrower's Behavior

- Borrower's behavior and their demand for credit
- Borrower's loss likelihood and loss severity

Pillar 3: Supply of Credit: Modeling Lender's Behavior and Business Strategies

- Lender's behavior and supply of credit
- Regulatory landscape and public policies
- Business and environmental commitments, funding and investment strategies, and reputational risk

The three pillars framework discussed here can provide a better understanding of the risks faced by the institution, which can allow management to determine strategic objectives based on the identified vulnerabilities. This can also help the Board members and senior managers to gain insight into the full implications of climate risk on credit, market, operational, and reputation risks and, ultimately, the franchise value.

REFERENCES

Bessis, J. (2005). *Risk Management in Banking*. Wiley, Chichester, UK, pp. 11–74.

Case, J. (2007). *Competition: The Birth of a New Science*. Hill and Wang, New York, pp. 115–144.

Day, G.S. (2013). Is It Real? Can We Win? Is It Worth Doing? *Innovation*. HBR's 10 Must Read Series. pp. 59–81.

Day, G.S., and Schoemaker, P.J.H. (2000). *Wharton on Managing Emerging Technologies*. John Wiley & Sons, New York, pp. 125–149.

Dicken, J. (2007). *Global Shift: Mapping the Changing Contours of the World Economy*. The Guilford Press, New York, pp. 73–136.

FSB (Financial Stability Board). (2021). *FSB Roadmap for Addressing Climate-Related Financial Risks*. Financial Stability Board, pp. 1–29.

IEA (International Energy Agency). (2021). Net Zero by 2050; A Roadmap for the Global Energy Sector. IEA Special Report, pp. 1–222.

IPCC (Intergovernmental Panel on Climate Change). (2018). Global Warming of 1.5°C. IPCC Special Report. www.ipcc.ch/srl5/

Keenan, S.C. (2015). *Financial Institution Advantage and the Optimization of Information Processing*. Wiley, Hoboken, NJ, pp. 27–100.

McCarthy, J.J. (2009). Reflections on Our Planet, Its Life, Origins, and Futures. *Science* 326 (December): 1646–1655.

Mokyr, J. (1990). *The Lever of Riches: Technological Creativity and Economic Progress*. Oxford University Press, New York, pp. 81–191.

NGFS (Network for Greening the Financial System). (2020). NGFS Climate Scenarios for Central Banks and Supervisors, (June), pp. 1–39. https://www.ngfs.net/sites/default/files/meelias/documents/820184-ngfs scenarios final version v6.pdf

Plimer, I. (2009). *Heaven and Earth: Global Warming, the Missing Science*. Taylor Trade Publishing, Lanham, MD.

Porter, M.E. (1985). *Competitive Advantage*. The Free Press, New York, pp. 164–314.

Rogers, E.M. (1995). *Diffusion of Innovations*. The Free Press, New York, pp. 252–280.

Ruddiman, W. (2005). *Plows, Plagues and Petroleum*. Princeton University Press, Princeton, New Jersey, pp. 177–189.

Smith, P. (2020). The Climate Risk Landscape: A Comprehensive Overview of Climate Risk Assessment Methodologies. United Nations Environmental Programme Finance Initiative, pp. 1–48. www.unepfi.org/wordpress/wp-content/uploads/2021/

Sobehart, J.R., and Keenan, S.C. (2005). New Challenges in Risk Credit Risk Modeling and Measurement. In *Risk Management: A Modern Perspective*. Ed. M. Ong. Elsevier, Oxford.

USTD (U.S. Department of the Treasury) (2021). Fossil Fuel Energy Guidance for Multilateral Development Banks. https://home.treasury.gov/system/files/136/Fossil-Fuel-Energy-Guidance-for-the-Multilateral-Development-Banks.pdf

CHAPTER 7

Pillar 1: Competitive Landscape and Climate Risk Scenarios for Stress Testing

ASSESSING THE INDUSTRY AND REGULATORY LANDSCAPE, SYNERGIES, AND COMPLEXITIES OF COMPETITIVE BUSINESS ENVIRONMENTS

The growing interest in understanding climate risk and ESG has increased the demand and complexity of periodic stress-testing exercises across risk types, which provides a strong incentive for financial institutions to develop and leverage advanced approaches and best practices to measure risk. While commercially available tools can be used as components of an overall solution for individual institutions, most institutions recognize that, ultimately, development and implementation of climate risk must be tailored to their portfolios. In many cases, whether climate risk analytics are based on commercial models or internally developed methodologies, loss models still rely on key statistics, such as obligor default rates, risk rating migration rates, loss severity rates, and their relationship to economic conditions to determine losses across risk types, changes in the portfolio credit quality to forecast balances and risk measures, such as expected credit losses, credit stress loss estimates, or risk capital (Bessis, 2002; BCBS, 2004, 2005a, 2005b).

Here we discuss default and credit events, their relation to credit cycles and their severity, which can provide a practical approach to the estimation of plausible economic scenarios that can be translated into portfolio losses and can be extended to accommodate climate-related risk drivers. The structural regularities observed in the dynamics of credit events can be used to characterize changes in credit quality using a few relevant indicators of economic activity. These regularities can also be used to include climate-related drivers to construct plausible "what-if" and stress-test scenarios for credit migration that include the effects of credit cycles, economic activity, and climate risk drivers for different industries and geographies beyond the limitations of historical data. This approach can be used to generate a range of plausible scenarios consistent with given economic conditions and climate risk scenarios and the typical uncertainty observed through credit cycles in different economies (Breuer, 2002).

SCENARIOS FOR THE GLOBAL ECONOMY: ECONOMIC, BUSINESS, AND CREDIT CYCLES

We start our analysis by focusing on the United States (US) economy, which has a long history of business cycles. The framework can be extended to other economies following the same basic principles provided here.

Historically, the US economy has experienced economic growth in the long run, but in the short run, the economy has exhibited a series of regular business cycles with periods of economic expansion and contraction that led to recessions and downturns (prolonged severe recessions), which can develop into economic crises, reflecting severe conditions that can affect the whole economy and financial system (Fisher, 1932, 1934; Kindleberger, 2000; Hubbard and O'Brien, 2008). Although most economic recessions, downturns, and crises are the result of a wide variety of unique circumstances and complex triggering events, they usually exhibit some common characteristics. More precisely, they are often associated with the peaks of business activity cycles and, therefore, with the evolution of economic expansions and contractions, reflecting changes in supply and demand for products and services and changes in monetary and socio-economic policies, government spending, and entitlement programs. They are also associated with periods of increased divergence between fundamental risk assessment and the pessimistic or optimistic perception of risk by market participants as often reflected in the widening and narrowing of credit spreads between periods of credit expansion and contraction (Juglar, 1915; Silverling, 1919; Fisher 1932, 1934; Hostetler, 1936; TEC, 1957; Bernake and Lown, 1991; Akhtar, 1994; Harris et al., 1994; Mosser and Steindel, 1994; Taylor, 1998; Crone, 2000, 2006; Kindleberger, 2000; Rothbard, 2000, 2008; Gujarati, 2006; Cooper, 2008; Iyetomi, 2010; Cooper and Bachman, 2012; Zidong et al., 2018). Figure 7.1 illustrates a typical business and credit cycle.

Recessions are periods with a significant decline in economic, business, and credit activity, affecting industrial production, employment, real income, and wholesale and retail trade, and lasting more than a few months. In practice, recessions are usually defined as two consecutive quarters of declining real Gross Domestic Product (GDP).

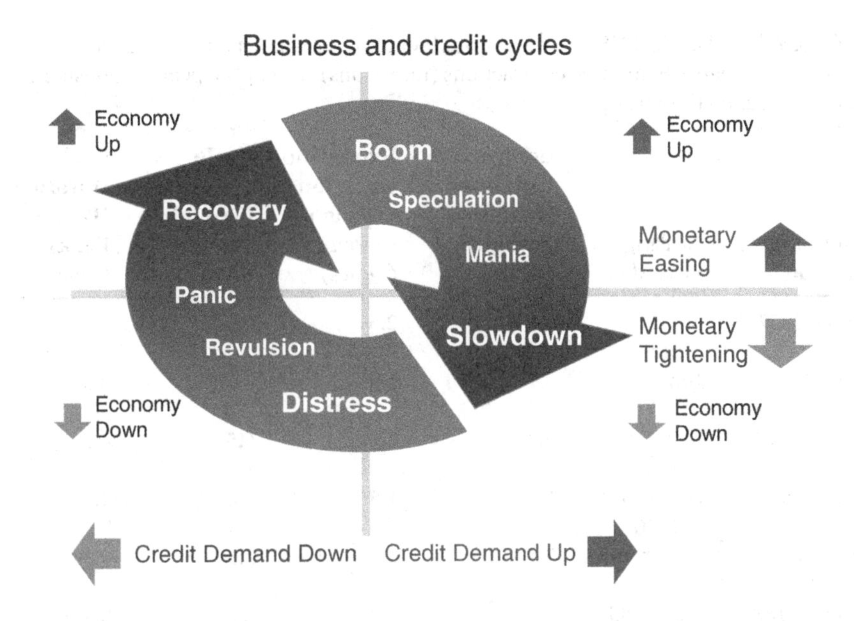

FIGURE 7.1 Simplified business and credit cycle stages and drivers.

In the US, business cycles and recessions are identified by the Business Cycle Dating Committee of the National Bureau of Economic Research (NBER) (Novak, 2008). Table 7.1 shows historical periods of economic expansion and contractions (recessions) and their duration. The duration of the recessions can be described as the period from the peak to the following trough. Figure 7.2 shows the peaks of the recessions identified by the NBER in the last decades (vertical lines), and the recovery and expansion periods between recessions. Figure 7.2 also shows the quarterly changes in GDP growth and (negative) changes in unemployment rate normalized with the standard deviation of changes over the period 1969–2023.

The US economy has experienced alternating periods of expansion and contraction of economic, business, and credit activity through time. In the late nineteenth century, the length of economic contractions was roughly about the same as the average length of economic expansions. This was driven by the cyclical nature of a commodities-based economy, an unstable financial system, and lack of adequate social and monetary controls to compensate for the economic swings. As the country moved away from a commodities-based economy toward an economy based on the production of goods and services, and new social programs and financial controls were introduced, the length of recessions generally decreased. However, it is only after the Great Depression in the 1930s that economists and policymakers gained better insight into

TABLE 7.1 Historical US business cycles including expansion and contraction periods (in years). Economic contractions (recessions) start at the peak of a business cycle and end at the trough

Starting Period	Ending Period	Contraction Duration (Peak to Trough) (Years)	Expansion Duration (Trough to Peak) (Years)	Cycle Duration (Trough to Trough) (Years)	Cycle Duration (Peak to Peak) (Years)
June 1857	December 1858	1.5	2.5	4.0	0.0
October 1860	June 1861	0.7	1.8	2.5	3.3
April 1865	December 1867	2.7	3.8	6.5	4.5
June 1869	December 1870	1.5	1.5	3.0	4.2
October 1873	March 1879	5.4	2.8	8.3	4.3
March 1882	May 1885	3.2	3.0	6.2	8.4
March 1887	April 1888	1.1	1.8	2.9	5.0
July 1890	May 1891	0.8	2.3	3.1	3.3
January 1893	June 1894	1.4	1.7	3.1	2.5
December 1895	June 1897	1.5	1.5	3.0	2.9
June 1899	December 1900	1.5	2.0	3.5	3.5
September 1902	August 1904	1.9	1.8	3.7	3.3
May 1907	June 1908	1.1	2.8	3.8	4.7
January 1910	January 1912	2.0	1.6	3.6	2.7
January 1913	December 1914	1.9	1.0	2.9	3.0
August 1918	March 1919	0.6	3.7	4.3	5.6
January 1920	July 1921	1.5	0.8	2.3	1.4
May 1923	July 1924	1.2	1.8	3.0	3.3
October 1926	November 1927	1.1	2.3	3.3	3.4
August 1929	March 1933	3.6	1.8	5.3	2.8
May 1937	June 1938	1.1	4.2	5.3	7.8
February 1945	October 1945	0.7	6.7	7.3	7.8

TABLE 7.1 (*Continued*)

Starting Period	Ending Period	Contraction Duration (Peak to Trough) (Years)	Expansion Duration (Trough to Peak) (Years)	Cycle Duration (Trough to Trough) (Years)	Cycle Duration (Peak to Peak) (Years)
November 1948	October 1949	0.9	3.1	4.0	3.8
July 1953	May 1954	0.8	3.8	4.6	4.7
August 1957	April 1958	0.7	3.3	3.9	4.1
April 1960	February 1961	0.8	2.0	2.8	2.7
December 1969	November 1970	0.9	8.8	9.8	9.7
November 1973	March 1975	1.3	3.0	4.3	3.9
January 1980	July 1980	0.5	4.8	5.3	6.2
July 1981	November 1982	1.3	1.0	2.3	1.5
July 1990	March 1991	0.7	7.7	8.3	9.0
March 2001	November 2001	0.7	10.0	10.7	10.7
December 2007	June 2009	1.5	6.1	7.6	6.8
February 2020	April 2020	0.2	10.7	10.8	12.2

Source: National Bureau of Economic Research (NBER) (US Business Cycle Expansions and Contractions).

the complex effects government policy could have on the economy and introduced new financial and monetary controls to stabilize the economy (Fisher, 1932, 1934). The impact of World War II (in the mid-twentieth century) on the global economy brought an era of additional financial controls and stabilization mechanisms.

Since then, the US federal government has actively used policy measures to reduce the length and severity of recessions and prolong the periods of economic expansion. This has created long periods of economic expansion and prosperity interrupted by relatively short recessions followed by periods of economic recovery leading to new economic expansions. Longer expansions and shorter contractions led to sustained growth of the US economy and other economies around the world through trade and investments.

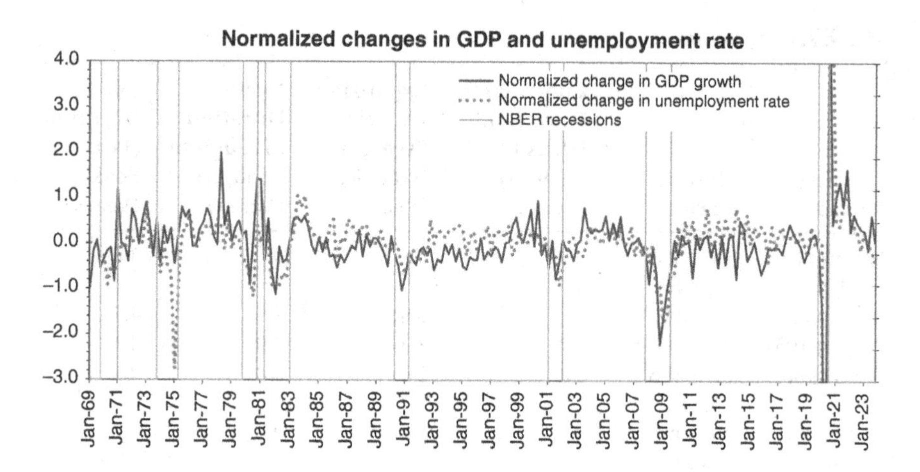

FIGURE 7.2 US recessions (vertical lines) and normalized quarterly changes in GDP growth and unemployment rate.

The fast pace of technological innovation and the globalization of financial systems and production and supply chains resulting from global economic growth led to a change in the industry composition of the developed economies, shifting from the manufacturing of goods to providing higher-value services (such as health care or financial investment). This changed the nature of the business cycles since the production of goods usually fluctuates more than the production of services during periods of economic expansion and contraction. Innovations in information, communications, manufacturing, and logistics contributed to the changes in economic activity by improving production, inventory controls, and distribution processes.

Business cycles in the US have become milder in recent decades, except for the Great Recession (2007–2009 financial crisis), which combined the effects of a regular business cycle with a prolonged preceding period of monetary and credit expansion and incentives to increase home ownership. Business activity expanded, fueled by cheap credit, and investments in riskier assets also expanded in the search for yield, creating economic bubbles in real estate and other asset classes. As the housing bubble collapsed, housing prices fell, and homeowners began to abandon their mortgages. The value of mortgage-backed securities declined dramatically, causing several financial institutions to collapse, or be bailed out (the subprime mortgage crisis). The sudden economic contraction that followed affected the financial sector and cascaded across all sectors of the global economy.

The primary triggering events, the length and severity of the expansion and contraction periods, and the sectors of the economy that are most affected are rarely the same in any two business cycles. However, business cycles and economic activity usually reflect common drivers:

1. the prevailing *monetary policy* used to control the availability of credit
2. the *fiscal policy*, which reflects government expenditures and revenue collection
3. the adopted *social policy*, which reflects the allocation of government expenditures among sectors of the economy, benefits, and entitlements.

These common drivers provide incentives and disincentives for economic activity, for example, through active market operations to control the money supply and interest rates, or by introducing carbon taxes to restrict high-carbon emission activities or tax breaks for alternative sources of energy and use of electric vehicles.

More precisely, prolonged periods of low interest rates and relatively inexpensive credit costs create incentives for investing across multiple sectors, accelerating economic activity during the expansion period of the business cycle. During this period, production, employment, and income are all usually increasing. Prices rise in response to increasing demand, potentially leading to inflation. As the economy reaches the end of an expansion, interest rates are usually rising and worker wages are often rising faster than prices, eroding profit margins. As the demand for funds increases, monetary policy may slow down the economic expansion by restricting the money supply and short-term borrowing (*monetary tightening*), effectively raising interest rates (i.e. increasing the cost of borrowing money). As interest rates rise, short-term interest rates start to align to long-term rates and the yield curve usually flattens. Costs increase and the demand for products, services, and credit slows down. As a result, profitability declines and the period of expansion ends with a business cycle peak. Toward the end of the expansion period, households and companies may have increased their debt to finance their spending and investments and are faced with the environment of higher financing costs.

At this stage, production, employment, and income decline as the economy enters a contraction period after reaching the business cycle peak. Economic contractions (or recessions) usually start with a decline in spending by households on new houses and consumer durables (such as cars, furniture, or appliances) and non-essential services (such as entertainment and leisure items or activities). Companies also show a pronounced decline in spending on capital goods (such as new factories, machinery, or equipment) and non-essential expenses (such as business travel and entertainment, supplemental employee benefits or staff training). The impact varies widely across different sectors of the economy. Durables tend to be affected more by the business cycle than non-durables (such as food or clothing) or essential services (such as health care). As overall consumer and corporate spending declines, sales for capital goods and consumer durables also decline. Revenues for the entertainment and hospitality service sectors also decline, reflecting decreasing demand. As revenues for products and services decline, inflation usually slows down, companies cut back on production and may begin to lay off workers, increasing unemployment (Keenan, Sobehart and Hamilton, 1999; Keenan and Sobehart 2004). Changes in the unemployment rate closely follow the decline in economic activity. The environment of high interest rates and financing costs, declining corporate profits, and rising unemployment reduces overall spending and income, and lowers the expectations of future economic growth. This creates a *positive feedback* mechanism (where changes drive more changes in the same direction) that leads to further declines in spending, income, and employment. This results in an increase in the number of consumer and corporate defaults that lags the increase in credit demand observed during the preceding expansion period.

As the economic contraction worsens, government usually reactivates the economy through monetary policy by increasing the money supply using open market operations and other mechanisms (*monetary easing*), effectively reducing interest rates (i.e. reducing the cost of borrowing money). Since long-term interest rates react at a different pace than short-term rates, the slope of the yield curve steepens. As loanable funds become available at low interest rates, the contraction phase comes to an end with a business cycle trough, after which another period of economic expansion begins, fueled by relatively inexpensive credit. Economic conditions gradually improve, and households and companies begin to reduce their debt or refinance them at lower cost, increasing their ability to spend and to borrow to finance new spending. Households begin to spend on consumer durables, entertainment, and leisure items and activities once again. Companies begin to spend on capital goods, research and development, and advertisements, as they anticipate future production needs. The fresh wave of spending reactivates other sectors of the economy and brings the recession to an end, creating the conditions for the next economic expansion to begin.

Even under these more favorable conditions, the unemployment rate may continue to rise or decline at a slower pace than expected well into the economic recovery period. One possible explanation is that companies may continue to operate below their capacity after a recession has ended and production has begun to increase. During this transition period between economic contraction and expansion, companies may still rely on existing workers to increase productivity and meet new demand rather than hiring

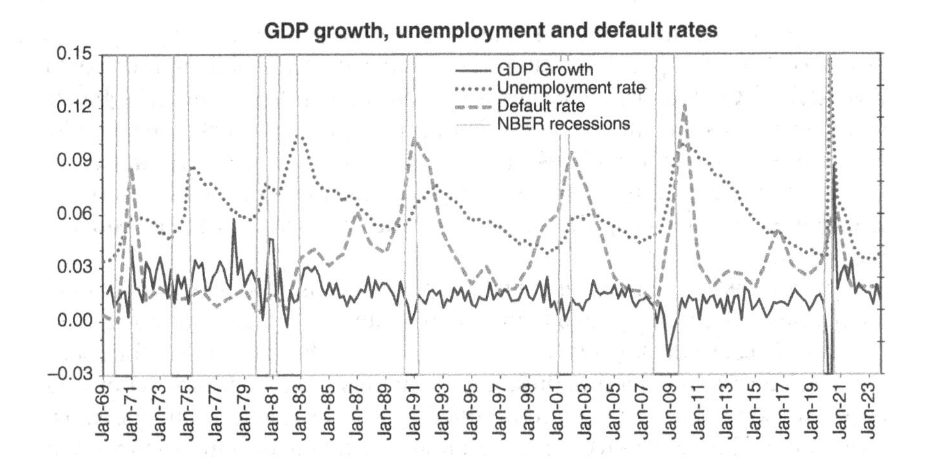

FIGURE 7.3 Relationship between changes in economic activity (GDP growth rate = solid line), credit deterioration and improvement of business activity (corporate default rates = dashed line), and changes in the workforce employment (unemployment rate = dotted line).

new workers. Companies may even continue to lay off workers based on their initial expectations of economic recovery and production needs instead of hiring back all the workers they have laid off during the contraction period. As a result, changes in unemployment rate will reflect the delayed response of the economy for hiring back workers, declining at a much slower pace than the fast increase observed during the contraction period.

Figure 7.3 shows the relationship between changes in economic activity (reflected in GDP), credit deterioration, and improvement of business activity (reflected in corporate default rates), and changes in the workforce employment (reflected in the unemployment rate). The observed changes follow the different phases of the economic and business cycle described in Figure 7.1.

BUSINESS CYCLES, CREDIT SUPPLY, AND DEMAND

The business cycle involves the interaction of different economic variables (Zamowitz and Boschan, 1975; Waud, 1989; Peracchi, 2001). We begin our discussion of the business cycles of expansions and contractions by focusing on the relationship between aggregate expenditures (demand) and the level of total production (supply).

The economy is never in perfect equilibrium. In practice, imbalances between supply and demand are the driving forces of the economy. When aggregate expenditure is greater than production, companies may sell more products than anticipated, which may result in a decrease of their anticipated inventories. Companies will increase their orders for additional goods, which translates into an increase in both production and worker hiring. The increase in production will cascade to suppliers, transportation, and other services in the economy, increasing overall production of goods and services and, therefore, increasing the Gross Domestic Product (GDP) and reducing unemployment.

In contrast, when the aggregate expenditure is less than production, companies will sell fewer goods than anticipated, and their inventories will increase. Because of slow sales, companies will cut back on orders for goods and services. The providers of these goods and services reduce production and lay off workers. The reduction in production will cascade to suppliers, transportation, and other services in the economy, reducing overall production of goods and services, reducing GDP growth, and increasing unemployment.

Government can affect the pace of economic activity and employment by changing spending through government purchases or by changing fiscal policy through taxes on households and companies. The use of fiscal policy as a steering mechanism of the economy can be traced back to the early work of Depression-era economist, John Maynard Keynes, who suggested that government could use its power of spending and taxation to stabilize the economy and lower unemployment during downturns. In Keynes's view, the principal cause of recessions or depressions was insufficient buying power or demand for goods and services (demand-side economics). That is, when supply and demand in the markets were out of balance and the economic activity was slowing down and unemployment increased, governments could steer the economy toward growth by facilitating consumer spending of goods and services, forcing businesses to start producing more. As a result, the economy grows, unemployment falls, and credit losses decline.

Keynesian economics was used as a guiding principle for driving fiscal policy for several decades until the US economy began to experience a combination of recession and high inflation (stagflation) in the 1970s. In the 1980s, Keynesian economics was rolled back by a change in fiscal policy to a supply-side economics and a move to free markets under government's regulatory supervision and a vision of smaller government. This approach to fiscal policy is based on the idea that economic growth is stimulated by investments on the supply side (production) and not from the demand side (consumer spending) as Keynes advocated. The change from a demand-side view to a supply-side view was also accompanied by changes to monetary policy, which increased monetary control of the financial system through market operations. This facilitated the more regular business cycles observed since the early 1980s, as shown in Figure 7.3.

Although monetary and fiscal policies play a critical role in driving economic activity, the third contributing factor is the social (or socio-economic) policy, which allocates government spending to different sectors, promotes, or discourages certain activities in the economy, and determines the level of entitlements and benefits. Social policy can provide incentives or disincentives for economic activity, for example, by restricting the use of internal-combustion engine (ICE) vehicles while promoting the use of electric vehicles (EV) for public transportation and private use; by restricting the use of fossil fuels as sources of energy for transportation, heating, or cooking; or by promoting low-carbon emission construction for housing projects. The use of social policy to drive large-scale changes to a low-carbon economy can also result in detrimental consequences for society due to the lack of a sound understanding of climate change, its drivers, and the interconnectivity across economic sectors, or due to politically motivated partisan agendas disguised as environmental concerns, or simply due to lack of competence to address these issues, disrupting economic activity and impacting the quality of life of its citizens without mitigating climate risk appropriately.

The discussion above highlights that having a better understanding of the relationships between business activity, the economic environment and competitive landscape and monetary, fiscal, and social policies may help refine credit, default, and loss severity models calibrated over multiple business cycles and their extensions for climate risk use. The idiosyncratic nature of some extreme recessions may require additional contextual information in order to understand the limitations of models developed for economic forecasting and stress-testing purposes. Note, however, that the discussion above also indicates that a few key economic indicators (such as GDP growth, unemployment rate, changes in market indices, the shape of the yield curve or changes in carbon-related taxes) may be able to provide valuable insight by capturing basic common features of credit cycles and downturns, which can provide us with a sensible means of modeling a wide range of hypothetical and forward-looking scenarios required for analyzing the impact of climate related events.

ECONOMIC DRIVERS

In the following we discuss key economic and market indicators and risk drivers for two main categories:

1. Indicators of economic activity reflecting the demand for credit.
2. Investment appetite from financial institutions and capital markets, reflecting the supply of credit.

Each category includes multiple economic and market indicators aligned to a common theme.

ECONOMIC ACTIVITY AND CREDIT DEMAND

The risks that financial institutions bear as lenders are reflected primarily in the risky activities being conducted by consumers, businesses, public institutions, and entrepreneurs in the marketplace and their demand for credit (Akhtar, 1994; Mosser and Steindel, 1994). Consumers, for example, are primarily attempting to fund consumption patterns based on their expectations of future income. However, for the lender, the risks include the uncertainty of future consumption and future income, as well as principal-agent problems associated with debt repayment. As a consolidator of consumer credit, a financial institution may be in a better position than individual consumers to evaluate the consistency of current consumption and expectations of future income. In contrast, corporate borrowers can be either small and medium enterprises (SMEs), middle market enterprises (MMEs), large commercial and industrial companies (C&Is), financial institutions (FIs), or government and public sector entities (PSs). These borrowers have different financing needs related to growth, cash flow management, and financial efficiency and may require more detailed credit assessments.

While individual financial institutions may seek to maintain a specific tolerance and strategy for risk acceptance, macroeconomic effects and monetary policies can change both the demand for credit and the supply of loanable funds. Significant increases in credit supply may lead to higher levels of risk-bearing and/or a fall in the market price for risk. This phenomenon can drive the credit cycle, helping to fuel higher future default and loss rates. Conversely, environments with high default rates and credit losses can lead to tightening the supply of credit and an increase in the price of risk. Therefore, financial institutions need to be sensitive to these systemic risks and manage their capital accordingly, even if their own credit policies are designed to constrain risk-taking within narrow limits.

Figure 7.4 shows the corporate default rate and the corporate bond market distress index (CMDI) (FRB, 2024). The CMDI provides a timely signal of corporate bond markets and identifies commonly accepted periods of market dislocation such as those around the global financial crisis peaking in 2008–2009, with the next largest peak during the COVID-19-related market stress in March 2020.

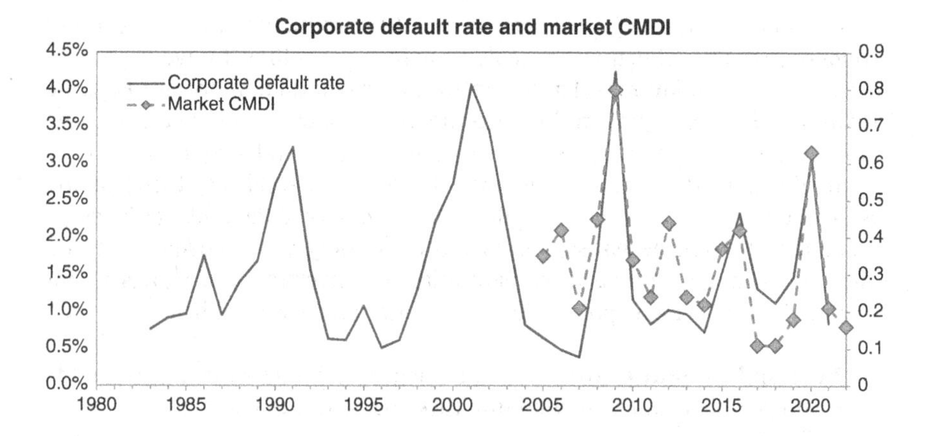

FIGURE 7.4 Credit cycle and economic activity: Corporate default rate and the corporate bond market distress index (CMDI). (Source: FRB, 2024.)

Measuring Total Production in the Economy: Gross Domestic Product (GDP)

One of the most widely used indicators of economic activity in credit models is the Gross Domestic Product (GDP), which measures the market value of all final goods and services produced in a country during a fixed period, typically one quarter or one year. A final good or service is one that is purchased by its final user and, therefore, is not included in the production of any other intermediate good or service. This eliminates potential double counting of goods or services in the total GDP value. However, the potential trade-off is the reduction in the information on specific sub-segments of the economy such as suppliers of intermediate products. GDP is an important concept in macroeconomics and plays a critical role in understanding business and credit cycles, and in determining economic activity in forecasting and hypothetical scenarios. Most macroeconomic scenarios for hypothetical and forward-looking views used for stress testing and climate risk analysis include estimates of GDP, which reflect basic economic activity for different sectors and geographies.

Although GDP measures the value of total production in the economy, it also measures income since the products and services covered in the estimation are consumed by households, companies, and governments. Notice, however, that GDP does not include two other types of production:

(1) household (internal) production; and (2) production in the *informal economy* (usually reflecting concealed buying and selling to avoid government taxes or regulations). These contributions could be important in some situations. For example, in low-income or less-developed countries, the informal economy could be a significant part of the overall economy.

In the United States, the Bureau of Economic Analysis (BEA) in the Department of Commerce compiles the necessary data and statistics to calculate GDP for different segments of the economy. GDP is usually divided into four major categories: (1) consumption, (2) investment, (3) government purchases, and (4) net exports. These categories are described below:

1. **Personal consumption** (C) – This category includes expenditures made by households, and is divided into three basic segments:
   a. Services
   b. Durable goods
   c. Non-durable goods.
2. **Gross private domestic investment** (I) – This category includes business purchases used for production, and is divided into three basic segments:
   a. Business fixed investment (new factories, office buildings, machinery, etc.)
   b. Residential investment (new residential housing)
   c. Changes in business inventories (goods produced but not yet sold).
3. **Government purchases** (G) – This category represents government consumption and gross investments and includes federal, state, and local government purchases of goods and services. Notice that transfer payments to households (social security and other entitlements and benefits) are not included in this category because they do not result in the direct production of goods and services. Government purchases are usually divided into two basic segments:
   a. Federal (Central Government) purchases
   b. State and local purchases.
4. **Net exports of goods and services** (NX) – This category is equal to exports to foreign countries minus domestic imports from other countries. Exports are goods and services produced domestically but purchased by foreign consumers, companies, or governments. Imports are goods and services produced in foreign countries but purchased by domestic households, companies, or government. GDP adds exports because it measures all spending on new goods and services produced domestically even if they are consumed elsewhere. In contrast, GDP deducts imports because these goods and services are produced elsewhere even when they are consumed domestically.

GDP is calculated by adding the four primary contributions discussed above:

$$GDP = C + I + G + NX \tag{7.1}$$

Historically, in the United States, consumption C represents roughly 70% of GDP, government purchases G is nearly 20%, while investments I is roughly 15% and, despite its relevance for measuring trade flows, net exports NX is only around −5%. Furthermore, consumer spending on services is usually greater than spending on durable and non-durable goods. The larger contribution of services in high-income/developed countries reflects the continuing economic trend away from production of goods and toward the production of services. As populations in high-income countries become older and wealthier, their demand for services, such as health care or financial services, outpaces their demand for goods. Also note that investments in fixed assets (new factories, office buildings, machinery, computers, etc.) are usually the largest component of investment for most companies. Changes in business investment tend to be cyclical (in response to changing economic conditions and expectations) and can play a significant role in the economy as they affect different industry sectors. In some situations, purchases made by state and local governments can be greater than purchases made by the US federal government because basic activities, such as education and law enforcement, usually occur at the state and local levels.

Finally, note that in the United States net exports provide a negative contribution to GDP since imports from other countries are significantly larger than exports to other countries. Both imports and exports are very sensitive to economic conditions and can change in different directions. For example, when the US dollar is strong relative to other currencies, foreign goods and services become cheaper and, therefore, imports may increase. In contrast, domestic goods and services become more expensive to foreign markets and, therefore, exports may decline. This situation is not unique to the United States and is experienced by other countries with unbalanced import and export flows with their trade partners. The net importer situation reflects a net outflow of financial resources to other countries that could be partially balanced by investments from other countries.

Notice that the net exports contribution (NX) can play a critical role in shaping the domestic economic activity in response to climate change, as inconsistent domestic policies and unilateral restrictions of certain high-carbon emission activities can be taken advantage of by countries with much larger high-carbon emission contributions but with fewer restrictions on manufacturing or energy generation activities (such as China or India). This provides some countries with a significant (unfair) competitive advantage that can result in a net transfer of wealth from countries that share

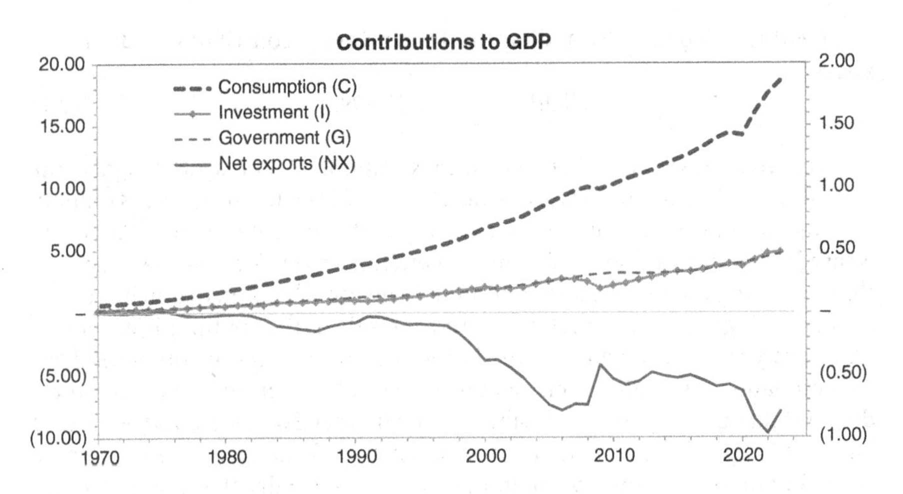

FIGURE 7.5 GDP contributions over time, including consumption (*C*), private investment (*I*), government (*G*), and net exports (*NX*) (right scale). (Source: Federal Reserve of St Louis.)

the economic and social costs of climate risk mitigation to those countries polluting the most and doing the least to mitigate climate risk effectively. Figure 7.5 shows the GDP contributions in the United States over time, including the next export component, which has been increasing over time with period changes in response to economic expansions and contractions. Figure 7.6 shows the largest contributions to carbon emissions and nominal GDP (from EDGAR, Emissions Database for Global Atmospheric Research). Note the disproportionate contributions of carbon emissions relative to the size of the economies. Figure 7.7 shows the change over time for relative contributions to global carbon emissions. Developed economies have reduced their carbon emission contributions while less-developed countries have increased their emissions, representing the largest contributions. Figure 7.8 shows the change in the amount of carbon emissions for the same countries and regions (in thousands of $MtCO_2$/year).

Because GDP measures production of goods and services in terms of nominal market value, it is sensitive to both the quantity produced and price changes over time. To separate price changes from quantity changes, economists introduced the concept of *real* GDP as an alternative to the *nominal* (price-sensitive) GDP. There are different approaches to remove the effects of price changes. Real GDP can be calculated by designating a particular year as the reference year, and then using the prices of goods and services for the selected reference year to calculate the value of goods and services for all other time periods at fixed prices. One drawback of this

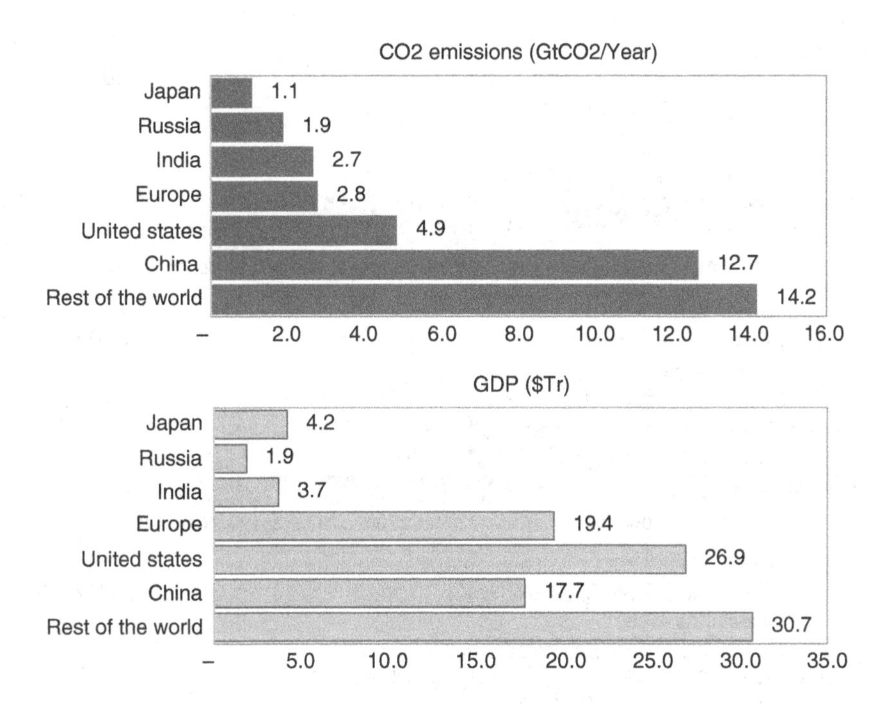

FIGURE 7.6 Largest contributions to carbon emissions and GDP.

approach is that prices may change relative to each other, distorting the GDP estimate over time. To mitigate this effect, economists use the concept of chain-weighted price, based on the relative price changes within consecutive periods referenced to the selected reference year.

Changes in price levels can be measured using the relative changes between real GDP and nominal GDP, or GDP deflator (nominal GDP/real GDP x100). As the nominal GDP and real GDP change over time, the GDP deflator provides the implied changes in price levels.

Real GDP is a more insightful measure of changes in production and economic activity than nominal GDP because it holds prices constant. Growth in the economy is usually measured as growth in real GDP (the relative percent change between consecutive periods). Increases in real GDP are usually related to increases in productivity and the amount of capital used to produce goods and services (e.g. machines, computers, warehouses, labor) and by the level of technology used for production. Innovation and changes in technology are critical to economic growth and are usually reflected in the purchase of new machinery, equipment, or software or changes to production processes.

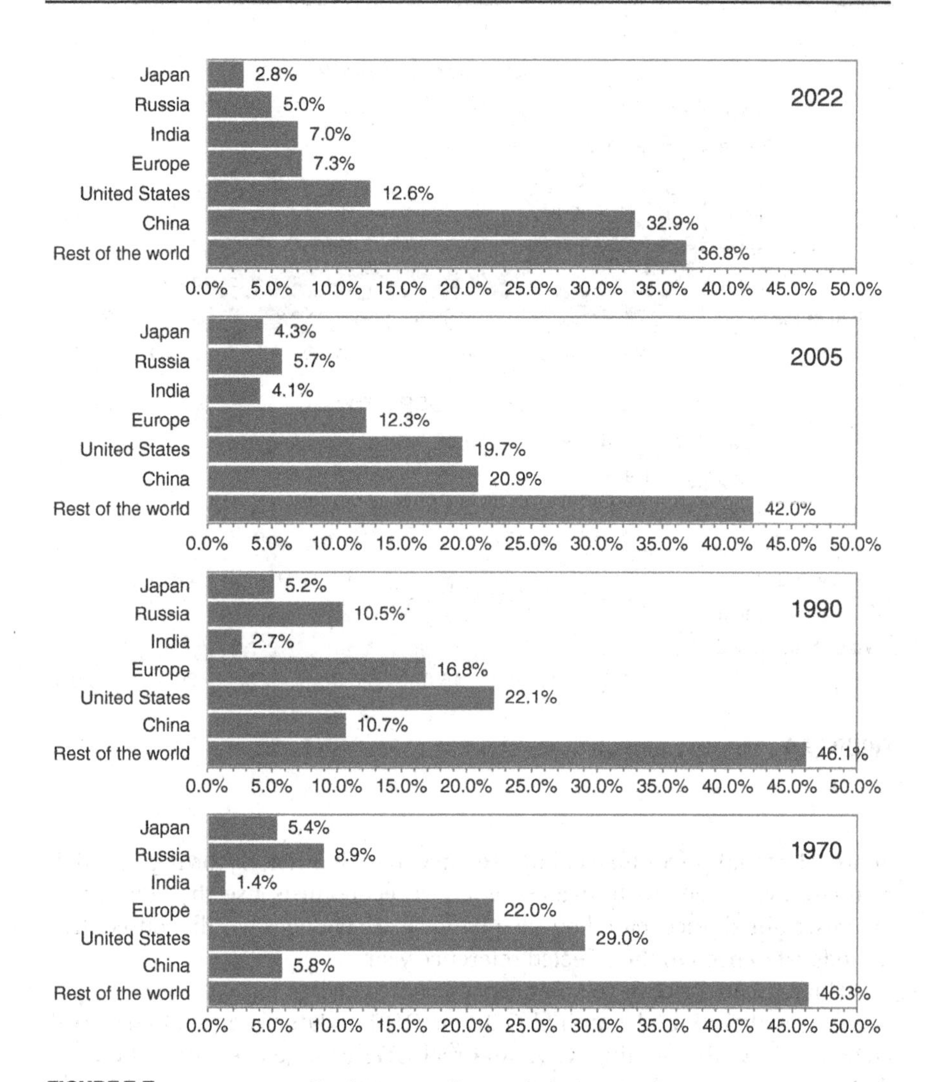

FIGURE 7.7 Largest contributions to carbon emissions over time.

Industries could contribute differently to GDP, depending on the nature of the goods and services they produce and their interconnectivity. To illustrate, the trucking business (ground-freight transportation and logistics) is a service activity, but it may be more related to the production of goods than to other services. As a result, the trucking industry could contribute to changes in GDP aligned to the production of goods rather than to changes in GDP for service industries, even when the trucking business is a service provider. A decline in the consumption of goods could affect the delivery of these goods,

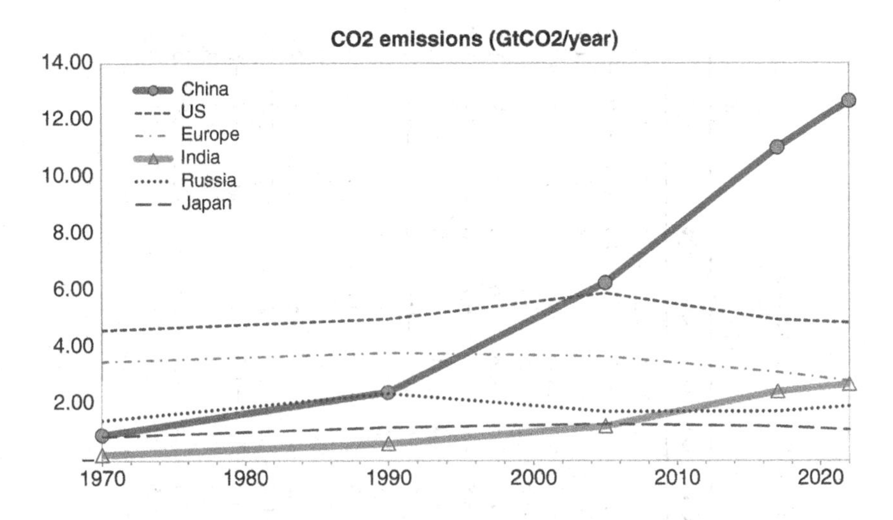

FIGURE 7.8 Contributions to carbon emissions for different countries and regions.

which may result in a decline in the service-component of GDP as observed in previous recessions. This interdependency of industries was also observed during the COVID-19 pandemic period (2020–2022), when logistics and transportation companies thrived as a result of increased on-line purchases, while physical shops struggled to stay afloat due to restrictions on customer interactions. Other industries could be more resilient to business cycles and severe economic conditions (e.g. food or health care), or less resilient to economic downturns due to the non-essential nature of the services they provide (e.g. tourism, lodging, or entertainment) as clearly reflected during the COVID-19 pandemic period. Table 7.2 shows the changes in US GDP growth for different industries for different recession and downturn periods, including the COVID-19 pandemic period.

Although GDP growth is an important indicator for describing changes in economic activity and in loss likelihood and loss severity estimates, the use of GDP statistics aggregated across several industry segments may limit the ability of loss models to reflect industry-specific patterns in great detail. Also note that definitions of GDP can vary across countries or regions, which can impact the reliability and performance of the loss models. Finally, GDP estimates are subject to revisions, which introduce modeling and parameter uncertainty into the relationship between the level of economic activity, credit quality migration, loss likelihood and loss severity at different points in the credit cycle or during severe downturn conditions. This uncertainty can impact the calibration of models and their sensitivity to climate-related risk drivers.

TABLE 7.2 Quarterly US (real) GDP growth rate for different industries including the COVID-19 pandemic period

	2019	2020				2021				2022				2023		
	Q4	Q1	Q2	Q3	Q4	Q1	Q2	Q3	Q4	Q1	Q2	Q3	Q4	Q1	Q2	Q3
Private industries	0.3	−5.4	−32.8	34.1	9.8	8.9	6.9	3.0	5.0	1.0	1.4	2.2	−0.4	3.7	1.3	3.5
Agriculture	5.6	5.7	−10.3	19.8	3.2	−3.4	−3.0	−2.5	−5.0	−8.1	−4.9	0.0	0.5	12.0	−1.0	−1.3
Mining	−9.0	−2.0	−66.9	7.4	14.1	21.8	32.7	−4.7	10.9	3.3	4.4	11.7	−3.2	8.0	−7.3	1.3
Utilities	−1.3	−11.2	−7.2	14.2	4.0	4.6	−0.5	−1.6	−7.5	9.9	20.5	−6.7	−6.3	−5.9	2.8	10.2
Construction	1.5	5.2	−12.3	6.2	10.6	8.6	−4.7	−7.5	−7.4	−1.0	−9.6	−10.7	−4.1	7.1	7.3	14.7
Manufacturing	−8.6	−7.6	−36.8	47.3	5.6	3.8	−4.4	−3.1	7.1	1.5	−4.8	4.6	3.0	−0.9	3.9	1.6
Wholesale trade	−3.1	−4.8	−40.0	59.1	14.6	15.8	9.7	−3.1	5.8	4.6	−6.7	−6.1	−7.6	−5.4	−8.9	−1.2
Retail trade	2.3	−10.7	−16.0	44.3	6.9	21.2	−2.9	−10.1	4.7	−0.7	4.1	1.2	1.0	5.8	−1.9	12.1
Transportation	2.3	−9.1	−66.5	68.2	27.0	12.1	9.0	13.7	10.9	6.6	10.6	−0.4	−6.4	−3.3	−0.3	1.9
Information	6.0	−0.3	−12.1	24.3	14.7	11.8	11.8	9.0	10.8	6.8	7.6	3.5	−0.3	6.1	1.7	8.4
Finance	3.1	−1.7	−6.4	8.9	6.0	4.4	4.1	4.0	1.8	−3.3	−1.0	4.6	−2.2	7.3	5.4	4.3
Services	5.4	0.1	−29.3	21.7	18.8	12.3	11.3	9.3	12.2	6.7	6.0	3.4	2.6	2.1	0.4	−0.4
Education, health care	0.4	−5.7	−39.6	40.4	11.5	0.6	9.4	7.4	5.3	−0.1	1.0	5.1	1.6	11.2	2.8	2.9
Entertainment	2.7	−29.8	−88.6	312.6	−1.2	33.6	77.4	31.9	−0.4	−6.4	30.8	2.2	4.0	7.4	−5.4	3.6
Other services	2.4	−12.7	−55.2	62.6	2.9	4.7	22.9	5.5	3.5	−3.5	−1.9	11.0	3.7	0.3	−2.7	−8.4
Government	2.9	0.1	−3.3	1.4	−1.4	7.6	−1.0	0.3	0.8	−1.0	1.6	2.3	1.7	2.6	1.8	3.3
Federal	−0.4	3.2	32.2	−12.7	−1.5	24.3	−8.9	−9.8	1.4	−7.1	0.1	3.1	1.1	2.5	1.0	3.4
State and local	4.3	−1.2	−16.1	8.6	−1.3	0.7	2.9	5.2	0.5	1.8	2.2	1.9	1.9	2.7	2.1	3.3

Unemployment Rate and Labor Force Participation

The unemployment rate, which is usually suited to a particular population or purpose, helps policymakers, economists, and market participants gain insight into the health of the economy. In the United States, the Department of Labor's Bureau of Labor Statistics (BLS) produces unemployment rate statistics. Similar statistics are provided by local entities around the world.

The unemployment rate measures the percentage of the labor force that is unemployed at the time of the unemployment survey, which includes people who did not work in the week before the survey and have been actively looking for work at some time during previous weeks. Although the unemployment rate is an insightful indicator of the economy, it is not a perfect measure of joblessness. Those who do not have jobs and are not actively looking are not considered part of the labor force (retirees, homemakers, full-time students, and active military personnel). Discouraged workers, who looked for a job at one time but stopped looking because they believe no jobs are available to them, are also excluded from the unemployment survey. Also, people who find part-time jobs while looking for full-time jobs are counted as employed, which may also lower the unemployment rate statistics. Furthermore, during recessions and severe business downturns, an increase in the population of discouraged workers may occur, as people who have a hard time finding jobs may stop looking for them. Because these workers are not counted as unemployed, the unemployment rate may understate the true degree of joblessness in the economy.

The labor force participation rate, that is the percentage of the population that contributes to the labor force (or workforce), is also important for estimating the impact of different economic scenarios because it determines the amount of labor that will be available to the economy from a given population (labor supply). The workforce is the sum of the civilian employment population and the unemployed population at a given point in time. The higher the labor force participation rate, the more labor will be available to generate higher levels of production. Changes in social behavior and demographics affect the participation rate, which shows declining trends for men and increasing trends for women in recent decades, with a noticeable impact after the 2008–2009 Great Recession and the 2020–2022 COVID-19 period. Some of the observed decline for men participation could be due to older men retiring earlier and younger men remaining in school longer. The observed changes in employment patterns (Figure 7.9) may affect unemployment rate figures by introducing non-stationary and non-cyclical components, which may impact the relationship between unemployment statistics and credit cycles and, therefore, can impact loss projections for different unemployment

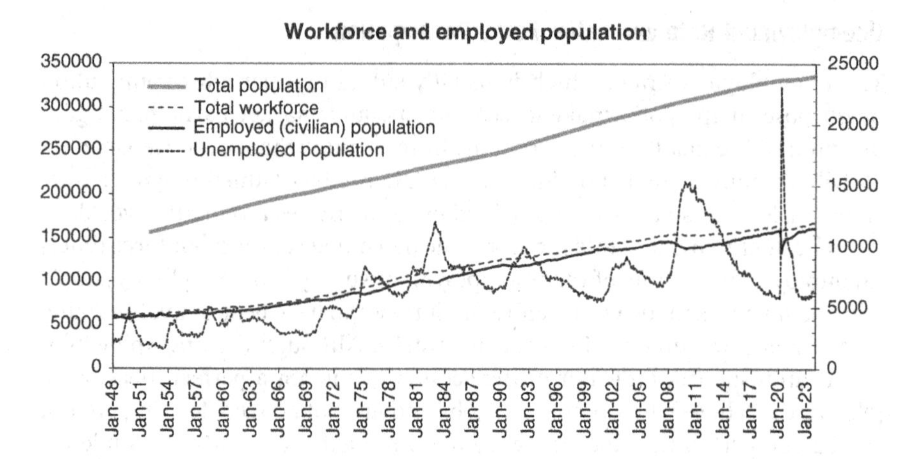

FIGURE 7.9 Changes in US population, total workforce, employed civilian population, and unemployed population (right scale). (Scale is in thousands of people.)

rate forecasts. Note, however, that there is still a significant regularity in the observed changes in unemployment rates over time, as shown in Figure 7.10.

Figure 7.10 shows that the unemployment rate closely follows the business cycles, rising during recessions and downturns, and falling during expansions but never reaching zero (full employment). The economic view on these changes is that there are three distinct types of unemployment:

1. Frictional unemployment
2. Structural unemployment
3. Cyclical unemployment.

Frictional unemployment is the short-term unemployment that arises from the process of matching workers with jobs. Some frictional unemployment may be unavoidable since the process of job searching and hiring takes time. Unemployment could be due to seasonal factors such as fluctuations in demand during different times of the year (Dickens and Triest, 2012; Diamond, 2013). Because seasonal unemployment can make the unemployment rate seem artificially high during some periods, the BLS usually reports seasonally adjusted and unadjusted unemployment rates.

In contrast, *structural* unemployment arises from a persistent mismatch between workers' job skills and the requirements of the job or its availability at compensation levels commensurable with domestic worker skills. While frictional unemployment is short-term, structural unemployment can last for

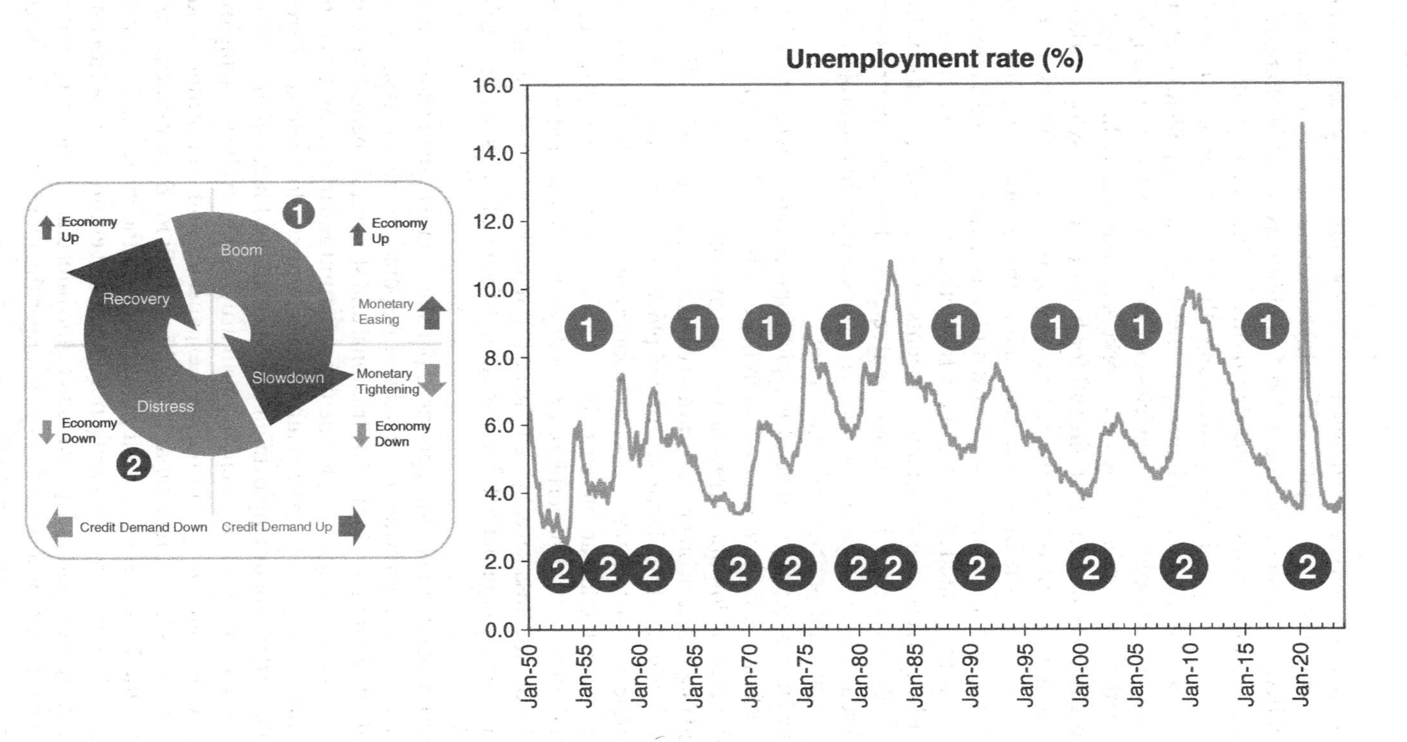

FIGURE 7.10 US unemployment rate and business and credit cycles.

longer periods since workers need time to learn new skills. Many manufacturing industries in the United States and in high-income/developed countries have exhibited a persistent decline of job availability as result of changes in technology and production processes but primarily due to competition from foreign producers and offshoring production with lower worker costs. Governments can help reduce structural unemployment by creating policies that aid retaining workers. However, other policies can, in fact, increase the unemployment rate either by increasing the time workers devote to searching for jobs, creating restrictions that make it difficult for firms to fire or lay off workers, or by creating disincentives for firms, that may hire workers otherwise. For example, historically the unemployment rate in the United States has been lower than the unemployment rate in most high-income/developed countries, partly because the United States has more restrictive requirements for unemployed workers to receive government payments and more flexibility for hiring employees. Thus, changes in unemployment rates for different countries may show different relationships to local business cycles and credit loss projections, reflecting local labor policies and regulations.

Cyclical unemployment reflects the typical upturns and downturns of the economy. When the economy moves from the expansion phase of the credit cycle to the contraction phase (recession or downturn), firms find their sales falling and may cut back on production and expenses and start laying off workers. The magnitude of the increase in unemployment reflects the severity of the economic contraction phase. As the economy recovers from the recession and business production accelerates, firms start rehiring those workers, creating (aperiodic) cyclical patterns of unemployment that mimic changes in loss likelihood and loss severity as reflected in default rates and risk rating migration rates for commercial, industrial, and financial firms, as shown in Figure 7.3.

As the economy moves through the expansion phase of the business and credit cycle, the unemployment rate falls in response to the increasing economic activity but remains limited by frictional and structural unemployment. When the only remaining unemployment is structural and frictional, the economy is at full employment. The remaining level of unemployment is referred to as the *natural* rate of unemployment for the economy. Fluctuations around this natural rate of unemployment are primarily due to cyclical unemployment resulting from changes in business expansions and contractions and, therefore, changes in the demand and supply of credit for business activities. Notice that the jobs created and eliminated during the cycles of economic expansion and contraction are usually not distributed evenly across all sectors of the economy. The demand for workers in the information, finance, education, and health care sectors tends to be stronger than for other sectors, such as manufacturing, which tends to be more sensitive to economic cycles and competition from low-cost offshore factories.

A very different type of employment contraction occurred during the 2020 COVID-19 pandemic (Figures 7.9 and 7.10). In December 2019, a respiratory disease outbreak based on a novel strain of coronavirus was reported in Wuhan (Hubei), China. The outbreak spread quickly globally by unimpeded flows of travelers going overseas and misleading information on the outbreak severity. The highly contagious nature of the virus and its severity triggered a global chain reaction, infecting millions of people, disrupting social activities and paralyzing economies worldwide through lockdowns and business restrictions. This stopped business activity temporarily, which was critically needed for job creation for a large share of the workforce globally. The tourism, travel, lodging, and entertainment industries collapsed due to travel restrictions, closing of public places, and government mandates. The retail sector faced reductions in store hours or temporary closures in many countries. Businesses around the world coped by increasing sanitation procedures and limiting customer-facing activities to minimum levels. As the effects of the pandemic subsided in 2021 and 2022, a strong economic rebound followed the lockdowns and restrictions worldwide, surviving businesses reopened, and employment recovered. The COVID-19 pandemic event and its economic outcome provide valuable insight into the disruptive nature of global idiosyncratic events, which could help assess severe physical and transition risk scenarios for climate risk.

For practical purposes, risk and loss models and economic forecasts from financial institutions, commercial vendors, and economic organizations tend to focus primarily on total (aggregated) unemployment rate statistics, given the common limitations for forecasting sector-specific unemployment rates. Although unemployment rate is an important economic indicator for describing changes in loss likelihood and loss severity, the use of unemployment rate statistics aggregated across multiple industry segments limits the ability of loss models to reflect industry-specific patterns in great detail. Also note that definitions of unemployment rates can vary widely across countries or regions, which can also impact the reliability and performance of the risk and loss models and economic forecasts.

Retail Consumption, Disposable Income, and Personal Savings

Practitioners and academics find disposable (personal) income to be a sensible indicator of economic activity and periods of financial stress. Disposable personal income is usually defined as total personal income minus personal current taxes and reflects the available income for consumption. That is, the income left after paying all the taxes. Subtracting personal outlays (which include the major categories of personal (or private) consumption expenditure) from disposable income yields the personal (or private) savings. This reflects the personal savings available for investment and later consumption that will also impact economic activity. Figure 7.11 shows the changes in disposable income and personal savings over time and their relation to economic activity. A significant change in both disposable income and personal savings can be observed during the COVID-19 period, which limits the ability of these variables as predictors of economic expansions and contractions.

Inflation and the Consumer Price Index

As the demand for products and services outpaces supply, prices increase, driven by the excess demand generating *inflation*. Prices could also increase when the value of a country's currency devalues relative to other currencies, reflecting a weak demand for the currency. In this case, the price increase may reflect higher production costs related to expenses paid in other currencies if the country is a net importer of goods or raw materials.

The inflation rate is the percentage increase in the prices of goods and services from one year to the next. A commonly used measure of inflation is the GDP deflator, which includes the price of final goods and services in the economy. However, these prices do not necessarily reflect the typical purchases of consumers and households since they also include economic activity in the corporate and government sectors. A closer measure of the costs of goods and services purchased by households is reflected in the Consumer Price Index (CPI), which is calculated by the Bureau of Labor Statistics. The CPI is an average of the prices of goods and services based on surveys of a wide range of household products purchased by a typical urban family. Figure 7.12 shows the changes in the inflation rate in the United States (as measured by CPI) over time and their relation to economic activity.

Although the inflation rate measures described above provide directional information on the economy, these measures need to be combined with other indicators of economic activity to gain a more comprehensive view. Note, however, that the way inflation impacts economic activity can be leveraged to gain insight into the potential impact of climate risk effects for both physical risk, where the price of insurance and replacement costs of assets could rise as a result of more severe weather events, and transition risk, where emission-based taxes or regulations could discourage economic activity by increasing costs.

Economic Activity and Fiscal Policy

The primary goal of *fiscal policy* (i.e. the government policy related to taxation and expenditures of public funds) is to secure funds and carry out

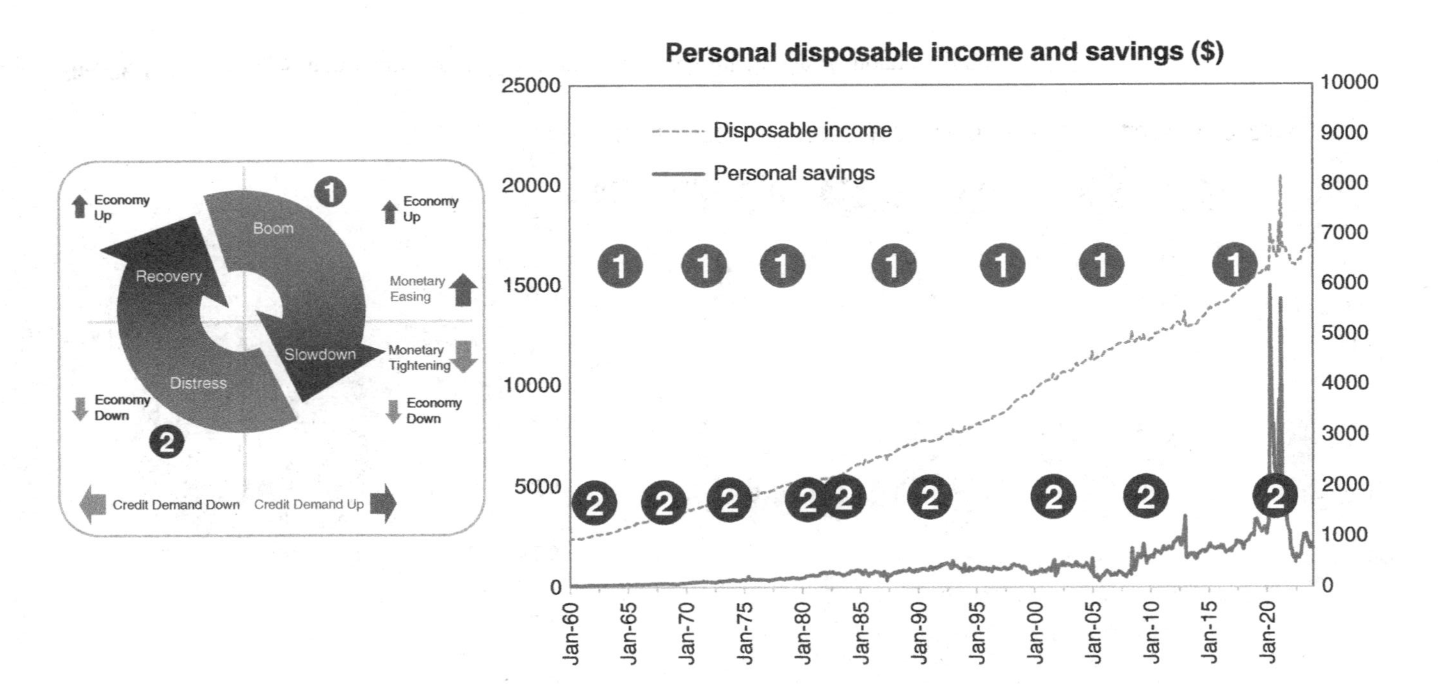

FIGURE 7.11 Changes in disposable income and personal savings over time and their relation to economic activity.

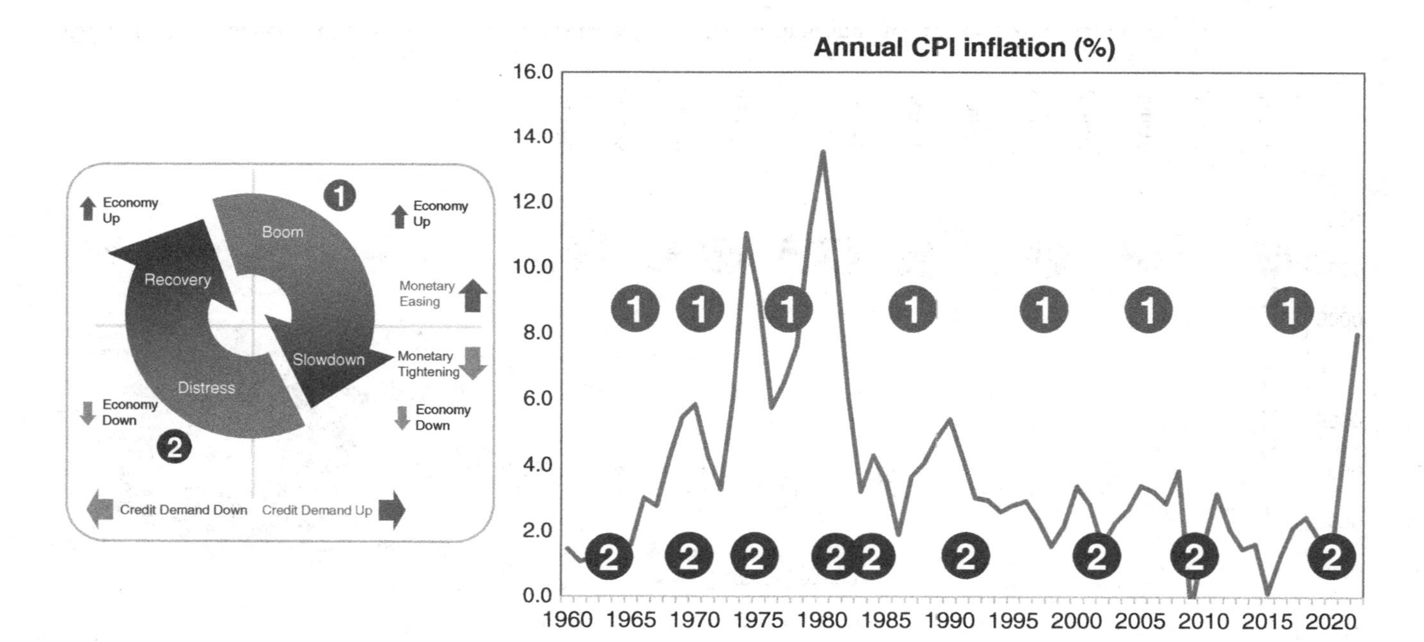

FIGURE 7.12 Changes in inflation rate/CPI over time and their relation to economic activity.

the appropriate spending to provide government services. In the United States, roughly about one-third of the national income generated annually is channeled toward public programs. Government decisions on taxation and spending on public programs can have a significant influence on economic activity. Government can deliberately influence the pace of economic growth by changing spending (government purchases) or the amount of taxes on households and companies.

This is particularly relevant in the context of climate change as governments may combine fiscal policy with socio-economic policy to promote or discourage different business activities and drive their strategy toward a low-carbon economy. This could have significant consequences for different sectors of the economy as the industry landscape is reshaped, driven by fiscal and socio-economic policies. Other countries may take advantage of misguided local policies, leading to a transfer of wealth or an increase in offshore production to countries that do not abide by the same low-emission rules or have a vested interest in their own internal economic growth. Figures 7.6, 7.7, and 7.8 show the relative contributions to global carbon emissions for major economies and their impact. Note the disproportionate regional contributions to carbon emissions and responses in terms of emission reductions, which result in significant economic impact.

As described in earlier sections, the use of fiscal policy as a steering mechanism for the economy can be traced back to the work of John Maynard Keynes. In Keynes's view, government could use its power of spending and taxation to steer the economy, lowering unemployment during downturns. In this demand-side view of economics, the reason for economic downturns is insufficient buying power or demand for goods and services. More precisely, when supply and demand in the markets are out of balance, economic activity slows down and unemployment increases. In this view, governments could steer the economy toward growth by facilitating consumer spending of goods and services. This increase in demand drives businesses to produce more, resulting in economic growth and the decline in unemployment and credit losses.

Government spending is one driver of fiscal policy. The other driver is *income taxation*. Lower-income households have relatively low disposable income for spending while upper-income earners can save or invest much of theirs. Keynes advocated that government spending could be financed by constructing a progressive tax structure, where higher-income earners pay higher taxes as a percentage of income than those with lower incomes. This tax structure would result in more disposable income in the hands of a broader sector of the economy that would lead to a larger demand for products and services. In theory, the progressive tax structure would create a mechanism for generating additional tax revenues from the higher earners that could be channeled to government agencies for more spending. This has significant consequences for the economy as many of the higher earners are small business owners, whose ability to hire workers, invest in, and grow their businesses is discouraged by taxation policy.

Keynesian economics remained a key driver of fiscal policy for several decades until the 1970s when the US economy began to experience *stagflation* (the combination of recession and high inflation periods). In the 1980s, fiscal policy shifted from demand-side economics to supply-side economics driven by a view of free markets under government's regulatory supervision and a smaller government. In this view, economic growth is stimulated by business investments on the supply side (production) and not only by the demand side (consumer spending) as previously advocated. By cutting taxes as opposed to increasing them, investors could provide companies with additional investment funds to increase production. The increase in economic activity cascades down to lower-income earners in the form of new jobs and higher salaries. In the United States, the change from a demand-side to a supply-side perspective was also accompanied by changes to monetary policy, which increased monetary control of the financial system through market operations to manage the money supply as opposed to direct lending through the discount window. This change helped create more regular business cycles as observed since the early 1980s, resulting in longer periods of economic growth followed by shorter recessions.

Because government decisions on taxation and spending can deliberately influence business activity and economic growth, having a sound and consistent fiscal policy can help steer the economy toward a more sustainable path that balances the society's need for economic prosperity and longer-term objectives to manage climate change and other emerging risks. In the United States and other developed countries, consistency in determining this delicate balance is difficult to achieve as fiscal policy typically shows periodic swings due to changes in the political affiliation of government administrations, switching from one governing party to its opponents and leading to significant fiscal deficits, inadequate taxation policies to support government spending, or inconsistent regulations to promote or limit business activities, such as fossil fuel extraction and power generation, which can have long-term consequences for the economy.

Over the years three basic arguments have been raised to suggest that the US has entered a period of overall economic decline. The first argument focuses on administrations predisposed to help politically connected rich donors at the expense of the middle class and low-income population through tax cuts skewed to the wealthy, the reduction of federal aid for homeownership, higher education, or transportation, and by facilitating the transfer of jobs abroad (offshoring and outsourcing) to satisfy the increasing demand for lower-wage labor required to boost corporate profitability. The second argument emphasizes the failure of government administrations to develop adequate technology and industrial initiatives that would enhance US competitiveness in the global economy. The third argument focuses on unbalanced fiscal budgets and excessive interest payments on the national debt as the greatest threat to the US economy. Large persistent deficits may eventually keep long-term interest rates high, reflecting investors' concerns about inflation. In this view, expensive borrowing not only would discourage private investment but also would limit public investment in the infrastructure needed for economic growth.

The persistent deficit gap view of economic decline discussed above has gained increasing credence in recent years as the last economic downturns (including the Great Recession of 2007–2009 and the COVID-19 period of economic slowdown in 2020–2022) moved fiscal deficits even deeper into the red with very anemic signals of long-term economic recovery and increasing levels of market volatility. Even the symbolic one-subgrade rating downgrade of the US government after the Great Recession assigned by one of the leading credit rating agencies highlighted the market perception of an overall economic decline. In the absence of an adequate long-term plan to regain global competitiveness, the US economy is primarily dependent on monetary policy and foreign capital investments to boost economic activity. This is relevant for modeling the long-term impact of transition risk for climate risk assessment. Figure 7.13 shows the changes in US fiscal deficits, interest payments, and their relation to the cycles of economic expansion and contraction (logarithmic scale in trillions of USD). Periods of net government savings (surpluses) are not shown (negative values). Note the significant changes in deficits as the gap between the government current receipts and current expenditures widens.

Money Supply

Money supply is the quantity of money (currency and demand deposits) in the financial system. Interest rate is the price borrowers are willing to pay for borrowing money today and repaying it later. These two concepts are inversely related. Ordinarily, as money supply in the economy increases, the availability of funds increases, and interest rates (the price of money) decrease. On the other hand, as money supply declines, the available funds become more valuable, borrowers are willing to pay more for accessing these funds and interest rates increase. Changes in the demand for funds exhibit similar behavior for a given level of money supply. Measures of the money supply have exhibited close relationships with economic variables such as nominal Gross Domestic Product (GDP) and the price level (inflation) and usually reflect the

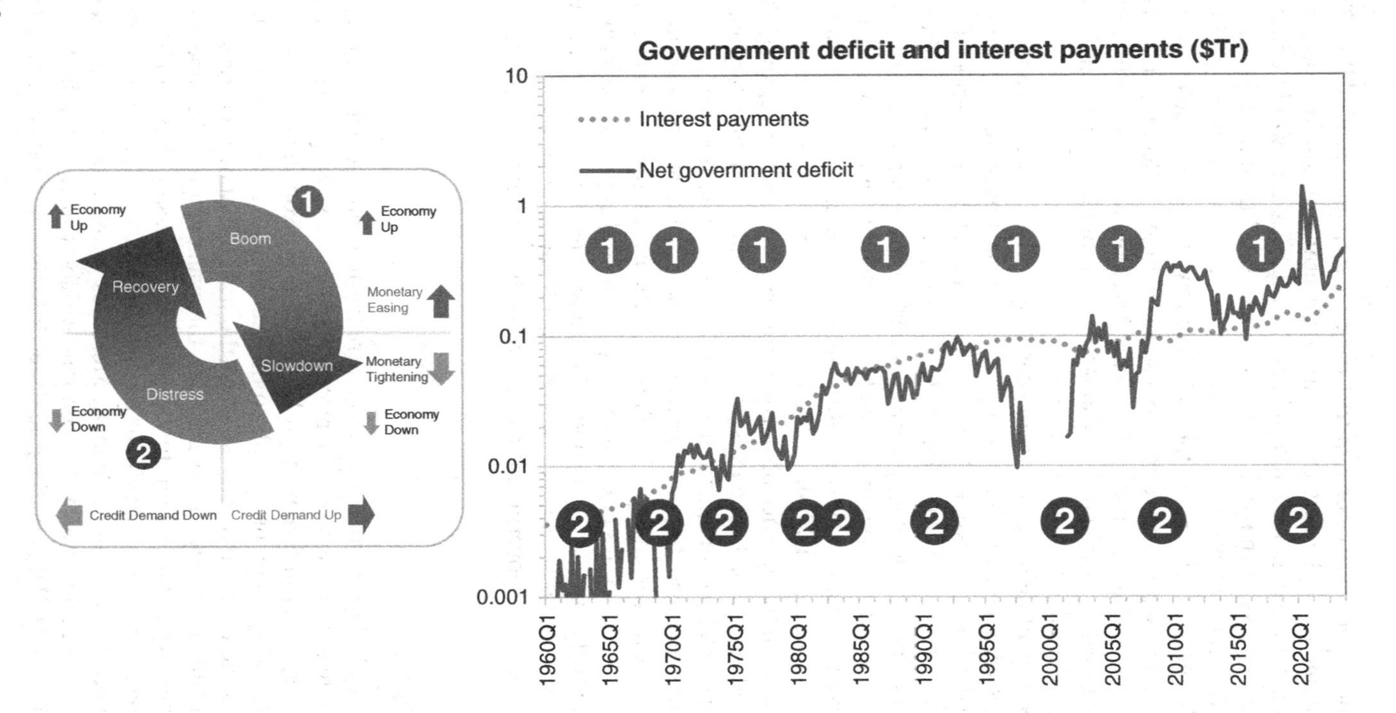

FIGURE 7.13 US fiscal deficits, interest payments, and their relation to economic cycles.

monetary policy response to economic expansions and contractions. However, the relationships between various measures of the money supply and economic variables such as GDP growth and inflation have been less stable in recent times after a significant expansion of the money supply following the Great Recession (2007–2009) and the COVID-19 pandemic downturn (2020). Understanding the relationship between the amount and price of money is important for assessing the impact of policies and incentives for transitioning to a low-carbon economy, which may require massive investments and reallocation of funding across sectors.

There are different definitions of money supply that reflect decreasing levels of liquidity of money substitutes. For example, the first type includes physical currency plus accounts at the central bank that can be converted into physical currency. The sum of all physical currency in circulation and reserve balances (deposits held by banks and other depository institutions in their accounts at the Federal Reserve) is the *monetary base* (MB). This is the base from which other forms of money are created.

M1 is the sum of currency held by the public and transaction deposits at depository institutions (which obtain their funds mainly through deposits from the public, such as commercial banks, savings and loan associations, savings banks, and credit unions). Before May 2020, M1 consisted of (1) currency outside the US Treasury, Federal Reserve Banks, and the vaults of depository institutions; (2) demand deposits at commercial banks (excluding amounts held by depository institutions, the US government, and foreign banks and official institutions) less cash items in the process of collection; and (3) other checkable deposits, consisting primarily of negotiable orders of withdrawal, and automatic transfer service, accounts at depository institutions, share draft accounts at credit unions, and demand deposits at thrift institutions. Since May 2020, M1 consists of (1) currency outside the US Treasury, Federal Reserve Banks, and the vaults of depository institutions; (2) demand deposits at commercial banks (with the same exclusions listed above) less cash items in the process of collection; and (3) other liquid deposits, consisting of other checkable deposits, savings deposits, and money market deposit accounts. Seasonally adjusted M1 estimates include additional changes.

The ratio of M1 to MB is known as the *money multiplier* and reflects the amount of money in financial accounts to the amount of physical money available for borrowing. The money multiplier reflects the leverage of the financial system to provide credit to borrowers without having to transfer physical money between parties.

M2 is M1 plus savings deposits, small-denomination time deposits (those issued in amounts of less than $100,000), and retail money market mutual fund shares. Before May 2020, M2 consisted of M1 plus (1) savings deposits (including money market deposit accounts); (2) small-denomination time deposits less individual retirement accounts and balances of tax-deferred retirement plans for self-employed individuals (Keogh plans) at depository institutions; and (3) balances in retail money market funds (with some exclusions). Since May 2020, M2 consists of M1 plus (1) small-denomination time deposits less individual retirement accounts and Keogh balances at depository institutions; and (2) balances in retail money market funds. Seasonally adjusted M2 estimates include additional changes. Other forms of money, such as M3, are no longer published. Figure 7.14 shows the contribution of physical currency (C), MB, M1, and M2 to the money supply in the financial system in the United States.

Interest Rates

Previously we discussed the relationship between supply and demand for credit and credit losses and the impact of monetary and fiscal policies, we now turn our attention to the role of the financial system and market for loanable funds, which encompasses a broad definition of financial institutions and capital markets.

The interaction between borrowers and lenders in the market for loanable funds in response to changes in economic activity determines the price of money (interest rate) and the quantity of loanable funds available to different borrowers (Rebonato, 1996; Cornyn, 1997; Rothbard, 2000, 2008; Ray, 2015). The demand for loanable funds is driven by the borrowers' willingness to borrow money to engage in new investment projects or refinancing old ones. These projects can cover a wide range of activities such as manufacturing products, building a factory, buying a house or a car. In determining whether to borrow funds, borrowers compare the return they expect to make on an investment with the interest and other costs they may pay to borrow the funds. Notice that both returns and costs can be impacted by climate risk and ESG issues, which could impact decisions on investments and costs of borrowing funds.

Ordinarily, the demand curve for credit (loanable funds) at a given interest rate is downward sloping as a function of the quantity of loanable funds because the lower the interest rate, the more borrowers of profitable investment projects can undertake, and the greater the quantity of loanable funds they will demand. On the other hand, the supply of loanable funds is primarily determined by the willingness of consumers, companies, and governments to save, making savings available to financial institutions for lending, or by the willingness of investors to provide funds at a given rate of return on their investment. The supply curve for credit (the quantity of loanable funds) is expected to be upwards sloping because the higher the interest rate, the greater the quantity of funds provided by financial institutions and investors.

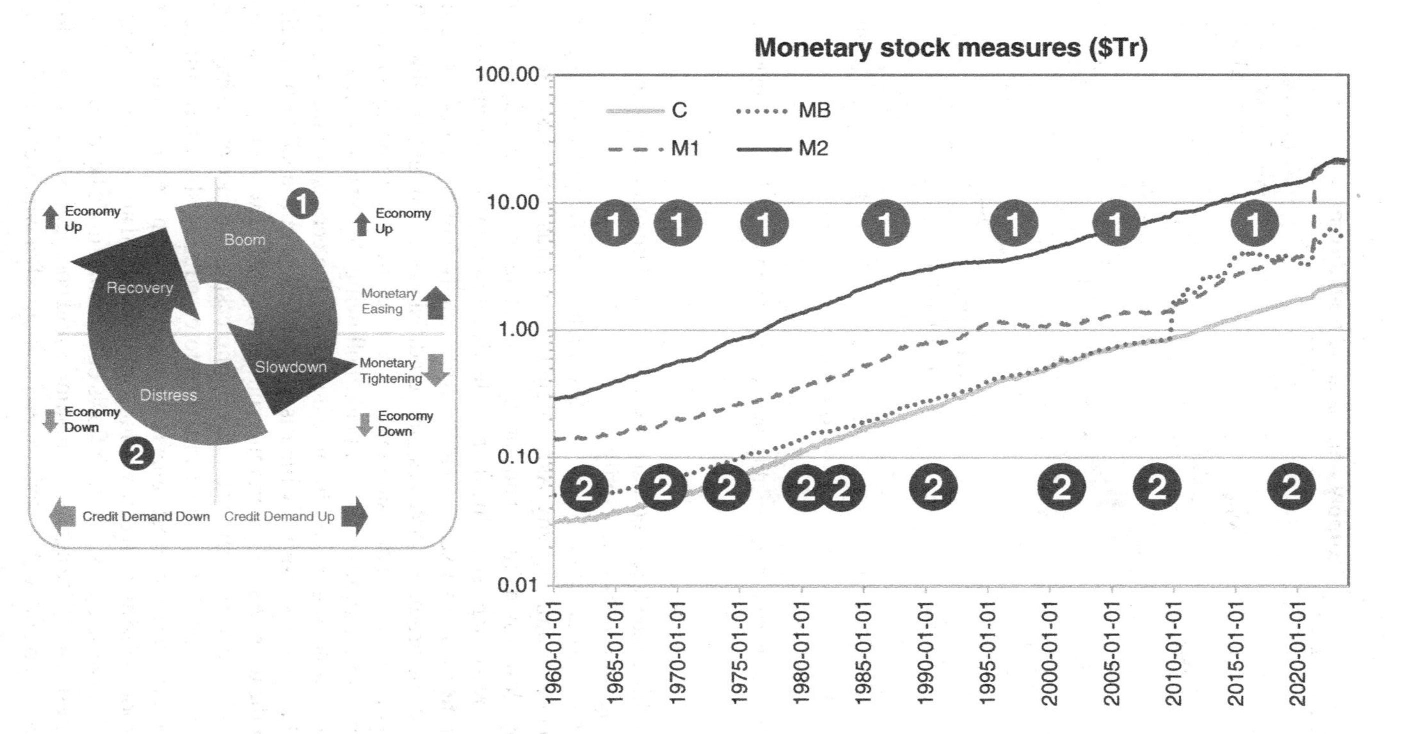

FIGURE 7.14 Relative contribution of C, MB, M1, and M2 to the total money supply. (Source: US Board of Governors of the Federal Reserve System, data retrieved from FRED.)

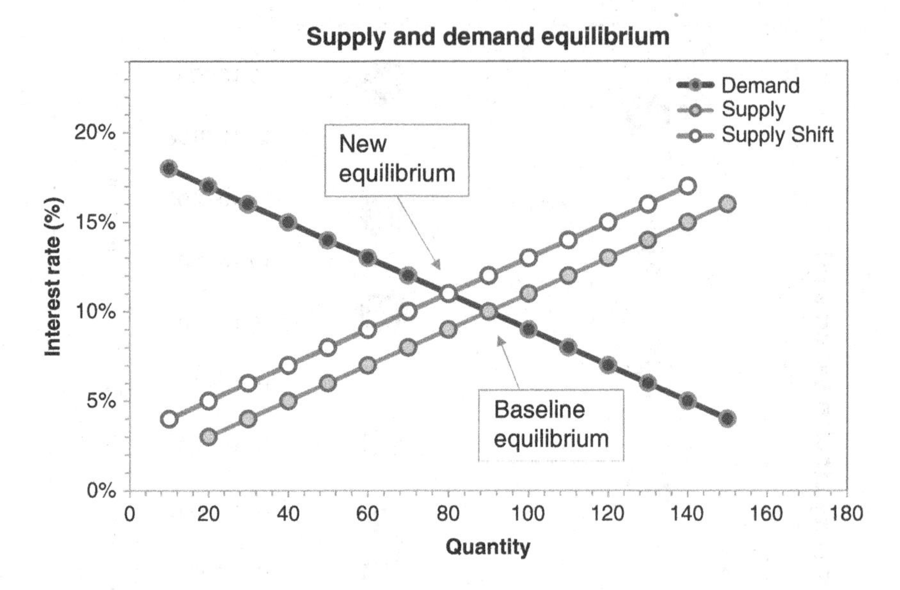

FIGURE 7.15 Credit supply and credit demand curves and shifts due to changes in economic activity.

Figure 7.15 illustrates the credit supply and credit demand curves and shifts due to changes in economic activity.

Persistent periods of low interest rates create incentives for investing and expansion of economic activity. As the demand for funds increases, monetary policy may slow down the expansion by restricting the money supply (e.g. by selling government securities) and effectively raising short-term interest rates. As interest rates rise, the yield curve usually flattens, costs increase, the demand for products, services, and credit slows down and profitability declines. The period of expansion ends with a business cycle peak, which is also reflected in an equity markets peak. Following the business *cycle peak*, production, employment, and income decline as the economy enters a contraction phase (or recession), which is also reflected in a decline in equity markets indicators. As the economic contraction phase worsens, monetary policy usually reactivates the economy by increasing the money supply (e.g. by buying government securities) and, therefore, reducing short-term interest rates. Since long-term interest rates react at a different pace than short-term rates in response to different expectations, credit demand and supply, the slope of the yield curve steepens. As fresh flows of loanable funds become available at low interest rates once again to consumers and companies, the contraction phase comes to an end with a business *cycle trough*, after which another period of expansion begins, fueled by relatively inexpensive credit. Figure 7.16 shows short- and long-term interest rates and the slope of the yield curve for several credit cycles. Notice both the overall declining trend in interest rates since the 1980s and the changes in interest rates driven by monetary policy in response to periods of economic expansion and contraction.

Capital Markets

Debt and equity capital markets are extremely important. Companies can raise capital for operations and business expansion by selling equity stocks (that is, shares of ownership of the company) in public markets or by issuing debt securities (bonds) to investors as opposed to borrowing directly from financial institutions. In the United States, most debt funding for large corporate clients is provided by capital markets.

The dynamics of prices of stocks and debt securities reflect economic activity and can influence or be an indicator of social mood and market perception of risk. The overall direction of the stock market is often considered a primary indicator of a country's economic strength and development. Rising equity prices ordinarily reflect increased business activity while declining equity prices reflect a reduction of business activity and periods of financial stress. Share prices also affect the wealth of households and their consumption patterns. The periods of increasing and decreasing prices can be observed in Figure 7.17 for equity markets indicators (S&P 500 Index).

The dynamics of stock markets and the reaction of market participants are relevant in the context of climate change as governments promote or discourage different business and consumer activities to steer the economy toward their climate-related goals. These goals can impact different sectors of the economy as the industry landscape is reshaped either toward an altruistic low-carbon economy or taken advantage of by governments with a vested interest in their own internal economic growth at the expense of the decline of economy activity elsewhere due to restrictions to reduce high-emission activities.

Credit Quality and Credit Spreads

Although the overall level of interest rates plays a critical role in determining investment activity in the economy, companies and consumers have different levels of credit risk. This is reflected in the incremental yield (or credit spread) required by investors to take the additional credit or liquidity risk. More precisely, the credit spread of a security is the difference in yield between two instruments: (1) the selected debt security, with a given credit quality and recovery expectations in case of a default and liquidity in the market, and (2)

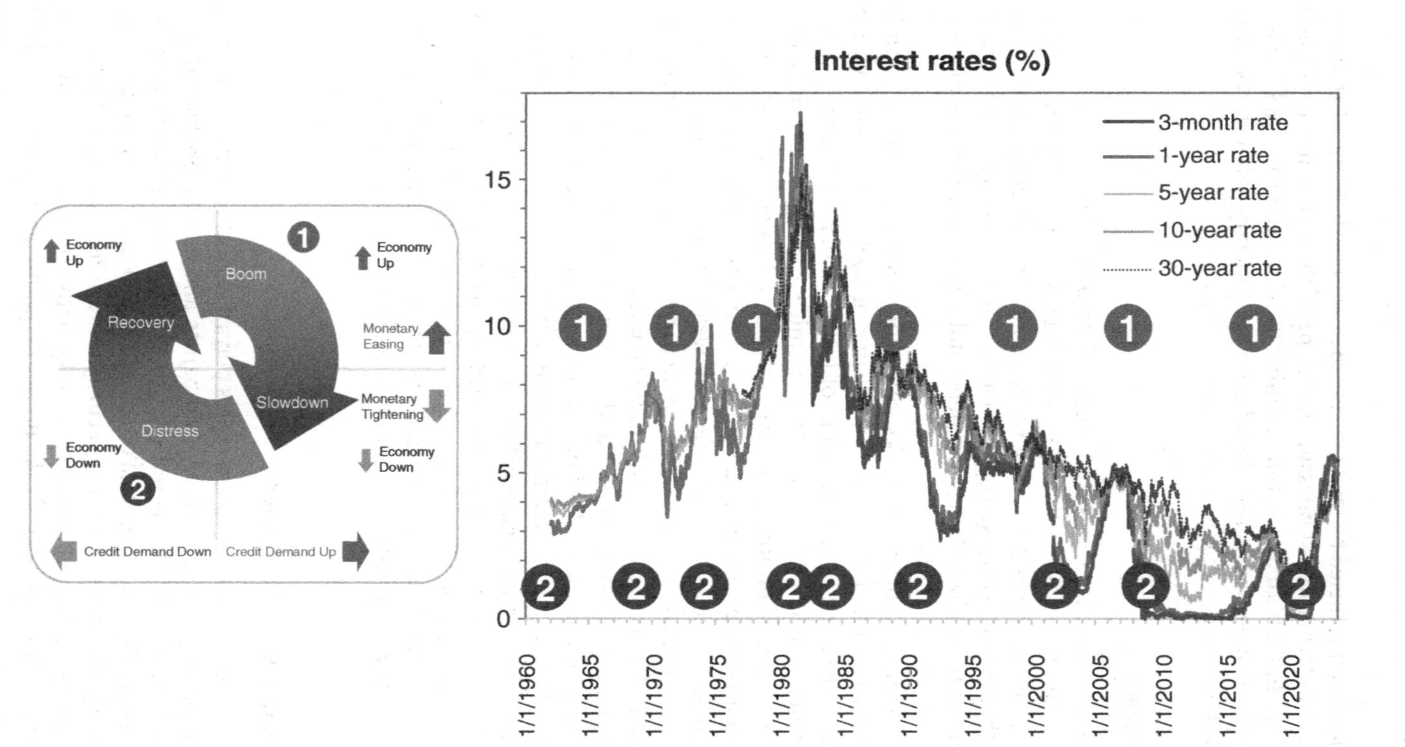

FIGURE 7.16 Interest rates for different credit cycles: 3 months, 1 year, 5 years, 10 years, and 30 years.

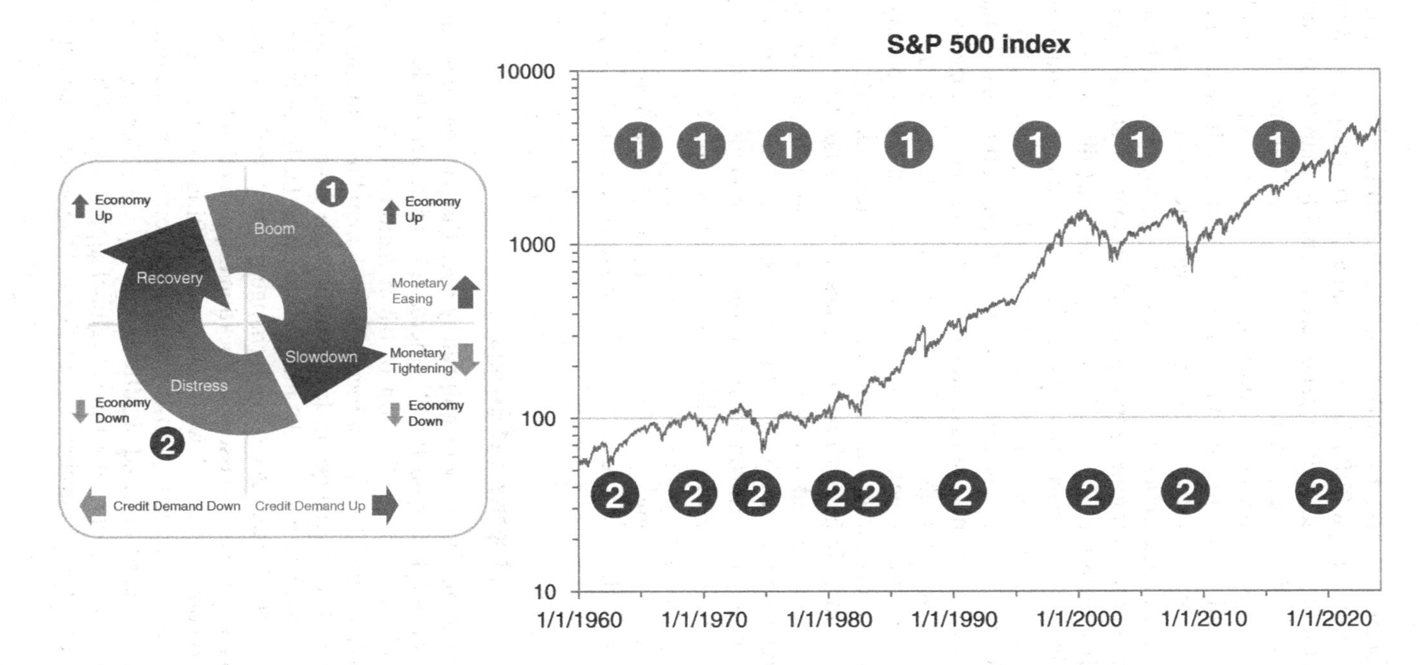

FIGURE 7.17 Evolution of equity market indicators (S&P 500) for different credit cycles.

a reference security of lesser risk. For example, the BBB credit spread reflects the difference between a BBB-rated bond and a US Treasury note of similar tenor (which is assumed to be a better credit quality than the riskier BBB bond). The credit spread reflects the additional net yield that investors demand from a security with more credit and liquidity risk relative to one with less credit and liquidity risk. Since credit spreads and related indicators reflect the dynamics of economic activity and investors' risk preferences, practitioners, policymakers, and academics tend to monitor the behavior of credit spreads as an indicator of economic activity and periods of financial stress. Figure 7.18 shows credit spreads for different risk rating categories (different credit quality) and their relation to periods of economic expansion and contraction.

Credit spreads are very important in the context of climate change as market participants will reflect their risk preferences in the yield demanded for companies that promote or discourage ESG and climate change activities. As investors demand higher levels of information from companies, the required credit spreads will reflect the companies' short- and long-term commitments to low-carbon emissions or net-zero strategies and their impact on the long-term viability of these companies. Over the last few years there has been increased interest in financing projects related to the development of alternative sources of energy or reduction of high-carbon emissions (green bonds and other debt instruments). However, market demand for green investments can also result in misleading information on the true nature of these projects, often referred to as "green washing." That is, dressing investments with the language of ESG and climate change to conceal their true nature and increase their appeal to investors looking for investment opportunities.

International Trade, Imports, and Exports

International trade has grown significantly over the past decades, affecting economies and markets globally. The increase in trade activity is the result of the overall reduction in the cost of labor, manufacturing, logistics and shipping costs around the world, increases in demand as countries become more developed, and changes in local and foreign government policies to promote the production of goods and services, and facilitate international trade (Root, 1990; Dicken, 2007). This change in the global supply chain of products has resulted in a massive reallocation of production from developed economies to countries with lower work wages, lifting countries exposed to decades of endemic poverty into the global economy. This reallocation of resources and wealth has had significant consequences for the economies involved. The trade-off of reallocating the global supply chain is the noticeable decline in production in developed countries and the systemic dependency of

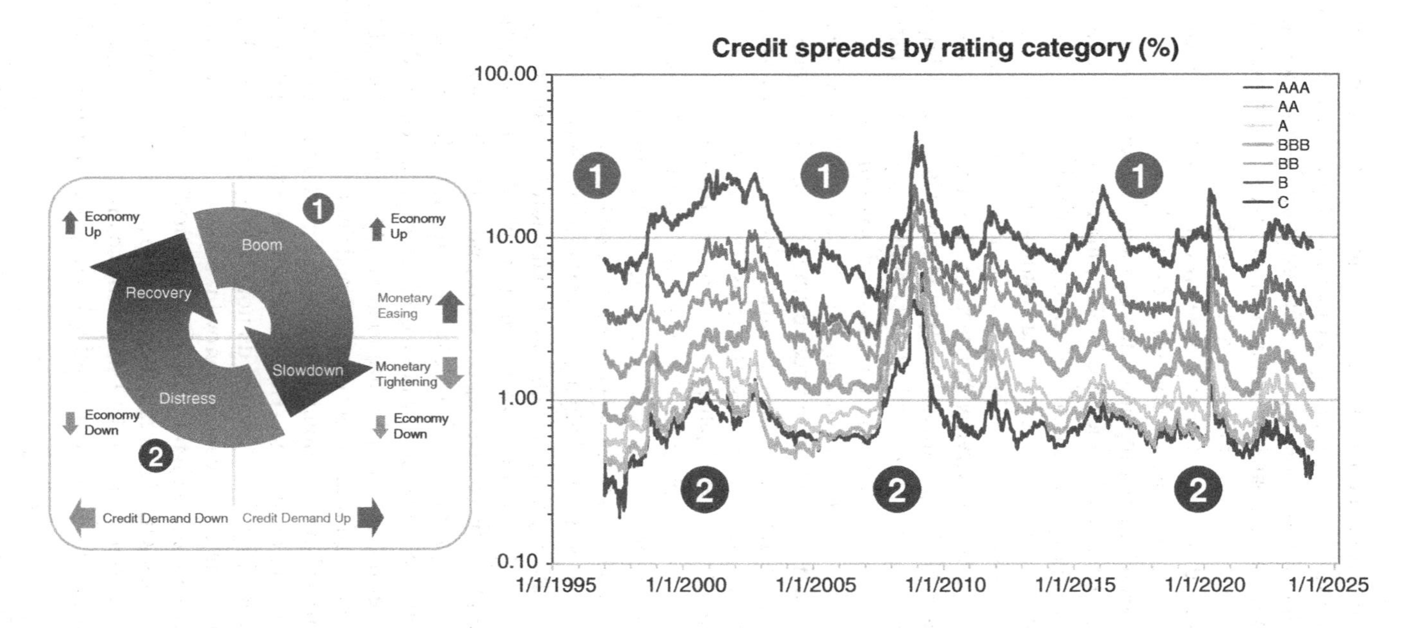

FIGURE 7.18 Credit spreads for different risk ratings and their relation to economic cycles.

their economies on offshore production, ranging from basic manufacturing to critical chemicals, medical and pharmaceutical products. This creates a significant vulnerability in the capacity of these countries to secure their supply chain locally and a means for low-cost suppliers and their governments to impact, or even control, other countries' supply chains.

In the United States and many developed economies, consumers buy increasing quantities of goods and services produced by other countries, while businesses sell increasing quantities of goods and services to consumers in other countries as well. Since the 1950s, imports and exports have been increasing as a significant fraction of US GDP, which reflects the value of all products and services produced in a country. The gap between imports and exports has been increasing in the last decades, reflecting differences in the cost of products and services between the US and other countries and the reallocation of the production supply chain. The gap in imports and exports exhibits cyclical swings reflecting periods of economic expansion and contraction, which can affect consumption and investments.

Not all sectors of the US economy are affected equally by international trade since certain products or services cannot be completely outsourced to other countries (e.g. health care or medical attention, defense, or basic services such as dining, plumbing, or barber haircuts). However, sectors such as the manufacturing of goods are highly affected by international trade. Although international trade can affect the US economy significantly, its impact has been viewed by policymakers as less significant than for most other countries, whose economic stability depends strongly on their ability to export. This short-sighted view is slowly changing as clearly shown during the COVID-19 pandemic in 2020–2022, where offshore production brought to light severe dependencies on other countries that led to dramatic shortages of critical medical supplies and products affecting the wellbeing of US citizens.

The flow of international trade is very relevant for climate change as governments promote or discourage different business and consumer activities to steer the economy toward their climate-related goals. This could have consequences on trade as government policies and regulations aimed at addressing physical or transition risks could affect the ability of companies to remain profitable or even viable. These local policies and regulations could be taken advantage of by countries that do not abide by the same rules or have an interest in their own economic growth, supporting or even promoting high-emission activities with the additional benefit of a net transfer of production from countries with tight regulations and no penalties for contributing to high-carbon emissions. Figure 7.19 shows the changes in international trade and consolidation of production creating a severe dependency on the global supply chain.

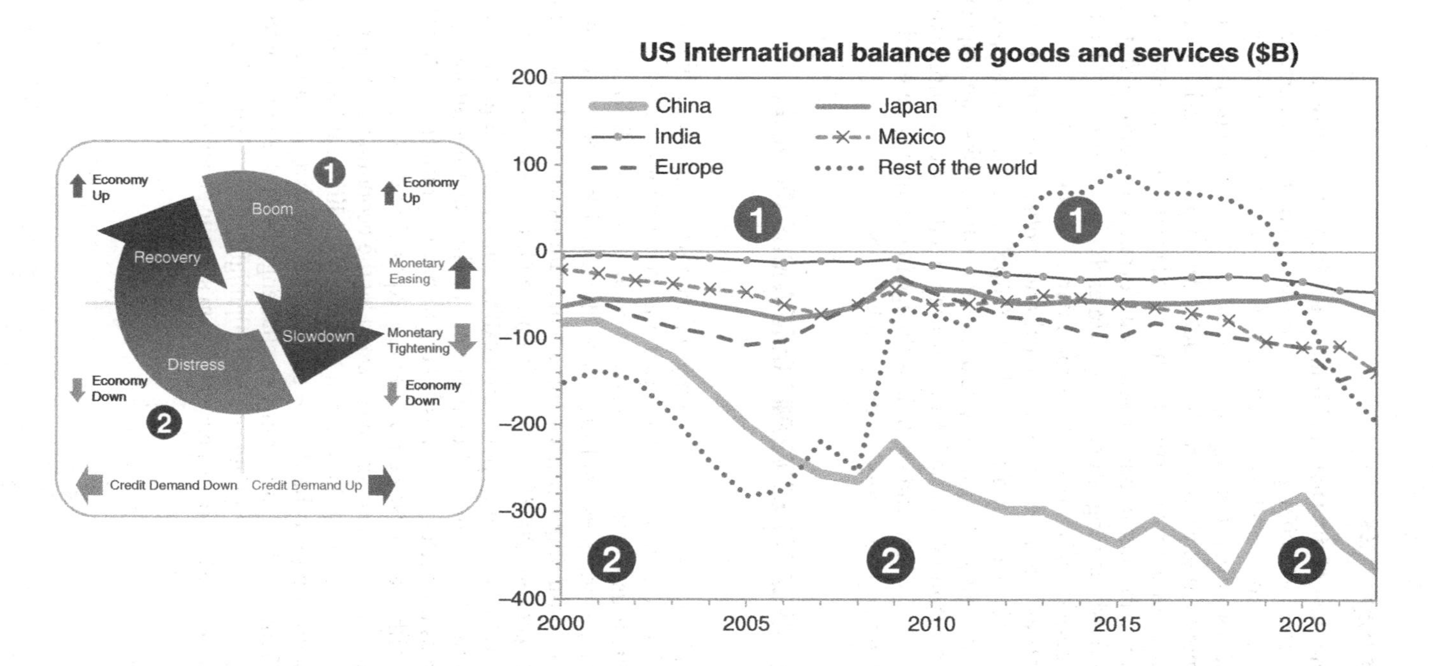

FIGURE 7.19 Changes in international trade flows.

Currency Exchange Rates

Related to the flow of international trade is the impact of currency exchange rates. The currency exchange rate is the price at which two currencies can be exchanged against each other (Figure 7.20). Ordinarily, exchange rates are classified as either floating or fixed. In a floating exchange rate, the daily movements in exchange rates are determined by the market, based on relative demand of the two currencies. In a fixed exchange rate, governments intervene in the market to buy or sell their currency to balance supply and demand at a fixed exchange rate. Changes in exchange rates reflect economic activity and periods of financial stress for different economies.

As countries are affected by the local and global responses to climate change and physical and transition risks, the supply and demand for products and services are also affected. Physical and transition risks can impact the supply and demand of products and services directly, or can impact market perception of risk, which can result in a change in the market demand for the country's currency, potentially affecting trade flows.

THE IMPACT OF LOCAL ECONOMIC CONDITIONS ON GLOBAL OBLIGORS

In previous sections we discussed a range of economic indicators that describe local conditions for different industries and geographies. These economic indicators allow us to forecast losses and create scenarios for a wide range of industries, geographies, and drivers of climate risk. Note, however, that most large multinational firms conduct operations in multiple countries (Root, 1990; Dicken, 2007). Large firms in high-emission sectors such as automotive, aviation, energy, petroleum extraction and refining, chemicals or large-scale manufacturing and distribution, or financial institutions financing high-emission sectors are most likely to have significant multinational operations. Although these companies are exposed to changes in local economic conditions, they can also be affected by the economic environment in other countries, which could result in sensitivity to different local economic scenarios, depending on the degree of international diversification in their operations. This presents a significant challenge from the point of view of modeling the impact of climate risk and highlights the potential limitations of models for forecasting losses based only on specific geographies.

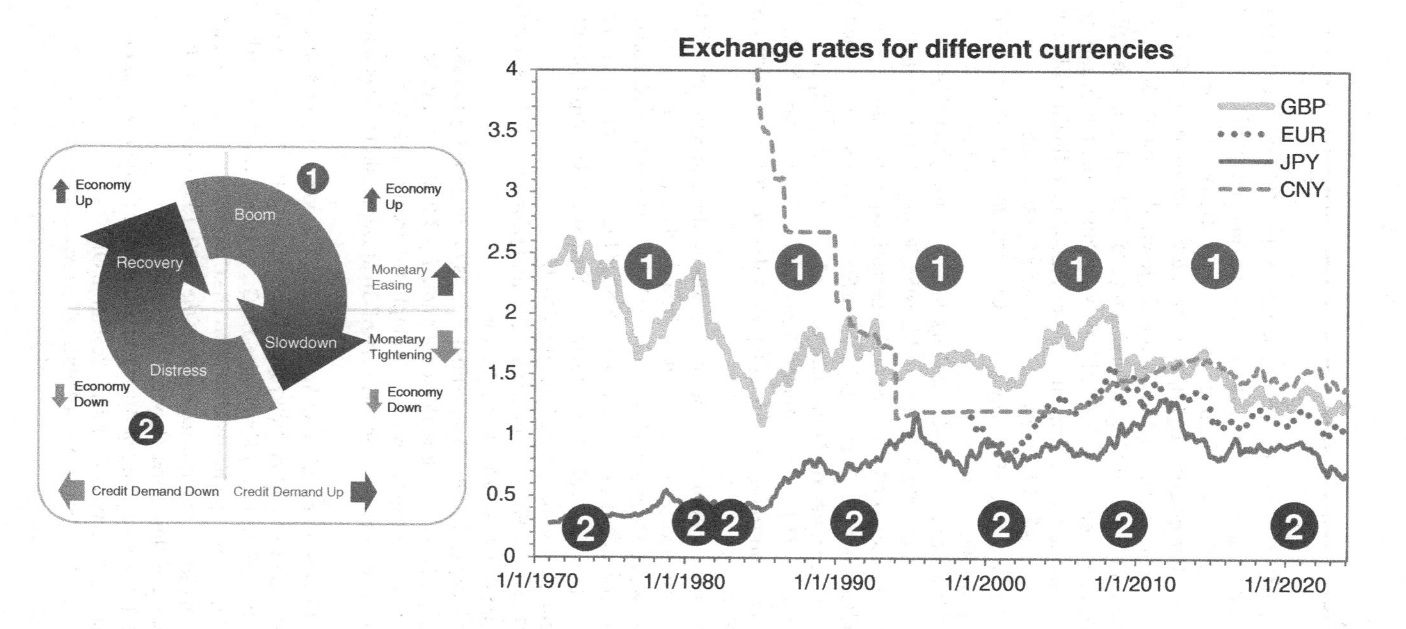

FIGURE 7.20 Currency exchange rates (in USD per currency unit) for selected currencies: GBP, EUR, JPY (x100), and CNY.

REFERENCES

Akhtar, M.A. (1994). Causes and Consequences of the 1982–92 Credit Slowdown: Overview and Perspective. In: *Causes and Consequences of the 1989–1992 Credit Slowdown,* Federal Reserve Bank of New York, February, pp. 1–38.

BCBS (Basel Committee on Banking Supervision) (2004). *International Convergence of Capital Measurement and Capital Standards,* Revised Framework, June.

BCBS (Basel Committee on Banking Supervision)(2005a). Studies on the Validation of Internal Rating Systems. AIG/RTF BIS Working Paper No. 14, February, pp. 1–120.

BCBS (Basel Committee on Banking Supervision) (2005b). Validation of Low Default Portfolios in the Basel II Framework. *Newsletter 6*, September.

Bernake, B., and Lown, C. (1991). The Credit Crunch. *Brooking Papers on Economic Activity* 2 : 206–247.

Bessis, J. (2002). *Risk Management in Banking*. Wiley Finance, New York.

Breuer, T., and Krenn, G. (2002). *What Is a Plausible Scenario? Fachhochschule Vorarlberg*, Working Paper, pp. 1–11. thomas.breuer@fh-vorarlberg.ac.at

Cooper, D. and Bachman, R. (2012). Cyclical and Sectoral Transitions in the US Housing Market. *Federal Reserve Bank of Boston, N 12-17*, Working Paper.

Cooper, G. (2008). *The Origin of Financial Crises*. Vintage Books, New York.

Cornyn, A. (1997). *Controlling & Managing Interest-Rate Risk,* New York Institute of Finance, New York.

Crone, T. (2000). A New Look at Economic Indexes for the States in the Third District. *Business Review. Federal Reserve Bank of Philadelphia*, pp. 3–14.

Crone, T. (2006). What a New Set of Indexes Tells Us About State and National Business Cycles. *Business Review. Federal Reserve Bank of Philadelphia*, Q1, pp. 11–24.

Diamond, P. (2013). Cyclical Unemployment, Structural Unemployment. *Federal Reserve Bank of Boston, N 13-5*, Working Paper.

Dicken, J. (2007). *Global Shift – Mapping the Changing Contours of the World Economy*. The Guilford Press, New York, pp. 73–136.

Dickens, T., and Triest, K. (2012). Potential Effects of the Great Recession on the US Labor Market. *Federal Reserve Bank of Boston, N 12-9*, Working Paper.

Fisher, I. (1932). *Booms and Depressions*. Adelphi Company, New York.

Fisher, I. (1934). The Debt-Deflation Theory of Great Depressions. *Econometrica*, 1(4): 337–357.

FRB (Federal Reserve Bank of New York) (2024). Corporate Bond Market Distress Index (CMDI). https://www.newyorkfed.org/research/policy/cmdi#/overview

Gujarati, D. (2006). *Essentials of Economics*. McGraw-Hill Irwin, New York.

Harris, E., Boldin, M., and Flaherty, M. (1994). The Credit Crunch and the Construction Industry. In: *Causes and Consequences of the* 1989–1992 *Credit Slowdown*. Federal Reserve Bank of New York, New York, February.

Hostetler, L.M. (1936). *75 Years of American Finance: A Graphic Presentation 1861 to 1935*. Western Reserve University, Cleveland, OH.

Hubbard, R.G., and O'Brien, A.P. (2008). *Macroeconomics*. Pearson Prentice Hall, Harlow.

Iyetomi, H., Nakayama, Y., Yoshikawa, H., Aoyama, H., Fujiwara, Y., Ikeda, Y., and Souma, W. (2010). What Causes Business Cycles? Analysis of the Japanese Industrial Production Data. Nigata University, Working Paper. 2010.

Juglar, C. (1915). *A Brief History of Panics*. Putnam's Sons, New York.

Keenan, S.C. and Sobehart, J.R. (2004). *Modeling Rating Migration for Credit Risk Capital and Loss Provisioning Calculations. RMA Journal*, October: 30–37.

Keenan, S.C., Sobehart, J.R., and Hamilton, D.T. (1999). *Predicting Default Rates: A Forecasting Model for Moody's Issuer-Based Default Rates*. Moody's Investors Service, August.

Kindleberger, C. (2000). *Manias, Panics and Crashes: A History of Financial Crises*. John Wiley, New York.

Mosser, P.C. and Steindel, C. (1994). Economic Activity and the Recent Slowdown in Private Sector Borrowing. In: *Causes and Consequences of the 1989–1992 Credit Slowdown*. Federal Reserve Bank of New York, New York, February, pp. 39–112.

Novak, J. (2008). *Marking NBER Recessions with State Data. Research Rap – Special Report*. Federal Reserve Bank of Philadelphia, Philadelphia, pp. 1–10.

Peracchi, F. (2001). *Econometrics*. John Wiley & Sons, New York.

Rebonato, R. (1996). *Interest-Rate Option Models*. John Wiley & Sons, New York.

Root, F.R. (1990). *International Trade and Investment*. South-Western Publishing, Cincinnati, OH, pp. 518–608.

Rothbard, M. (2000). *America's Great Depression*. Mises Institute, Auburn, AL, pp. 3–36.

Rothbard, M. (2008). *The Mystery of Banking*. Mises Institute, Auburn AL, *2008*.

Silverling, N. (1919). British Financial Experience 1790–1830. *Review of Economic Statistics*, pp. 282–297 (FRASER, fraserstlouis.ed.org).

Taylor, J.B. (1998). *Monetary Policy and the Long Boom. Business Review*. Federal Reserve Bank of St Louis, St Louis, November/December.

TEC (Tension Envelop Corp.) (1957). *Business Booms and Depressions since 1775*, Century Press, Cleveland, OH.

Waud, R. (1989). *Macroeconomics*. Harper & Row Publishers, New York.

Zamowitz ,V. and Boschan, C. (1975). *Cyclical Indicators: An Evaluation and New Leading Indexes*. Federal Reserve Bank of St Louis. Working Paper.

Zidong, A., Tovar, J., and Loungani, P. (2018). *How Well Do Economists Forecast Recessions?* WP 18/39, IMF, Working Paper.

CHAPTER 8

Pillar 2: Demand for Credit: Modeling Default Risk and Loss Severity

SUPPLY AND DEMAND FOR CREDIT: EXCESS CREDIT DEMAND

Broadly speaking, economic growth is driven by the ability of companies to expand their operations, buy equipment, build factories or improve services, train workers, and adopt new technologies in a timely and efficient way. Ordinarily, companies finance some of these activities from their retained earnings, which are profits reinvested in the company rather than redistributed to shareholders. However, when retained earnings are not sufficient to finance growth and new investments, companies must raise capital through the issuance of debt securities or equity in financial capital markets, or from financial intermediaries such as banks and investment funds. Financial markets and intermediaries play a fundamental role in the ability of companies to expand and acquire new technologies, and in providing funding for household purchases and investments, local and central governments, and other public sector entities. Companies, consumers, governments, and public sector entities drive the *demand for credit*, while the financial system (capital markets and intermediary financial institutions) provides the *supply of credit*, which comes from savings of other domestic or foreign households, companies, investors, or governments (Jones, 1991; Galbraith, 1994; Jones and Mingo, 1998; Kindleberger, 2000; Malkiel, 2007).

The relationship between supply and demand for credit determines the dynamics of prices and volume of goods and services for different economic sectors. When credit markets are in equilibrium and only credit demand or only credit supply shifts, economic theory can be used to estimate the effect on equilibrium prices and quantity. But when both supply and demand for credit shifts over time through the credit cycle at a different pace or with different lags, the dynamics of credit markets, borrowers' ability to access credit, credit quality, and potential losses become difficult to predict

(Cantor et al., 1999; Murray et al., 2000; Hamilton et al., 2001; Dembo et al., 2002; Mahoney, 2002; Stumpp, 2002; Hamilton and Cantor, 2004). Other relevant variables that drive the mechanics of credit demand and subsequent credit losses are the regulatory environment, fiscal and monetary policies, and the socio-political environment, which can facilitate or discourage the availability and appetite for different credit products through the promotion of credit programs, direct market intervention, or through different forms of taxation and financial restrictions.

Understanding the dynamics of supply and demand for credit over time can provide us with a sensible framework for predicting how changes in the actions of borrowers, credit providers, and the credit environment can cause changes in the quantity of credit demanded and the subsequent credit losses (Bucay and Rosen, 2000; Bessis, 2002; Ch'ng, 2002). This is relevant in the context of climate risk and ESG, which are driven by regulatory restrictions and policy changes as well as by social influence on consumers and companies. While climate change can lead potentially to costly damages caused by physical changes in the frequency and severity of storms, sea-level changes, or wildfires, it can also impact long-term consumer behavior and can shape policies for transitioning to a low carbon-emission economy and, consequently, the supply and demand for credit across multiple sectors.

Early work on understanding business and credit cycles (Keenan, Sobehart, and Hamilton, 1999; Keenan and Sobehart, 2004; Sobehart and Mogili, 2007) identified two key economic indicators: (1) a broad measure of aggregated excess credit demand, which should reflect all types of lending to the private sector by banks and non-banks, and (2) GDP growth, which is a measure of economic activity. Excess credit demand and economic activity are relevant for understanding the dynamics of credit supply and demand and credit quality because the timing of the flow of new entrants into the credit market affects the probability of changes in credit quality, and these flows are themselves affected by economic drivers and the overall credit conditions. Broad measures of excess credit demand are not often readily available and need to be constructed from different economic variables.

In the United States, the aggregate debt of all nonfinancial businesses (both large and small) has increased consistently over time, supported by issuance in the corporate bond and syndicated loan markets. Figure 8.1 shows the ratio of total debt of nonfinancial businesses over gross domestic product over multiple business cycles (FRB, 2024). The demand for credit shows both an increase relative to the growth in GDP and cyclical variations that reflect changes in business activity and periods of economic expansion and contraction. The regularity of the changes was interrupted by the COVID-19 pandemic. Although financial conditions deteriorated significantly in mid-2020 at the onset of the COVID-19 pandemic, they returned

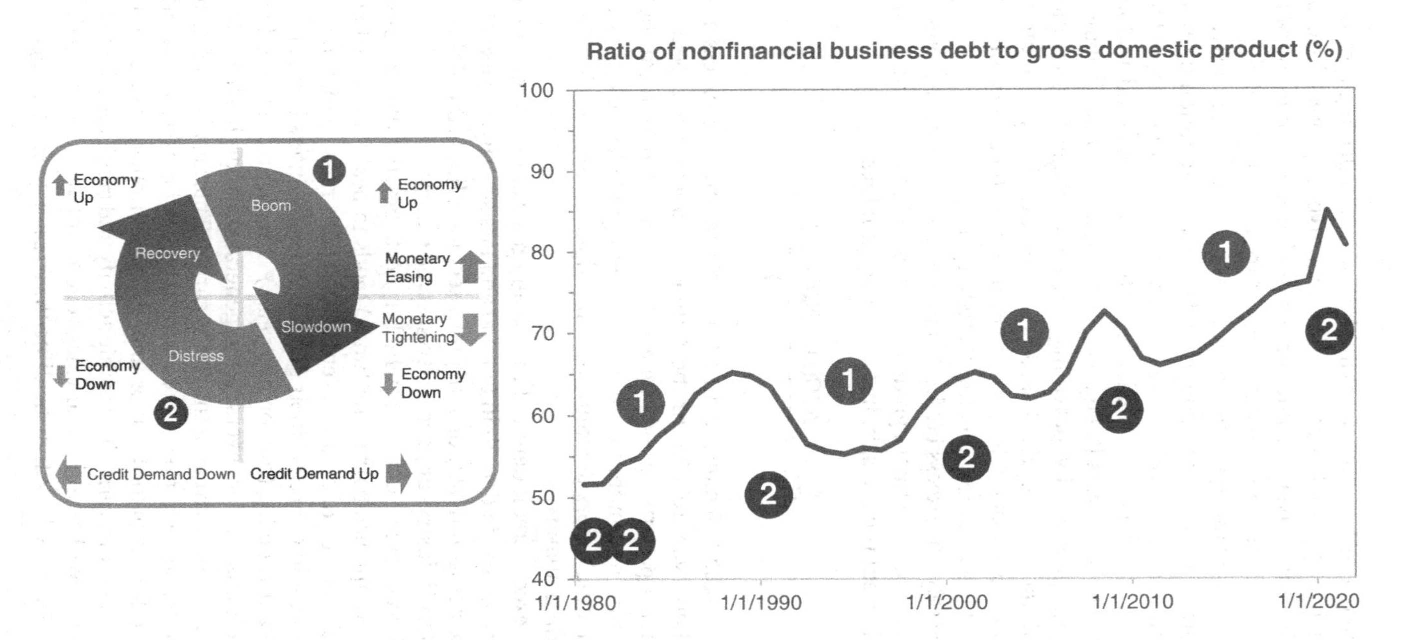

FIGURE 8.1 The ratio of total debt of nonfinancial businesses over gross domestic product over multiple business cycles.

to pre-pandemic levels for the surviving businesses, reflecting a recovery in economic activity supported by government initiatives to protect the economy during the downturn. Overall equity prices rose, although interest rates and inflation also rose in response to fiscal, monetary, and social policies.

Ideally, models of the dynamics of credit supply and demand driving business and credit cycles must include a fundamental description of monetary and fiscal policy, business and economic activity, and changes in preferences of investors and financial intermediaries. These models could include the effects of climate risk drivers in changes in credit supply and demand. Here we simplify the problem by providing phenomenological descriptions of credit demand and credit cycles based on the excess of lower credit quality borrowers above its expected long-term average, which has a credit demand component and is a leading indicator of credit cycles.

More precisely, the excess credit demand is described by the ratio of the total number of non-investment grade borrowers (NIG) to its long-term trend (NIG_{LT}) for different products, industries, or geographies:

$$E_{CD} = \frac{NIG(t) - NIG_{LT}(t)}{NIG_{LT}(t)} \tag{8.1}$$

The excess credit demand defined above captures the excess number of corporate borrowers relative to the number of borrowers expected from normal long-term patterns, providing a measure of potential credit bubbles. Figure 8.2 shows the number of non-investment grade issuers (that can be used for estimating excess demand) over multiple business cycles. Note the swings in credit demand.

Although excess credit demand can be an important economic indicator for describing changes in default likelihood and loss severity, the use of statistics aggregated across industry or geography segments limits the ability of loss models to reflect industry- or geography-specific patterns.

Because excess credit demand (which reflects broadly the excess of low credit quality companies borrowing funds) tends to peak much earlier than the observed frequency of default events (or default rates, which reflect the companies that could not repay their debt), it typically enters in credit models with a significant time lag (one or two years). This lag reflects the fact that these low credit quality borrowers fail to generate enough business to repay their obligations after borrowing the funds earlier. This lag also makes model forecasts more sensitive to previous historical values for excess demand, and less sensitive to forecasts for GDP or unemployment rates, which tend to be more coincidental indicators of economic activity and credit deterioration. For loss forecast windows beyond the time lags of one or two years, credit demand needs to be projected, which may be difficult. Thus, the use of a lagged excess

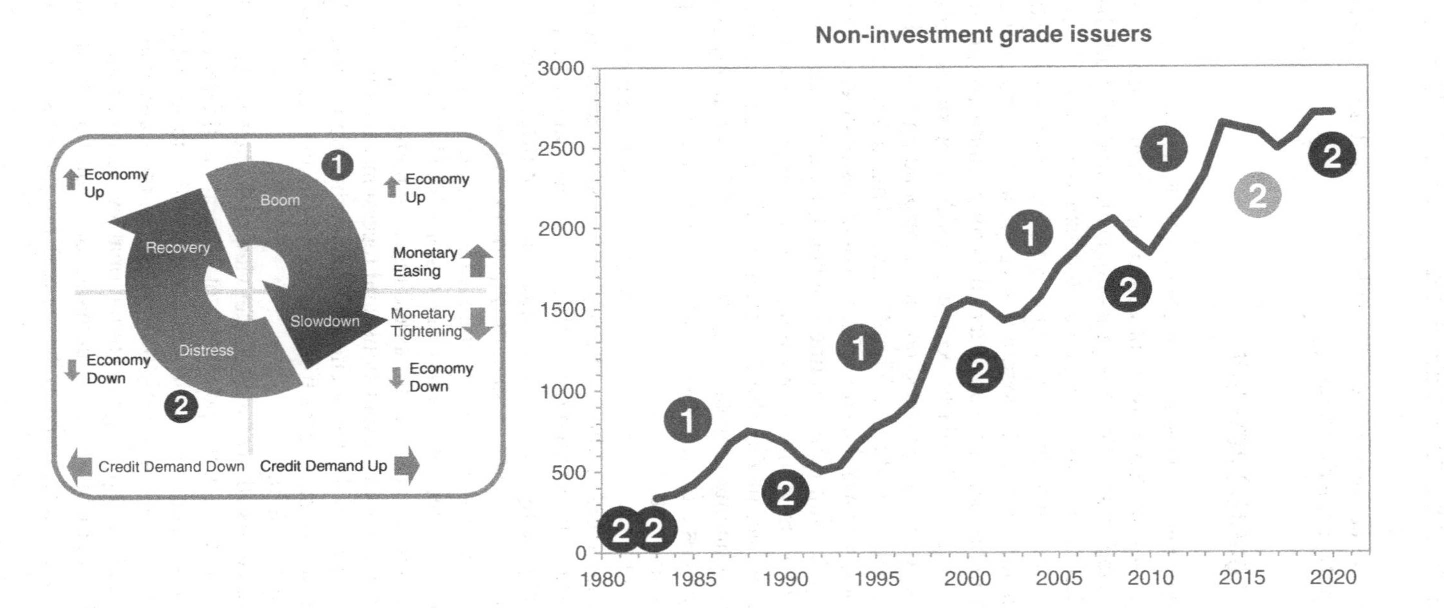

FIGURE 8.2 Number of non-investment grade issuers over multiple business cycles. The light circle reflects an episodic economic expansion and contraction in the energy sector.

demand variable as a key driver of credit models for default likelihood or loss severity may require further research.

ANALYSIS OF ECONOMIC ACTIVITY

The analysis of business and credit cycles is usually based on models whose explanatory variables are key drivers of economic activity such as country and government statistics (changes in gross domestic product, trade balance, government deficit or surplus, foreign and domestic debt, level of reserves), fiscal policy (tax rates, interest rates, money supply), labor market statistics (employment, labor force), demographics (population, households, and migration), and industrial and construction activity (industry production, housing starts and sales). Similarly, economic models for states, provinces, or counties can be constructed based on the corresponding regional or local drivers, and then aggregated at the country level. This approach recognizes that country economies are the sum of sub-economies of diverse size (state or provinces), which require more refined models linking the sub-economies through capital flows, migration flows and unemployment rates among other variables. For example, in the United States, workers tend to move from states where the economy is weak to states where the economy is strong and business activities are encouraged. Sub-economy models must explicitly account for these interaction effects.

As described in Chapter 7, there is a wide range of indicators of economic activity and business and credit cycles. Some of these variables are leading indicators of changes in economic activity (e.g. changes in credit quality of issuers with agency ratings), others are coincident with these changes, while others are lagging indicators (e.g. default rates or unemployment statistics). Although lagging indicators have limited value for forecasting, they are valuable to confirm changes in trends or recessions. The leading, coincident, and lagging indicators can be used for assessing current and future economic trends, particularly economic expansions, and recessions. Usually, indicators are grouped according to their tendency to change direction before, during, or after the general economy turns from a recession to an expansion or from an expansion to a recession.

This type of analysis allows for a broader view of the key drivers influencing the economy and offers the possibility of generating different scenarios by forecasting or stressing the main drivers of credit models (Breuer and Krenn, 2002; Sorge, 2004). Changes in fiscal policy, monetary policy, social policy, or demographics at the regional or national level can be translated into economic activity that can affect the demand for credit (i.e. the amount of funding and use of credit required by borrowers). From the risk management viewpoint, the ability to forecast the need for funds, use of credit lines and credit loss statistics (defaults, bankruptcies, and general business failures, asset write-offs, delinquency, and recovery rates) is critical to understanding the demand for credit.

DEFINING KEY CONCEPTS: DEFAULT, LOSS LIKELIHOOD, AND LOSS SEVERITY

Before we can discuss credit risk modeling techniques and measurement, we need to define basic concepts of loss likelihood and loss severity, which are the fundamental building blocks upon which credit risk management systems are built (Nishiguchi et al., 1998; Wilson, 1998; Nuxoll, 1999; Heitfield, 2003; BCBS, 2004, 2005a, 2005b).

Loss likelihood relies on a precise determination of a default event on a contractual agreement and is reflected in the *probability of default* (PD) over a fixed time horizon (e.g. 12 months). The magnitude of the potential loss resulting from the default event is reflected in the *exposure at default* (EAD), or amount at stake when the default event occurred, and the *loss given default* (LGD), which measures the amount lost as result of the default event. Depending on the purpose and use of the estimates, LGDs could reflect accounting losses (i.e. the amount disclosed in financial reports in accordance with accounting principles) or economic losses (i.e. the total loss incurred due to collection costs and legal expenses, replacement costs, time value of money, etc.). Conceptually, the expected loss *EL* is:

$$\mathrm{EL} = PD \times LGD \times EAD \tag{8.2}$$

Consistency in the application of these definitions is crucial. Institutions involved in developing quantitative PD, EAD, and LGD models need loss data to construct and validate their models. Loss data is most useful when based on consistently applied definitions and standards of interpretation. More precisely, PD models are typically of the binary response type and are estimated from binary flags (zero or one) to represent the default and non-default outcomes. But, in fact, the binary nature of the outcome is an artifact of the way we have chosen to observe the process and determine that a credit event has occurred. The definition of default is the filter through which heterogeneous information is converted into data. Therefore, the default definition applied in the modeling process should be the same as that used to classify and write off assets in the portfolio. Clearly, if the PD estimates from these models are used as inputs to an integrated portfolio management system, the definition of default and the calculation of loss applied to actual exposures in the portfolio must be the same as used in the modeling process or the model estimates may not reflect the expectation of losses accurately.

Default Events and Probability of Default

Although the basic concept of default seems straightforward (failure to repay in accordance with the terms of the lending agreement), it requires careful review (La Porta et al., 1998). For example, many indentures and loan agreements include a grace period for interest (and sometimes principal) payments. While taking advantage of the grace period does not violate the terms of the agreement, it does mean a failure to make a scheduled payment, and many financial institutions and rating agencies define default to include use of the permitted grace period. In other situations, financial institutions may allow for default status to be applied on the subjective basis that repayment is highly unlikely. If, for example, a company files for bankruptcy, one may not want to wait until a scheduled payment date to assign default status to the borrower. More subtle and complex issues may also arise when a borrower defaults on selected debt obligations or on one set of creditors, which can trigger cross-default events across other obligations. Many loan agreements and indentures contain language establishing default status if the obligor defaults on other obligations. These cross-default provisions are intended to facilitate the coordination of creditors in restructurings and bankruptcies, but they are ultimately contractual arrangements that can themselves be contested, introducing additional sources of uncertainty in the determination of default events. There are other relevant questions related to the determination of default and credit events. What sorts of liabilities are or should be included as triggers under the cross-default component of the overall definition? Should the definition cover only tradable securities and loans? What if a company fails to pay its suppliers or employees? What if an insurance company fails to honor its clients' policies? Or what if a bank fails to allow clients unlimited access to their deposits?

Another subjective category of default is that of *distressed exchange*. A default occurs when an obligor exchanges one set of debt obligations for another to reduce its debt burden, interest expense burden, or both. Typically, to qualify as distressed exchange, creditors must either accept uncompensated extensions of maturity, uncompensated reductions in seniority, or reduced interest rates on outstanding debts. But how should compensation be measured? If the obligor seeks to replace a portion of the debt with equity, how should the equity be valued? If a lender of a distressed borrower sells loans back to the borrower or to third parties at distressed market prices, should such transactions be considered a default event?

The definition of default also needs to be aligned to regulatory, legal, and accounting standards in the jurisdictions in which the financial institutions operate. For example, default events may reflect business failures due to court proceedings such as bankruptcy, foreclosure, and receivership, and businesses that closed voluntarily either through compromises with creditors or by leaving unpaid debts.

To illustrate, the US bankruptcy code[1] contains five *Chapters* (Chapters 7, 9, 11, 12, and 13) under which bankruptcy petition may be filed:

- **Chapter 7** provides for "liquidation": the sale of a debtor's nonexempt property and the distribution of the proceeds to creditors. In a Chapter 7 filing, a trustee is appointed to collect and liquidate the assets and distribute the proceeds to creditors in accordance with set priorities.
- **Chapter 9** provides for reorganization of municipalities, which includes cities and towns, as well as villages, counties, taxing districts, municipal utilities, and school districts.
- **Chapter 11** provides for reorganization, usually involving a corporation or partnership. Chapter 11 is a reorganization where a debtor seeks to reorganize its financial structure without liquidating all its assets to keep its business alive and pay creditors over time. The reorganization usually takes one to three years, and the percentage of recovery could vary significantly. In a pre-packed Chapter 11, a company in financial distress reaches an agreement on the terms of the reorganization plan with its creditors prior to filing for bankruptcy.
- **Chapter 12** provides for adjustment of debts of a family farmer or a family fisherman. This chapter seeks to reorganize and rehabilitate the financial structure of the debtor. Normally it allows a debtor to propose a plan to pay creditors.
- **Chapter 13** is known as the wage earners' chapter and provides for adjustment of debts of an individual with a regular income. Chapter 13 allows debtors to keep property, propose plans to pay all or part of their debt, and repay debts over time, usually three to five years.

Data on business failures can be obtained from court records for bankruptcies, foreclosures, receivership, attachment, etc. Data for non-court failures can be obtained from local credit management groups and boards of trade, sales notices, and auction sales. Here we focus mainly on debtors of financial institutions, for which this information is critical for developing credit risk models. Note that for portfolios including high investment grade/high quality obligors or assets and specialized structured transactions, historical data on defaults may be rare or non-existent. In such cases, financial institutions may need to infer default risk using indirect methods such as extrapolating from statistics for other risk ratings or borrowing techniques for predicting rare events from other disciplines.

For loss modeling and data management purposes, it is not only critical to establish the precise date on which a default occurred but also the date for the termination of the default episode and resolution of the event. Once an obligor's status shifts from "in default" to "not in default," the risk of future default resurfaces and must be evaluated anew. The time frame for the calculation of realized losses (LGD) is determined by the *default date* and the *resolution date*. Different conventions are used to define a "default episode" and to establish what circumstances constitute a termination of such an episode. For example, if an interest payment is delayed, the obligor may be considered "in default" for a period of six months or a year, even if payment is made within that period. Such practices recognize the escalating nature of credit events associated with a default, with delayed payments often cascading to distressed restructurings and, perhaps, followed by a bankruptcy filing. A bankruptcy emergence date may be used as a termination date for the default episode, or the borrower may be required to file at least one set of new financial statements before being considered a viable firm again.

Estimating Default Rates

Related to the estimation of probabilities of default is the calculation of default rates. Default rates reflect realized/observed frequencies of (*past*) default events and can be leveraged for the estimation of probabilities of default, which reflect *future* possible events. Here we discuss primarily the estimation of corporate default rates and the implications of different definitions for measuring, understanding, and forecasting credit losses. We focus on data issues, definitions, and a methodology for calculating default rates for companies that can vary by credit quality (ratings), industry, and geography. Similar approaches are also applicable to the calculation of consumer default rates for retail portfolios.

Default rate statistics are important for portfolio managers and investors seeking to establish a bullish or bearish stance on credit. Changes in the actual or perceived state of the credit cycle can have an impact on portfolio performance that can go beyond simple statistics on defaults. Turning points in the credit cycle can appear gradually or abruptly, creating risks and opportunities for those able to anticipate these trend changes correctly. Furthermore, because climate risk and credit stress scenarios are usually tied to historical default rates in periods of economic downturn or financial crises to be able to translate the impact of the scenario drivers into credit losses, it is important to have a clear understanding on how default rates are calculated and how they change in response to economic activity.

Figure 8.3 shows US corporate default rates over the last few decades, exhibiting multiple distinct episodes of high default rates closely associated with different events: the railroad crisis sparked by the Penn-Central default in 1970, the leveraged buy-out (LBO) debacle in 1989, and the savings and loans (S&L) crisis in 1990, the bursting of the tech-bubble in 2000–2001, the financial crisis of 2008–2009 (the Great Recession) and more recent periods that reflect changes to the global economy and credit environment as a result of the 2020 COVID-19 pandemic. The observed changes in default rates reinforced the idea of credit cycles aligned to changes in economic and business conditions. The relationship between economic activity and credit events can be leveraged to introduce the impact of economic drivers of climate risk and ESG and translate them into credit losses for different scenarios.

A natural question to ask is, which factors cause these dramatic swings in default rates? The way we choose to observe and describe these events is a critical determinant of how we ultimately understand them. More precisely, we need to carefully review how defaults are defined and how default rates are calculated to evaluate the impact on the measurement of credit cycles.

For bond investors whose portfolios consist primarily of rated instruments, the default rates published by the rating agencies provide relevant information on credit quality. However, the population of companies holding credit ratings from the major agencies is only a small fraction of credit borrowers, including bank loans, private placements, or other forms of borrowing. Statistics that represent a broader universe of borrowers might help to provide a different view of credit cycles and may ultimately lead to a different understanding of their drivers that can provide a better framework for modeling climate risk and ESG effects.

We begin by reviewing the concept of default rate. More precisely, let f_t be the default rate (frequency of default events) in month t, D_t be the number of defaulters in month t, and N_t be the number of companies in month t. The default rate over period T is the sum of all default events D_s occurring over the selected period $t - T \leq s \leq t$, divided by the number of companies N_{t-T} that could default during the same period, adjusted for the number of companies W_E that exit the population.

$$f_t = \frac{\sum_{s=t-T}^{t} D_s}{N_{t-T} - W_E} \tag{8.3}$$

Note that companies can exit the population during the observation period for multiple reasons such as cessation of operations or moving their banking relationship to another institution. Because companies can leave the population for a variety of reasons, the estimation of W_E requires supplemental assumptions about the default characteristics of these companies.

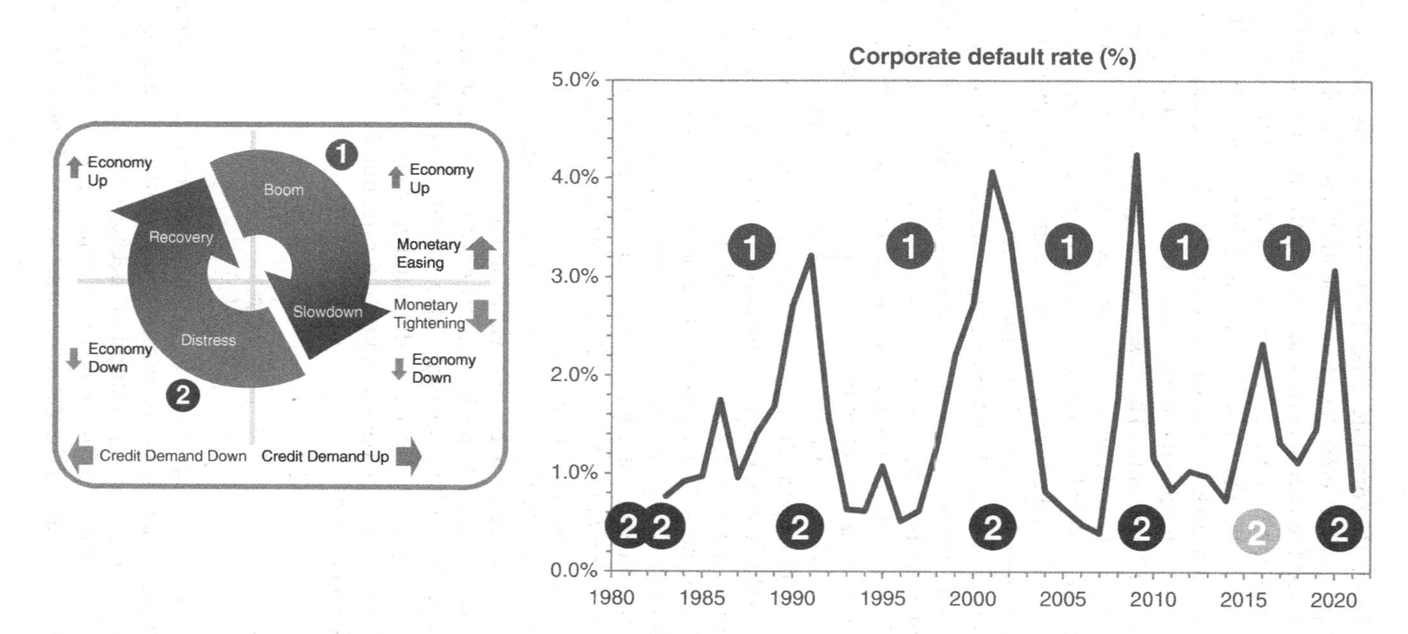

FIGURE 8.3 Stylized US corporate default rates over time. The light circle reflects an episodic economic expansion and contraction in the energy sector.

To illustrate, if these companies could not default after leaving the population (e.g. due to cessation of operations), the effective adjustment would be $W_E = 0$. In contrast, assuming these companies could default at the same rate of the remaining companies after leaving the population (e.g. moving to a different bank), the effective adjustment would be $W_E = \sum_{s=t-T}^{t} W_s$, that is, the sum of all exiting companies W_s over the selected period $t - T \leq s \leq t$. Another common adjustment is $W_E = \frac{1}{2}\sum_{s=t-T}^{t} W_s$, under the assumption that companies exit uniformly through the period T.

Note also that company subsidiaries are usually excluded from Equation (8.3) since few subsidiaries have independently traded equity, reported financials, or act independently from their parent companies. The inclusion of subsidiaries can lead to changes in default rates due to the correlation of parents and subsidiaries, which should be considered when determining the confidence level of model parameters calibrated with historical default rates.

Obtaining timely default information to calculate the numerator in Equation (8.3) is critical. A default event is deemed to have occurred whenever:

1. A company fails to make a timely payment of interest or principal on bank loans or public debt obligations, including bonds, notes, or commercial paper. This includes debt waivers by creditors and delayed payments, even when grace periods are provided for in the indenture or loan agreement.
2. A company files for bankruptcy protection (or its equivalent) or is forced into bankruptcy by its creditors.
3. A company engages in distressed exchange to reduce its debt burden, interest expense burden, or both. Distressed exchange occurs when creditors must either accept uncompensated extensions of maturity, uncompensated reductions in seniority, or reduced interest rates on outstanding debts. When the obligor is clearly distressed, the selling of loans back to the obligor or to third parties at "market rates" or the acceptance of debt-for-equity swaps constitute distressed exchange defaults.
4. A company makes a formal announcement that it has insufficient funds to repay creditors, and seeks either debt forgiveness, new equity injections, or both.
5. A distressed company is provided with below-market loans from a government or government agency, specifically to avert or forestall bankruptcy.

Because the rating agencies cover primarily debt securities and stay in close contact with the issuers of these securities, they can identify default events for the issuers they rate. In contrast, constructing default statistics for a broader population of companies requires additional sources of information on companies and default events to capture the situations listed above.

The issues discussed above highlight that the calculation of default rates reflects a range of supplemental assumptions that can impact the default rate values and their relationship to economic drivers and, therefore, can impact the estimation of probabilities of default as discussed in later sections.

Exposure at Default

Although the calculation of the exposure amount when a default event occurs (EAD) may seem simple at first blush, the existence of credit products with amortization schedules, discretionary drawdown options, and market-sensitive exposures (e.g. derivative products) creates uncertainty as to what the exposure amount to the borrower or counterparty will be at any point in the future. In cases where access to cash could potentially help a company to avoid bankruptcy, one would expect this company to make full use of discretionary credit lines, hoping for a favorable outcome. However, in cases where bankruptcy is inevitable, a company may wish to limit its debt to present a more favorable picture for restructuring its liabilities and may choose not to pay other obligations rather than to draw on existing credit lines. For derivatives products, the exposure amount will tend to fluctuate with market conditions making predictions difficult. The term *derivatives* refers to multiple types of over-the-counter and exchanged traded transactions including interest and currency swaps, swap options, interest rate caps, collars, floors, weather derivatives, and other financial products. Derivatives contracts often include margin agreements so that the actual exposure amount depends on the behavior of the underlying assets or cashflows as well as the specific details of the margin agreements. The International Swap Dealers Association (ISDA) developed standard contract definitions, provisions, and master agreements that helped parties to streamline the documentation of derivatives transactions. EAD for these products is usually based on complex models and numerical simulations that reflect basic features of these contracts.

Let's turn our attention first to the calculation of EAD for simple products and then discuss more complex products. Let's assume we have a simple debt obligation (e.g. term loan) with a principal amount V_0 outstanding at the time $t = 0$. The debt obligation provides interest and amortization payments at regular time intervals $t = 0, 1, 2, \ldots, T$, where T is the remaining tenor of the obligation measured in payment periods (e.g. monthly, quarterly, or annual interest payments). Let the contractual cash flows CF_t be paid at regular time periods t

$$CF_t = C_t + W_t \quad 0 \leq t \leq T \tag{8.4}$$

The contractual cash flow CF_t includes interest payments (coupons/fees) $C_1, C_2, \ldots, C_T$ and amortization payments $W_1, W_2, \ldots, W_T$ of the total principal amount. As a result of the amortization of the principal amount, the principal amount value V_t outstanding at the beginning of each payment period t is simply:

$$V_t = V_0 - \sum_{s=0}^{t-1} W_s = V_{t-1} - W_{t-1} \tag{8.5}$$

To illustrate, for a simple term loan/bond paying fixed fees/coupons with a single balloon payment of principal $W_T = V_0$ at expiration T, the cash flows are:

$$CF_t = \begin{cases} C_t & 0 \leq t \leq T \\ C_T + W_T & t = T \end{cases} \tag{8.6}$$

If the obligor defaults during the period $[t-1, t]$, there are no further interest or amortization payments. At this point, the exposure at default EAD is defined in terms of the remaining principal amount value V_t outstanding at the time of default t plus any accrued interest C_t^A that remained uncollected before the default event:

$$EAD_t = V_t = V_0 - \sum_{s=0}^{t-1} W_s + C_t^A \tag{8.7}$$

Equation (8.7) describes the EAD for funded loans and funded banking products (that is, funds were already provided to the borrower). This EAD also applies to debt securities that are simple direct obligations of the issuer (e.g. bonds with no complex features).

Now let's discuss the EAD for unfunded loans with *unused commitments* and repayment amortization schedules for drawn amounts. Unused commitment refers to the amount of funds that financial institutions agree to lend to a borrower but has not yet been drawn (e.g. revolver or line of credit). An unused commitment represents a potential liability for the lender since the exposure-at-default depends on the borrower's future drawdowns on the available funds.

The contractual cash flows for unused commitments include two fees/interest rate components:

1. interest rate payments on the drawn amount V_t (with interest rate r_v)
2. interest fees on the undrawn amount U_t available for future drawdowns (with fee/rate r_u)

That is, the cash flows for interest, fees and amortization can be described as follows:

$$CF_t = r_v V_t + r_u U_t + W_t \quad 0 \leq t \leq T \tag{8.8}$$

As a result of the amortization of the principal amount W_t and incremental drawdowns d_t, the principal amount value V_t outstanding at the beginning of each payment period t is simply:

$$V_t = V_0 + \sum_{s=0}^{t-1} d_s - \sum_{s=0}^{t-1} W_s = V_{t-1} + d_{t-1} - W_{t-1} \tag{8.9}$$

Similarly, the unused commitment amount available is in each period is:

$$U_t = U_0 - \sum_{s=0}^{t-1} d_s + \sum_{s=0}^{t-1} W_s = U_{t-1} - d_{t-1} + W_{t-1} \tag{8.10}$$

Here the incremental drawdown d_t is determined by the difference in *incremental utilization* IU_t in the period $[t-1, t]$

$$d_t = IU_t - IU_{t-1} \tag{8.11}$$

The incremental utilization IU_t represents the expected drawdown of the unused commitment in each period t.

Each contractual cash flow CF_t is based on the amount outstanding at that point in time. The expressions for expected losses are the same equations used for funded exposures except that EAD_t includes both the funded amount and the *expected* funded amount at default from the available unused committed exposure as reflected in the Credit Conversion Factor (CCF), which represents the expected fraction of the unused commitments at time t that will be drawn in the event of default:

$$EAD_t = V_t + CCF_t U_t \tag{8.12}$$

Note that CCF_t reflects the expected drawdown in the event of default (usually defined over a fixed 12-months period prior to default). CCF_t could be higher than the normal drawdowns reflected in the incremental utilization IU_t. The effects of incremental utilization are included in the definitions of V_t and U_t. An alternative approach is to estimate the full term-structure of drawdowns for every period until the default event, which would reflect the combined effects of the fixed window CCF_t and the cumulative effects of incremental utilization IU_t. Climate risk and ESG issues can affect the borrower's needs for funds (e.g. to cover factory damages or upgrades for physical risk mitigation, or to upgrade technologies to lower carbon emissions in response to transition risk). This can affect the normal utilization of credit facilities (as reflected in the incremental utilization IU_t) and the drawdowns in default (as reflected in the credit conversion factor CCF_t).

Similar EAD concepts apply to *contingent exposures* such as trade, performance, or standby letters of credit, which are fund commitments provided by a bank or financial institution to a beneficiary on behalf of the obligor. Contingent exposures allow the obligor to make payments or provide a guarantee to a beneficiary substituting its own credit quality with that of the institution and, therefore, represent (future) contingent credit liabilities for the institution. The EAD for contingent exposures reflects the expected exposure amount due to a beneficiary or paid to the beneficiary but that remains uncollected by the financial institution from the obligor in the event the obligor defaults on its obligations.

For derivatives transactions, the estimation of EAD is often based on complex models and numerical simulations of basic features of these contracts. Derivatives contracts often include margin and netting agreements so that the actual exposure amount depends on the behavior of the underlying assets and market conditions as well as the specific details of the margin and netting agreements.

When financial institutions engage in derivatives transactions, such as interest rate swaps, currency swaps, credit default swaps, options, or weather derivatives, they hedge out as much of the risk in the underlying as possible. More precisely, they hedge their positions to reduce the sensitivity of their interest rate swap book to changes in interest rates, and the sensitivity of their equity options books to changes in the underlying stock prices. This is not always possible or cost effective, so some residual risk often remains unhedged. The residual risk can be mitigated further through the transaction agreement since derivative contracts often require counterparty performance clauses, including honoring margin agreements over the life of the contract and complying with the terms of the contract at maturity or if contract termination triggers are reached.

The calculation of losses for derivatives transactions requires estimating both the likelihood that the counterparty will default on its contractual obligations and the exposure amount at stake. The potential exposure amount of a counterparty is contingent on the market factors affecting all transactions as well as any legally enforceable agreements, such as netting and margin agreements. The circumstances under which the counterparty will default are usually highly specific and tied to market factors that also affect the exposure amount, potentially creating wrong way risk, where the exposure amount increases as the likelihood of default increases. Modeling counterparty risk for derivatives products often requires complex simulations including future market conditions, and hypothetical counterparty responses to each scenario. The exposure amount for derivatives exhibits a distribution of possible outcomes that reflects the nature of the contract and the underlying assets or cash flows. The "expected" exposure amount can be positive (money the counterparty owes) or negative (money owed to the counterparty) and is often determined at different confidence levels to reflect the uncertainty in outcomes.

Two exposure amount measures aligned to the EAD concept for derivatives are the expected positive exposure (EPE) and expected negative exposure (ENE).

$$EPE_t = E_X[NM_t^+ + N_t^+ + UN_t^+] \tag{8.13}$$

$$ENE_t = E_X[NM_t^- + N_t^- + UN_t^-] \tag{8.14}$$

Here $E_X(y)$, is the expectation of variable y for all possible values (paths) of the market factors X. For each individual path of the market factors, NM_t^+ is the sum of positive exposures over all netting and margin agreement sets, N_t^+ is the sum of positive exposures over remaining netting sets, and UN_t^+ is the sum of positive exposures over un-netted transactions. Similarly, NM_t^- is the sum of negative exposures over all netting and margin sets, N_t^- is the sum of negative exposures over remaining netting sets, and UN_t^- is the sum of negative exposures over un-netted transactions.

$$\begin{array}{ll} NM_t^+ = \sum_m\left(\left(\sum_j PV_{t,jm}\right)^+ - M_{t,m}\right)^+ & NM_t^- = \sum_m\left(\left(\sum_j PV_{t,jm}\right)^- - M_{t,m}\right)^- \\ N_t^+ = \sum_n\left(\sum_j PV_{t,jn}\right)^+ & N_t^- = \sum_n\left(\sum_j PV_{t,jn}\right)^- \\ UN_t^+ = \sum_s (PV_{t,s})^+ & UN_t^- = \sum_s (PV_{t,s})^+ \end{array} \tag{8.15}$$

Here $M_{t,m}$ is the margin posted for netting/margin set m, $PV_{t,k}$ is the present value of the cash flows for transaction k at time t, $(x)^+ = \max(x, 0)$ and $(x)^- = \min(x, 0)$. For derivatives involving products with multiple parties (such as collateralized debt obligations (CDO) and basket products), the estimation of correlations in credit quality for these parties is also critical in evaluating the risk of these transactions as they impact the cash flows $PV_{t,k}$.

Loss Given Default

Depending on the purpose and use of the *loss given default* estimates, LGDs could reflect accounting losses (i.e. the amount disclosed in financial reports in accordance with accounting principles) or economic losses (i.e. the total loss incurred due to collection costs and legal expenses, replacement costs, time value of money, etc.) (Araten et al., 2004; Li et al., 2009; Mora, 2010;

Yashkir and Yashkir, 2013). There are essentially two different approaches that an institution can use to determine loss given default: (1) the mark-to-market approach, which reflects losses based on market valuation, and (2) the workout approach, which reflects ultimate losses incurred through the workout process.

The mark-to-market approach seeks to estimate loss from the market value observed for a company's obligations once that company has defaulted. For traded obligations such as bonds or other securities, prices are often quoted for some time after default has occurred. One common convention is to calculate losses as the percentage of the par value at which a bond trades one month after the default event. To illustrate, if a \$100 bond is trading at \$60 after one month, then the loss is \$40 and the LGD is 40%. In practice, historical samples of defaulted obligations can be used to produce a distribution of losses that can be used to estimate LGD for defaults that have not occurred yet and may materialize in the future.

The mark-to-market approach is simple and objective, but it has some drawbacks. As the uncertainty over the prospects of the distressed company diminishes over time, prices on the company's obligations will tend to cluster around par value for surviving companies, around some residual asset value for liquidated companies, and somewhere in between these values for companies with court-mandated settlements that go bankrupt and later re-emerge from bankruptcy. During periods of rising (or falling) interest rates, fixed income assets will often trade at significant discounts (or premiums) to par value. When this happens, the percentage of par value becomes a poor measure of loss due to default. The use of traded assets may also induce biases in LGD estimation since companies with less liquid obligations are not represented, and even liquid instruments may stop trading in extreme loss situations.

In contrast, the workout approach measures the value of assets recovered over the resolution period of the workout process (post default), discounting all cash flows back to some appropriate date (usually the default date) at some appropriate discount rate (usually the contractual lending rate). This approach accounts for the timing of payments and expenses, such as legal costs and interest drag (lost or delayed interest) and reflects the ultimate loss experience by the lender or investors.

In both cases, market-to-market LGD or workout LGD, realized losses often show a distribution of outcomes with wide dispersion. The distribution of LGDs usually differs from a bell-shaped distribution around a mean value. The distribution tends to be flatter over a wide range of LGD values, reflecting uncertainty in losses and recovery process. Depending on the nature of the debt obligation (e.g. bonds or loans), the distribution could have an L-shape or slight U-shape, with relatively higher probability of low losses reflecting risk mitigation factors such as seniority, collateral or guarantees, or reflecting default events with no materialized losses.

Figure 8.4 illustrates the difference in the distributions of outcomes for market-based and workout LGD. Furthermore, realized losses depend on characteristics, such as product type, sector and geography, and risk mitigation factors such as debt seniority, collateral, and guarantees. These characteristics exhibit different losses under stress and benign economic conditions observed during business contractions and expansions.

The estimation of LGD requires collecting recovery information and discounting the recovery amounts appropriately to reflect realized economic or accounting losses using present value calculations. The present value of the recovery amounts is:

$$RPV = \sum_{t=1}^{T} \frac{C_t}{(1+r)^t} \tag{8.16}$$

The received cash flows C_t during the resolution period T are discounted to the time of default ($t = 0$ in Equation (8.16)) using the appropriate discount rate r. The recovery rate is defined as $R = RPV/EAD$, where EAD is the exposure at default. The loss given default is simply the difference between the amount outstanding at default and the realized recovery amount as a percentage of EAD.

$$LGD = 1 - R = 1 - RPV/EAD \tag{8.17}$$

An important challenge in modeling LGD for different product segments and risk drivers is that the LGD values often have unusual distributions. Ordinarily LGD values are bounded between 0% and 100%, although LGD could be higher than 100% due to collection costs or legal costs, and the LGD distribution tends to be L-shaped or U-shaped (slightly bimodal). These distribution characteristics make the use of standard statistical regressions, such as ordinarily least squares (OLS), or normality assumptions limited or inappropriate for modeling LGD. Due to the bounded nature of the LGD values, logistic-type transformations or similar approaches are common practice to model LGD as a combination of economic and risk drivers $\{X_1, \ldots, X_m\}$ and product characteristics $\{Y_1, \ldots, Y_n\}$ (i.e. product type, seniority, collateral, support, guarantees).

$$\log\left(\frac{LGD}{1-LGD}\right) = \sum_{k=1}^{m} \alpha_k X_k + \sum_{k=1}^{n} \beta_k Y_k + \gamma \tag{8.18}$$

More fundamental LGD models may be based on a deeper understanding of the recovery cash flows and their drivers. Ordinarily, the recovery cash flows are discounted with the interest rate of the defaulted loan or debt obligation

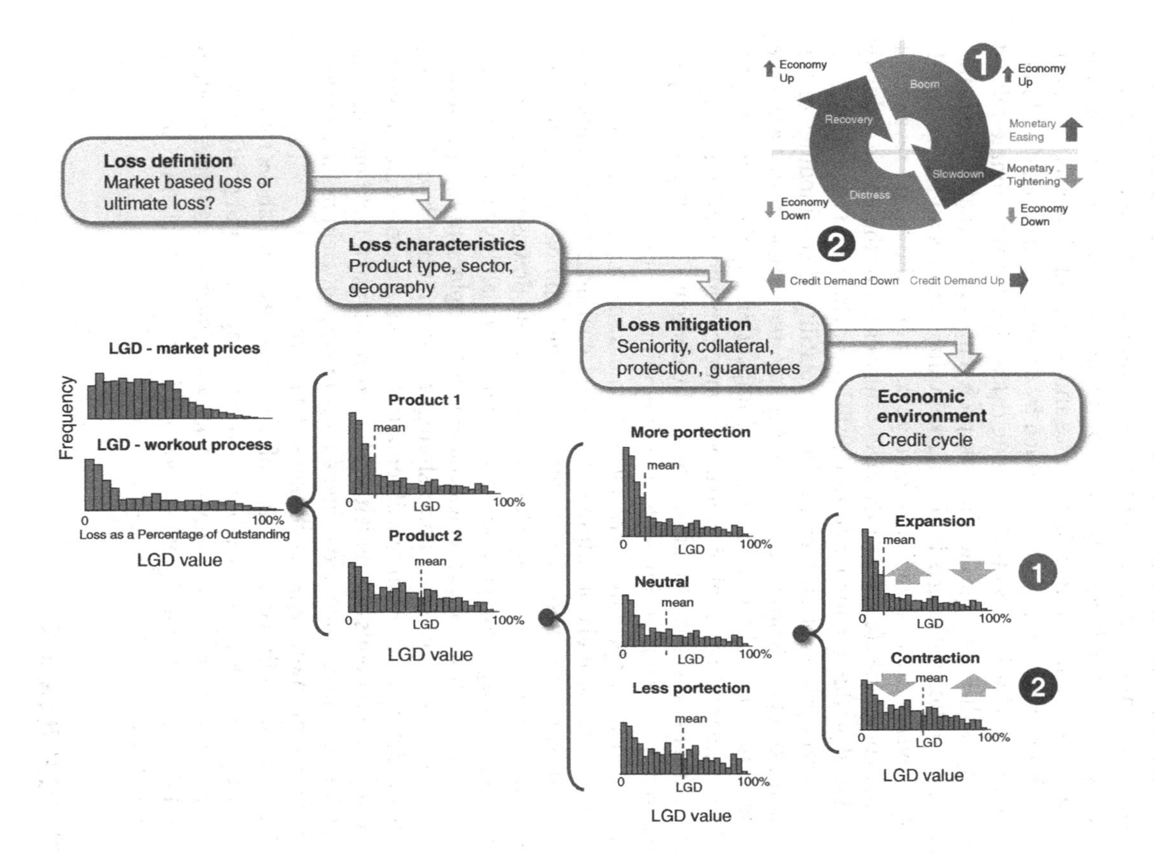

FIGURE 8.4 Differences in the distribution of LGDs for different characteristics.

(or some average discounting rate for similar assets) effective at the time of default. Note, however, that the estimated discounting rates reflect economic conditions close to the default event while the recovery process can be measured in years, with economic conditions very different from those at default.

The recovery cash flows may reflect payment collection through the resolution period from the company's revenues and the liquidation of assets or collateral, whose values could be determined by changes in economic and market conditions. This is relevant for climate risk and ESG since climate-related events can result in changes to the risk preferences of market participants, leading to significant changes in the market value of assets and their volatility. Acute events driven by climate change can affect the future revenue of individual companies (e.g. due to a severe hurricane) or the entire industry (e.g. reduced production for agriculture or fishing). This could lead to a reduction of market prices, or an increase in the volatility of commodities or the value of the company's productive assets. Climate-related events that are more chronic in nature could lead to sustained impact on supply chains, import or export activities, regional or global economic conditions affecting capital markets and interest rates that could affect the company's ability to repay its obligations. Changes in technologies and migration to a low-carbon emission economy can also impact the market valuation of *stranded assets* and revenue generation (e.g. thermal coal mining, oil and gas, coal-fired power plants) and, consequently, the recovery process for the affected companies. Accounting for stranded assets impacted by climate-related write-downs is an important aspect of estimating recoveries and LGD. Limited availability of data and analytics to address these issues creates challenges to implementing appropriate loss estimation and risk mitigation strategies. Transition risk and loss and recovery estimation uncertainty will remain elevated in sectors that rely heavily on high-carbon emission activities or carbon-intensive technologies that need significant investments to become more energy-efficient or have limited availability of insurance against climate-related events.

Understanding the consequences of the issues described above for physical and transition risks is critical due to the growing social and political pressure to divest from greenhouse emission activities, such as fossil fuel production and thermal coal mining, where stranded assets create significant credit risk for the loss recovering process.

CREDIT CORRELATION

Central to the estimation of credit losses is the problem of credit correlation (default and rating migration correlation). Conceptually, there are two basic types of phenomena commonly referred as to "correlation" in the literature:

1. **temporal correlation**, which reflects the co-movement of the credit quality of the companies
2. **conditional correlation** of credit events, which reflects the (conditional) inter-obligor correlation at any point in time.

Because default events are usually rare and, given there is only one history of changes in risk ratings and default events (realized path) for each individual borrower, models based on temporal correlation are easier to test and calibrate than models based on conditional correlations. Most models found in the literature that attempt to address the issue of conditional correlation are usually calibrated using temporal data under strong and debatable assumptions on stationarity and other effects.

Why is credit correlation important? Credit correlation is important for estimating losses when multiple parties are involved. For example, default correlation for two obligors could arise in the context of counterparty risk for credit derivatives, letters or credit, or parent company or third-party support. Credit correlation with respect to multiple parties is relevant for portfolio management of loans, bonds, and other debt instruments, and for pricing and risk management of structured instruments such as collateralized debt obligations. The shape of the distribution of losses in a credit portfolio is driven by correlation effects (Lucas, 1995; Sullivan, 1995; Burgisser et al., 1999; Duffie and Singleton, 1999; Embrechts et al. 1999; Ertork, 1999; Finger, 1999, 2000; Li, 1999; Gersbach and Lipponer, 2000; Kim, 2000; Kim and Finger, 2000; Li, Song, and Ong, 2000; Lindskog, 2000; Nyfeler, 2000; Zhou, 2000; Das, Fong, and Geng, 2001; Erlernmaier, 2001; Nagpal and Bahar, 2001; Zeng and Zang, 2001, 2002; Das and Geng, 2002; Das et al., 2002; De Servigny and Renault, 2002; Mashal and Zeevi, 2002; Neftci, 2002; Perez-Barrio and Garcia-Cespedes, 2002; Yu, 2002; Pluto and Tasche, 2005).

Let's illustrate the relevance of credit correlation for a simple portfolio of loans, one for each of N obligor. Each obligor has a probability of default p on its loan of size E and $LGD = 100\%$. Let's also assume that the credit (default) correlation for any pair of obligors in the portfolio is ρ. The total loss variability σ of the portfolio (per loan) is:

$$\sigma^2 = E^2 p(1-p)\left(\frac{1}{N} + \left(1 - \frac{1}{N}\right)\rho\right) \tag{8.19}$$

Even small values of credit correlation can have a significant impact on the variability of losses in the portfolio, reducing diversification effects. For example, for $N = 100$ loans and correlation $\rho = 0\%$, diversification effects reduce the variability of portfolio losses per loan to an amount $\sigma = E\sqrt{p(1-p)}/10$. However, if the correlation is as low as $\rho = 0.9\%$, we need, at least, $N = 1{,}000$ loans to achieve the same level of loss diversification.

Common assumptions for the estimation of credit correlation are:

- Correlations are often long-term averages or stress values representing periods of stress economic conditions.
- Correlations are usually estimated from credit quality proxies (e.g. equity prices or implied asset prices).
- Correlations often include linear and symmetric effects.

For example, if companies A and B are correlated and company A defaults, then company B can also default. If company B defaults, company A can default driven by the same correlation.

In practice, there are several shortcomings for correlation estimates:

- Correlations are variable, dynamic, and uncertain, and have a distribution of outcomes.
- Correlations may depend on industry and economic conditions and can change significantly through business and credit cycles.
- The distribution of correlations may not peak around a mean value but can have a wide range of outcomes.
- Contagion effects may exhibit non-linear and asymmetric correlation effects.

For example, if obligor A is a large company and obligor B is a smaller supplier of company A with no other clients, when obligor A defaults, it may drag obligor B also into default but not necessarily the other way round.

The impact of correlation on credit losses can be calculated using different approaches:

- Under the assumption of joint log-normality, joint probabilities of default can be calculated analytically.
- Under other distributions assumptions, numerical simulations are usually required.
- Imposing an exogenous relationship between variables driven by a *copula function*, where all variable dependencies are summarized in the copula function structure.

Copula functions are an alternative means to formulate correlations (Schonbucher, 2000; Schonbucher and Schubert, 2000). In principle, the correlation of credit events can be described in terms of copula functions, which are multivariate cumulative distribution functions of key risk factors for which the marginal probability distribution of each variable is uniform in the interval [0,1]. More precisely, given random variables $x_1, \ldots, x_N$ with multivariate cumulative probability distribution and multivariate probability density functions:

$$F(x_1, \ldots, x_N) = P(X_1 < x_1, \ldots, X_N < x_N) \text{ and } f(x_1, \ldots, x_N) = \frac{\partial^N F}{\partial x_1 \ldots \partial x_N}, \tag{8.20}$$

and (univariate) marginal distribution and density functions

$$F_k(x_k) = F(\infty, \ldots, x_k \ldots, \infty) \text{ and } f_k(x_k) = \frac{\partial F_k}{\partial x_k}, \tag{8.21}$$

we define the vector $(u_1, \ldots, u_N) = (F_1(x_1), \ldots, F_N(x_N))$. Each component u_k exhibits a marginal uniform distribution in the interval [0,1]. The copula function of random variables $x_1, \ldots, x_N$ is defined as the joint cumulative distribution function of variables $u_1, \ldots, u_N$

$$\begin{aligned} &C(u_1, \ldots, u_N) = F(F_1^{-1}(u_1), \ldots, F_N^{-1}(u_N)), \quad \text{or} \\ &F(x_1, \ldots, x_N) = C(F_1(x_1), \ldots, F_N(x_N)) \end{aligned} \tag{8.22}$$

There is a wide range of copula approaches that can be used when describing correlated events with different properties. Note, however, that selecting a copula function with a structure unrelated to the original variables may impose an exogenous solution on the correlation of events.

In practice, common alternatives when describing the correlation of defaults and rating migration are usually aligned to two basic approaches:

1. the structural asset-based approach based on extensions of the Merton model
2. the factor-based reduced form approach.

The structural approach is based on the assumptions that defaults occur when the (unobservable) market value of the company's assets falls below the value of the company's liabilities at a given point in time. In this framework, the correlation of credit quality is driven by the correlation of asset returns (Vasicek, 1987, 1991; Wehrspohn, 2003). The dynamic of the company's assets is described in terms of common market and industry factors and idiosyncratic components. This basic assumption simplifies the description of the borrower's dynamic and makes the simulation of correlation of default events and rating migration computationally easier. Although the literature on factor-based correlation estimates is broad, sound studies on the validation of factor-based correlations are limited, with very few studies indicating the reliability of the estimates based on equity or debt price information.

In the structural framework, given company asset values A_1 and A_2 with asset returns μ_1 and μ_2, volatilities σ_1 and σ_2 and asset correlation ρ_{12}, the correlation structure is described as follows:

$$\begin{aligned} dA_1 &= A_1(\mu_1 dt + \sigma_1 dZ_1) \\ dA_2 &= A_2(\mu_2 dt + \sigma_2 dZ_1) \\ \mathrm{Cov}(dZ_1, dZ_2) &= E(dZ_1 dZ_2) = \rho_{12} dt \end{aligned} \tag{8.23}$$

Since in practice there is only one credit history to draw from for any company, the expectation of any pair of variables f and g can be calculated using historical (temporal) averages over a fixed period T (*ergodic* assumption)

$$E(fg) \approx \langle f, g \rangle = \frac{1}{T}\int_0^T f(t)g(t)dt \tag{8.24}$$

In the structural framework, "default correlations" reduce to decompose the asset returns ΔZ_j for company $1 \leq j \leq N$ into M regional and industry factors $f_1, \ldots, f_M$

$$\begin{aligned} \Delta Z_j &= \sum_{n=1}^{M} \beta_{jn} f_n + \varepsilon_j \qquad 1 \leq j \leq N, 1 \leq m \leq M \\ \beta_{jm} &= \langle \Delta Z_j, f_m \rangle \quad \langle \varepsilon_j, f_m \rangle = 0 \end{aligned} \tag{8.25}$$

The variance $\tilde{\sigma}_j^2 = \sigma_{jj}$, covariance σ_{jk}, and correlation ρ_{jk} for companies j and k are described below:

$$\begin{aligned} \sigma_{jj} &= \langle \Delta Z_j, \Delta Z_j \rangle = \sum_{m=1}^{M} \beta_{jm}^2 + \varepsilon_j^2 \\ \sigma_{jk} &= \langle \Delta Z_j, \Delta Z_k \rangle = \sum_{\substack{m=1 \\ n=1}}^{M,M} \beta_{jm}\beta_{kn}\langle f_m, f_n \rangle \\ \rho_{jk} &= \frac{\sigma_{jk}}{\sqrt{\sigma_{jj}\sigma_{kk}}} \qquad \langle \varepsilon_j, \varepsilon_k \rangle = 0 \end{aligned} \tag{8.26}$$

Note that the regional and industry factors $f_1, \ldots, f_M$ can be constructed as an orthogonal set $\langle f_m, f_n \rangle = \delta_{mn}$ in the period T, which simplifies the covariance calculations.

The description of the company's asset returns in terms of regional and industry factors can improve the efficiency of portfolio loss simulations significantly. If the asset returns for the N companies could be described using only $M << N$ credit factors, the number of correlation parameters would be $(M + 1)N$, which would be much smaller than the $(N + 1)N/2$ pairwise correlations required to describe all pairs of companies. To illustrate, for $N = 10{,}000$ obligors and $M = 10$ factors, the number of correlation factors would be $(M + 1)N = 110{,}000$ as opposed to the $(N + 1)N/2 = 50{,}000{,}000$ required when simulating all pairwise correlations. The selection of a parsimonious representation in terms of a reduced number of factors results in a 500-fold increase in computational efficiency to simulate correlation effects. Note, however, that the selected credit factors need to exhibit sensible explanatory power and predictability of observed correlations of credit events to reduce spurious diversification effects due to poor representation of correlations.

Because structural models are tractable only under very strong technical assumptions and do not allow for defaults as surprise events, academics and practitioners have adopted an alternative framework known as "reduced form models," where defaults occur with a prescribed intensity of events (Kenney, 1939; Giesecke, 2002; Saunders and Allen, 2002; Jaynes, 2003). Most implementations of reduced form models focus on the temporal correlation of single-company default intensity rates as opposed to the conditional dependence of joint default events (e.g. two or more firms defaulting simultaneously). Temporal correlation is reflected in the correlation of the fundamental variables of the model and common factors similar to the ones used in structural models. Although there is extensive literature on these models, detailed reliability analyses of the estimated default correlations and validation against statistically significant samples of defaults are sparse.

More precisely, the survival function of two companies until times t_1 and t_2 can be described as follows (Thompson, 1969):

$$\begin{aligned} S(t_1, t_2) = {} & \text{Prob(no default events for company 1 during time } t_1) \\ & \times \text{Prob(no default events for company 2 during time } t_2) \\ & \times \text{Prob(no joint default events for companies 1 and 2} \\ & \quad \text{during time } \max\{t_1, t_2\}) \end{aligned} \tag{8.27}$$

Let the single default event intensities for the two companies be λ_1 and λ_2, and let the joint intensity of default events be λ_{12}. The joint probability function of random survival times τ_1 and τ_2 for companies 1 and 2 is simply:

$$\begin{aligned} S(t_1, t_2) &= \text{Prob}(\tau_1 \geq t_1, \tau_2 \geq t_2) \\ &= \exp\left(- \int_0^{t_1} \lambda_1(s)ds - \int_0^{t_2} \lambda_2(s)ds - \int_0^{\max\{t_1,t_2\}} \lambda_{12}(s)ds \right) \end{aligned} \tag{8.28}$$

The marginal survival probabilities are:

$$S_1(t_1) = \text{Prob}(\tau_1 \geq t_1) = \exp\left(-\int_0^{t_1} (\lambda_1(s) + \lambda_{12}(s))ds\right)$$

$$S_2(t_2) = \text{Prob}(\tau_2 \geq t_2) = \exp\left(-\int_0^{t_2} (\lambda_2(s) + \lambda_{12}(s))ds\right) \tag{8.29}$$

When the default intensities λ_1 and λ_2 and joint intensity λ_{12} are constant, we can estimate basic summary statistics for the survival times τ_1 and τ_2

$$\begin{aligned} E(\tau_1) &= \frac{1}{\lambda_1 + \lambda_{12}} \quad \text{Var}(\tau_1) = \frac{1}{(\lambda_1 + \lambda_{12})^2} \\ E(\tau_2) &= \frac{1}{\lambda_2 + \lambda_{12}} \quad \text{Var}(\tau_2) = \frac{1}{(\lambda_2 + \lambda_{12})^2} \\ E(\tau_1\tau_2) &= \frac{1}{\Lambda}\left(\frac{1}{(\lambda_1 + \lambda_{12})} + \frac{1}{(\lambda_2 + \lambda_{12})}\right) \\ \text{Cov}(\tau_1\tau_2) &= \frac{\lambda_{12}}{\Lambda(\lambda_1 + \lambda_{12})(\lambda_2 + \lambda_{12})} \\ \Lambda &= (\lambda_1 + \lambda_2 + \lambda_{12}) \end{aligned} \tag{8.30}$$

Finally, the correlation of survival times is:

$$\rho_t = \text{Corr}(\tau_1, \tau_2) = \frac{\lambda_{12}}{\Lambda} \tag{8.31}$$

Note that $0 \leq \rho_t \leq 1$. From the survival function in Equation (8.28) we can calculate the conditional probability that the survival time for company 1 is $\tau_1 > t_1$ when time $\tau_2 = t_2$ for company 2.

$$\begin{aligned} P(\tau_1 > t_1 \mid \tau_2 = t_2) &= \lim_{\varepsilon \to 0} \frac{S(t_1,t_2) - S(t_1,t_2 + \varepsilon)}{S(0,t_2) - S(0,t_2 + \varepsilon)} \\ &= \begin{cases} e^{-\lambda_1 t_1} & t_1 \leq t_2 \\ \dfrac{\lambda_2}{\lambda_2 + \lambda_{12}} e^{-\lambda_1 t_1 - \lambda_{12}(t_1 - t_2)} & t_1 > t_2 \end{cases} \end{aligned} \tag{8.32}$$

The pairwise correlation structure for the reduced form framework discussed above can be extended to a multi-variate correlation structure for reduced form models:

$$S(t_1, \ldots, t_N) = \text{Prob}(\tau_1 \geq t_1, \ldots, \tau_N \geq t_N) = \exp\left(- \sum_{p(1,\ldots,N)} \int_0^{T_p} \lambda_{k_1 k_2 \ldots k_p}(s)ds \right) \tag{8.33}$$

Here $T_p = \max\{t_{k_1}, t_{k_2}, \ldots, t_{k_p}\}$ and $p(1, \ldots, N)$ denotes the selection of a particular combination of p obligors among the N companies available, aggregated over all the possible combinations of companies. This generalization of the correlation of multiple companies is known as the Olkin model (Thompson, 1969).

Capturing the correlation between credit events is critical for climate risk and ESG loss estimation since correlation determines the shape of the distribution of losses in a portfolio, making severe losses more or less likely to occur. Both physical risk and transition risk can lead to higher correlations. For example, higher severity of floods or hurricanes can lead to more correlated losses across real estate properties, or changes in policies in response to transition risk can result in a decline of revenues or an increase in the volume of stranded assets in a high-carbon emission sector, which can lead to lower estimated recoveries across multiple companies.

NOTE

1. Source: Administrative Offices of U.S. District Courts. Personal bankruptcy data are compiled by the Administrative Office of the U.S. Courts from the reports of the various circuits of the U.S. Bankruptcy Courts. Bankruptcy Basics | United States Courts (uscourts.gov).

REFERENCES

Araten, M., Jacobs, M., and Varshney, P. (2004). Measuring LGD on Commercial Loans: An 18-Year Internal Study. *RMA Journal*, May: 28–35.

BCBS (Basel Committee on Banking Supervision) (2004). *International Convergence of Capital Measurement and Capital Standards*. Revised Framework, June.

BCBS (Basel Committee on Banking Supervision) (2005a). Studies on the Validation of Internal Rating Systems. *AIG/RTF BIS Working Paper No. 14*, February, pp. 1–120.

BCBS (Basel Committee on Banking Supervision). (2005b). Validation of Low Default Portfolios in the Basel II Framework. *Newsletter 6*, September.

Bessis, J. (2002). *Risk Management in Banking*. Wiley Finance, New York.

Breuer, T., and Krenn, G. (2002). *What is a Plausible Scenario? Fachhochschule Vorarlberg*, Working Paper, pp. 1–11. thomas.breuer@fh-vorarlberg.ac.at

Bucay, N., and Rosen, D. (2000). Applying Portfolio Credit Risk Models to Retail Portfolios. *Algorithmics Research Quarterly* 3(1): 45–74.

Burgisser, P., Kurth, A., and Wolf, M. (1999). *Integrating Correlation.* UBS Technical Report, Working Paper.

Cantor, R., Fons, J. S., Mahoney, C.T., and Pinkes, K.J.H. (1999). *The Evolving Meaning of Moody's Bond Ratings.* Moody's Rating Methodology, August.

Ch'ng, E. (2002). Consultative Paper on Credit Stress Testing. *Market Infrastructure and Risk Advisory Department, Monetary Authority of Singapore, Working Paper, January* 31, pp. 1–53. www.mas.gov.sg

Das, S.R., Fong, G., and Geng, G. (2001). Impact of Correlated Default Risk on Credit Portfolios. *Journal of Fixed Income,* December: 9–19.

Das, S.R., Freed, L, Geng, G., and Kapadia, N. (2002). *Correlated Default Risk.* Santa Clara University, Working Paper.

Das, S.R., and Geng, G. (2002). *Modeling the Processes of Correlated Default.* Santa Clara University, Working Paper.

Dembo, A., Deuschel, J.D., and Duffie, D. (2002). Large Portfolio Losses. *Stanford University Working Paper, March* 29, pp. 1–17.

De Servigny, A., and Renault, O. (2002). *Default Correlation: Empirical Evidence.* S&P Risk Solutions, London, pp. 4–27.

Duffie, D., and Singleton, K. (1999). *Simulating Correlated Defaults.* Graduate School of Business, Stanford University, Working Paper.

Embrechts, P., McNiel, A., and Straumann, D. (1999a). *Correlation and Dependence in Risk Management: Properties and Pitfalls.* Department of Mathematics, ETHZ, Switzerland, Working Paper.

Embrechts, P., McNiel, A., and Straumann, D. (1999b). Correlation: Pitfalls and Alternatives. *Department of Mathematics,* ETHZ, Switzerland, Working Paper.

Erlernmaier, U. (2001). *Models of Joint Defaults in Credit Risk Management: An Assessment.* University of Heidelberg, Germany, Working Paper.

Ertork, E. (1999). Is Default Risk Unsystematic? Investigating Default Correlation Among Investment-Grade Borrowers. Asset-Backed Securities Research, S&P Structured Finance Special Report, pp. 2–7.

Finger, C. (1999). Conditional Approaches for Credit Metrics Portfolio Distributions. *Credit Metrics Monitor,* April: 14–33.

Finger, C. (2000). A Comparison of Stochastic Default Rate Models. *Risk Metrics Journal,* November: 49–74.

FRB (Federal Reserve Board) (2024). Statistical Release Z.1, Financial Accounts of the United States. Total Debt and Equity of Nonfinancial Businesses, 1980–2021.
https://www.federalreserve.gov/data.htm

Galbraith, J.K. (1994). *A Journey through Economic Time.* Houghton-Mifflin, Boston.

Gersbach, H., and Lipponer, A. (2000). *Default Correlation, Macroeconomic Risk and Credit Portfolio Management.* University of Heidelberg, Germany, Working Paper.

Giesecke, K. (2002). *An Exponential Model for Dependent Defaults.* Humboldt University, Germany, Working Paper.

Hamilton, D., James, J., and Webber, N. (2001). *Copula Methods and the Analysis of Credit Risk.* Moody's Investors Service, Risk Management Service, Working Paper.

Hamilton, D.T., and Cantor, R. (2004). *Rating Transitions and Defaults Conditional on Watchlist, Outlook and Rating History*. Moody's Special Comment, February.

Heitfield, E. (2003). *Rating System Dynamics and Bank-Reported Default Probabilities Under the New Basel Capital Accord. Federal Reserve Board of Governors*, Working Paper, March.

Jaynes, E.T. (2003). *Probability Theory: The Logic of Science*, Cambridge University Press, Cambridge.

Jones, D. (1991). *The Politics of Money: The Fed Under Alan Greenspan*. New York Institute of Finance, New York.

Jones, D., and Mingo, J. (1998). Industry Practices in Credit Risk Modeling and Internal Capital Allocations: Implications for a Models-Based Regulatory Capital Standard. *FRBNY Economic Policy Review*, October: 53–60.

Keenan, S.C. and Sobehart, J.R. (2004). *Modeling Rating Migration for Credit Risk Capital and Loss Provisioning Calculations. RMA Journal*, October: 30–37.

Keenan, S.C., Sobehart, J.R,. and Hamilton D.T. (1999). *Predicting Default Rates: A Forecasting Model for Moody's Issuer-Based Default Rates*. Moody's Investors Service, August.

Kenney, J.F. (1939). *Mathematics of Statistics*. Van Nostrand, New York, pp. 153–161.

Kim, J. (2000). Hypothesis Test of Default Correlation and Application to Specific Risk. *Risk Metrics Journal* 1: 36–47.

Kim, J., and Finger, C. (2000). A Stress Test to Incorporate Correlation Breakdown. *Risk Metrics Journal* 1: 61–75.

Kindleberger, C. (2000). *Manias, Panics and Crashes: A History of Financial Crises*. John Wiley, New York.

La Porta, R., Lopez-de-Silanes, F., Shleifer, A., and Vishny, R. (1998). Law and Finance. *Journal of Political Economy* 106(6): 1113–1155.

Li, D. (1999). On Default Correlation: A Copula Function Approach. *Credit Metrics Monitor*, April: 4–13.

Li, D., Bhariok, R., Keenan, S., and Santilli, S. (2009). Validation Techniques and Performance Metrics for Loss Given Default Models. *Journal of Risk Model Validation* 3–6, Fall: 3–25.

Li, W., Song, Y., and Ong, M. (2000). Reconciling Credit Risk+ and Credit Metrics Models. *Paper presented at Modeling Credit Risk, Risk Conferences*, London, pp. 1–18.

Lindskog, F. (2000). *Linear Correlation Estimation*. RiskLab, D-Math ETH-Zentrum, Switzerland, Working Paper.

Lucas, D.J. (1995). *Default Correlation and Credit Analysis. Journal of Fixed Income*, March: 76–87.

Mahoney, C. (2002). *The Bond Rating Process: A Progress Report*. Moody's Investors Service, Global Credit Research, February.

Malkiel, B. (2007). *A Random Walk Down Wall Street*, W.W. Norton, New York.

Mashal, R., and Zeevi, A. (2002). *Beyond Correlation: Extreme Co-movements between Financial Assets*. Graduate School of Business, Columbia University, Working Paper.

Mora, N. (2010). *What Determines Creditor Recovery Rates?* Federal Reserve Bank of Kansas.

Murray, C., Cantor, R., Collins, T., Hu, C. M., Keenan, S.C., Nayar, S., Ray, R., Ruttan, E., and Zarin, F. (2000). *Promoting Global Consistency for Moody's Ratings.* Moody's Rating Methodology, May.

Nagpal, K., and Bahar, R. (2001). *Modelling Default Correlation.* Risk, April: 85–89.

Nishiguchi, K., Kawai, H., and Sazaki, T. (1998). Capital Allocation and Bank Management Based on the Quantification of Credit Risk. *FRBNY Economic Policy Review,* October: 83–94.

Neftci, S. (2002). Correlation of Default Events: Some New Tools. *City University of New York, ISMA Centre, University of Reading, UK, Discussion Papers in Finance* 2002-17.

Nyfeler, M.A. (2000). *Modeling Dependencies in Credit Risk Management.* ETH Ecole Polytechnique Fédérale de Zurich, PhD dissertation.

Nuxoll, D.A. (1999). Internal Risk Management Models as a Basis for Capital Requirements. *FDIC Banking Review,* 40.

Perez-Barrio, M., and Garcia-Cespedes, J.C. (2002). *On Asset Correlation Estimation.* Grupo BBVA Technical Report.

Pluto, K., and Tasche, D. (2005). *Thinking Positively.* Risk, August: 72–78.

Saunders, A., and Allen, L. (2002). *Credit Risk Measurements,* Wiley Finance, New York, pp. 107–120.

Schonbucher, P.J. (2000). *Factor Models for Portfolio Credit Risk.* Department of Statistics, Bonn University, Working Paper.

Schonbucher, P.J., and Schubert, D. (2000). *Copula-Dependent Default Risk in Intensity Models.* Bonn University, Department of Statistics, Working Paper.

Sobehart, J.R., and Keenan, S.C. (2005). New Challenges in Credit Risk Modeling and Measurement. In: M. Ong (Ed.), *Risk Management: A Modern Perspective.* Academic Press, New York.

Sobehart, J.R., and Mogili, D. (2007). Modeling Credit Migration Citigroup Global Markets. *Quantitative Credit Analyst,* April: 34–41.

Sorge, M. (2004) Stress-testing Financial Systems: An Overview of Current Methodologies. Monetary Economic Department, *BIS Working Paper* 165, December, pp. 1–41

Stumpp, P, Arner, R, Mahoney C., Perry, D., Hilderman, M., and Madelain, M. *(2002) The Unintended Consequences of Rating Triggers.* Moody's Investors Service, Special Comment, December 2001.

Sullivan, G. (1995). *Correlation Counts. Risk,* August: 36–37.

Thompson, W.A. (1969). *Applied Probability.* Holt, Rinehart and Winston, New York, pp. 125–129.

Vasicek, O.A. (1987). *Probability of Loss on Loan Portfolio.* KMV LLC Technical Report, February.

Vasicek, O.A. (1991). Limiting Loan Probability Distribution. *KMV Corporation, August* 9, pp. 1–6.

Wehrspohn, U. (2003). Analytic Loss Distribution of Heterogeneous Portfolios in the Asset Value Credit Risk Model. Alfred Weber Institute (Heidelberg, Germany), Working Paper, March, pp. 1–19.

Wilson, T.C. (1998). *Portfolio Credit Risk. FRBNY Economic Policy Review,* October: 71–82.

Yashkir, O., and Yashkir, Y. (2013). Loss Given Default Modeling: A Comparative Analysis, *Journal of Model Validation* 7(1): 25–59.

Yu, F. (2002). *Correlated Defaults in Reduced Form Models.* University of California, Irvine, Working Paper.

Zeng, B., and Zang, J. (2001). *An Empirical Assessment of Asset Correlation Models.* KMV LLC Technical Report, November.

Zeng, B., and Zang, J. (2002). *Measuring Credit Correlations: Equity Correlations Are Not Enough!* KMV LLC Technical Report, January.

Zhou, C. (2000). Default Correlation: An Analytical Result. *Paper presented at Modeling Credit Risk, Risk Training Conferences*, London, pp. 1–28.

CHAPTER 9

Pillar 2: Demand for Credit. Risk Assessment and Credit Risk Ratings

THE PATH TO BUSINESS FAILURE

The path to a company's business failure is very diverse as each company can collapse in its own way (Charan and Useem, 2002). Some companies may collapse suddenly, others linger for a while, and others fizzle out over many years until they disappear or are eventually absorbed by other companies. In some cases, they fail individually, and, in other cases, they fail in groups in response to similar events. Company failure and success are part of the natural business lifecycle: companies grow, mature and, eventually, fail and disappear, leaving room for a new generation of businesses. For example, the nifty-fifties companies in the 1960s, the dot-com companies in the 2000s or the overpopulation of brick-and-mortar retailers and entertainment companies competing for a limited market during the COVID-19 period in 2020s were showcases in whether certain businesses were resilient enough or even viable for their competitive environments. While in many cases, these companies were not strong enough to survive, in other cases, companies were simply struck down and disappeared for bad reasons (management incompetence, fraud, reputation issues, over-regulation, etc.).

Company executives often have a range of reasons and excuses for why companies fail: a weak economy, cost increases, and reduction of revenues due to inflation or reduction in purchasing power, market turbulence, domestic and international competitive pressure, inadequate products and services, climate change, a hundred-year flood, perfect storms, and other causes very much outside management control. In some situations, these arguments might be based on facts. But a close analysis of corporate failures suggests that often companies fail for one simple reason: *managerial error.*

Failure does not necessarily mean only a bankruptcy event. A dramatic fall from good standing as a respected company (referred to as "fallen angel") qualifies too. Often during severe economic contraction periods of the business cycle, companies lose a significant fraction of their market value with rare exceptions. This can also result in a deterioration of the company's revenues and ability to service its debt, ultimately, impacting its credit quality.

The speed at which these events can unravel varies widely. Companies that seem healthy just a few days earlier can suddenly turn for the worse as markets and lenders uncover the true standing of these companies and their managerial practices. For example, failure could occur due to poor managerial judgment, lax practices and ineffective controls may go unchecked for a long time. Large and complex institutions are more prone to this situation. These issues combined usually with unrealistic performance goals and financial targets, and people's reluctance to review objectively the limitations of their own competence and abilities can lead to managerial error and failure.

Managerial error leading to failure reflects denial, wishful thinking, poor communication, lax oversight, greed, deceit, and other common aspects of human behavior. In the end, this adds up to a failure to execute a sound business strategy. Of course, this is not an exhaustive list of corporate failures but behavioral issues and economic issues both require models and analytical tools to help assess risks. This is particularly important in the context of climate risk and ESG where social behavior impacts companies' management, customers, government, and regulatory entities (e.g. through regulations and policies).

RISK ASSESSMENT AND CREDIT RISK RATINGS

While financial institutions, lenders, and investors may rely on multiple approaches to assess the creditworthiness of debt securities issuers, bank borrowers and counterparties in transactions (or, more generally, *obligors*), in major capital markets the standards for fundamental analysis have been set historically by accredited rating agencies, which have decades of experience in financial risk assessments (Keenan et al., 2000; Murray et al., 2000; Zarin, 2000; Crouhy et al., 2001a, 2001b; Griep and De Stefano, 2001; Stumpp et al., 2001; Lyon, 2002).

In 1909, John Moody issued the first credit risk ratings for railroad-company bonds in the United States. The Poor's Co. began publishing investor's guides in 1860, followed by the Standard Co. in 1923. Standard and Poor's (S&P) merged in 1941. Fitch entered the industry in 1924. By 1930, US Federal regulators began to require that financial institutions evaluate the ratings on bonds in their portfolios. The business model of the rating agencies was primarily to receive revenue from their publications to investors.

As bond issuers became more eager to reassure investors about the quality of debt issues, companies started paying the rating agencies for the evaluation of their securities. In 1975, the use of risk ratings from accredited agencies extended further when the US Securities Exchange Commission (SEC) decided that those rating agencies that demonstrated reliable and unbiased opinions would receive the status of Nationally Recognized Statistical Rating Organization (or NRSRO). Over the years, the reliability of agency risk ratings was questioned following periods of economic expansion and contraction that resulted in significant credit events and losses. The overreliance on NRSRO's opinions and reliability of the risk ratings came into full question as the aftermath of the 2007–2009 global financial crisis (Great Recession), leading to the Dodd-Frank regulation (2010) in the United States that imposed restrictions on the use of NRSROs for US regulatory agencies and the financial institutions they supervise.

Note, however, that for decades capital markets relied on agency risk ratings as a gauge of the financial health of securities issuers. Rating agencies use different grading scales to rank obligors according to their ability to service their obligations. Rating scales are usually based on a combination of (capitalized and lowercase) letters such as: AAA/Aaa, AA/Aa, A, BBB/Baa, BB/Ba, B, CCC/Caa, where AAA/Aaa ratings are assigned to obligors with the highest credit quality, and CCC/Caa ratings or lower ratings indicate poor ability to service debt obligations. Some rating agencies that focus on default likelihood (e.g. S&P) include an extra symbol D to indicate that a firm is already in default. Other agencies (e.g. Moody's Investors Service) do not include a D rating as they had focused historically on loss severity.

Long-term corporate bond default studies by the leading rating agencies show a strong relation between ratings, the likelihood of default, and the severity of losses, with lower ratings generally linked to higher defaults rates and severe losses. Because the agencies have published extensively on their methodologies and performance statistics, which differ by sector and product, financial institutions often benchmark their internal rating scales against agency ratings.

Agency risk ratings and risk ratings assigned by credit analysts at banks and other financial institutions are expert *opinions* on credit quality intended to quantify the expected likelihood and severity of default events. Here we describe basic concepts on the quantification of credit risk through expert opinions and the process of rating assignments by credit analysts. These issues are particularly important within the context of climate risk and ESG where risk identification, measurement, monitoring, and management's response may play a key role in determining consistent risk assessment practices for financial institutions. Beyond providing transparency for investors and financial institutions, a more detailed understanding of risk ratings will help provide a more solid basis on which to evaluate quantitative and qualitative approaches. The latter are driven by how judgment is affected by contextual information and social influence.

The assessment of business portfolios includes the review of relationships and the exposure to the obligors in these relationships. The analysis must address the overall creditworthiness of the relationships, including the appropriateness of the overall structure of debt issues or loan facilities, while also assessing the ability of specific obligors to repay their debt. When the review includes conglomerate relationships with a variety of different and unrelated obligors, the analysis also extends to exposure to these obligors.

CREDIT RISK RATINGS

Credit risk ratings are assessments of the credit quality of a debt obligation (e.g. debt issue or loan facility) or an obligor (e.g. debt issuer or bank borrower). Risk ratings rank the creditworthiness of debt issuers and debt issues using categories ranked in terms of riskiness from the highest credit quality to the lowest credit quality. Risk ratings provide an ordinal scale as opposed to probabilities of default, which quantify the likelihood of default for a given time horizon. Risk ratings are usually identified by their source: (1) risk rating from rating agencies (usually referred to as "external" ratings), which focus on assessing the risk of debt issuers and issuers and provide this information to investors, or (2) risk ratings from financial institutions (usually referred to as "internal" ratings), which are based on proprietary scales and risk practices, and vary across institutions.

Risk ratings that apply to borrowers or debt issuers provide an assessment of the *likelihood of default* of the obligors. In contrast, risk ratings that apply to loan facilities or debt securities (debt issues) provide an assessment of the *severity of losses*, reflecting both the likelihood of default on the obligation and the expected loss (or, equivalently, the expected recovery). For example, ratings assigned to senior unsecured debt are usually close to the issue ratings since its probability of default is driven by the obligor, while loss expectation reflects loss during the recovery period. Secured debt, however, could benefit from various forms of collateral, guarantees or protections, which differentiate the risk of the debt from the pure/intrinsic default risk of the obligor. In this case, the risk rating of the debt can reflect the role of any supporting entity and the recoveries from guarantees or collateral attached to the debt.

More generally, risk assessments can be reflected in different types of risk ratings:

1. **Obligor risk ratings**, which assess the likelihood of a default event.
2. **Issue risk ratings or facility risk ratings**, which assess the loss severity for individual debt issues, facilities, or transactions.

3. **Portfolio risk ratings**, which reflect the overall risk of a pool of transactions.
4. **Portfolio trend ratings**, which reflect changes in the credit quality of the portfolio.
5. **Risk management ratings**, which reflect the risk of internal risk management practices to assess credit risk. That is, they reflect the risk of the institution's credit assessment process as opposed to the credit risk of the obligors.

Each individual risk rating provides a different perspective on the level of risk from individual assessments of clients to a portfolio-level view, including the quality of the risk management practices. In the following we focus primarily on risk ratings for obligors and their obligations.

Critical to the assignment of risk ratings is the identification of the period covered by the credit assessment. While rating agencies generally focus on a fundamental analysis of the debt and obligor using a long-term view of the obligor or through the life of the debt instrument, financial institutions may focus on a shorter-term or longer-term view aligned to the nature of the relationship and needs of the borrower.

Risk ratings are based on a range of criteria, from qualitative factors, such as strengths or weaknesses of management, to quantitative factors measuring financial flexibility and performance or capacity to borrow.

External (Agency) Risk Ratings

Credit risk ratings are assessments of credit quality. There are numerous organizations dedicated to providing credit assessments reflected into risk ratings. Among these organizations, Moody's, S&P, and Fitch are the most familiar of the risk rating agencies for their coverage and long-term histories of providing rating services to investors. Rating agencies use coded letters (such as Aaa/AAA, Aa/AA, A, Baa/BBB, etc.) combined with symbols (such as 1, 2, 3 or + and -) to categorize the riskiness of obligors and their obligations. These rating scales include roughly about 20 different rating levels that can be grouped into two broad categories: (1) investment grade ratings, which reflect high credit quality, and (2) non-investment grade ratings, with lower credit quality. Note that, while the population of issues or issuers may overlap across rating agencies and the assigned ratings could be similar, rating agencies have different rating definitions and practices. Some agencies may focus more on loss likelihood, while others focus more on loss severity. Table 9.1 illustrates typical risk rating levels and their descriptions.

Rating agencies rate primarily public debt issues, which could have different risk ratings due to seniority, guarantees, or other risk mitigants even when all of them share the same risk that the issuer could default. For example,

TABLE 9.1 Typical risk rating categories

Segment	Rating	Risk Level	Summary Description
Investment grade	AAA/Aaa	Highest credit quality	Highest credit standing, very stable and predictable company with excellent asset quality, debt capacity, liquidity, and management.
	AA/Aa	High credit quality	High credit standing, very stable company with very good asset quality, debt capacity, liquidity, and management.
	A/A	Good credit quality	Good credit standing, stable company with good asset quality, debt capacity, liquidity, and management.
	BBB/Baa	Acceptable credit quality	Good credit standing, company with moderate volatility, good or acceptable asset quality, debt capacity, liquidity, and management.
Non-investment grade	BB/Ba	Acceptable credit quality with higher risk	Acceptable credit standing with higher level of credit risk, company with moderate volatility, acceptable but lower asset quality, limited debt capacity due to leverage, and little excess liquidity.
	B/B	Moderate to low credit quality	Considerable credit risk, high volatility of earnings or company performance with low asset quality, limited debt capacity, little liquidity, or poor management. The credit situation often requires lender supervision.

TABLE 9.1 (*Continued*)

Segment	Rating	Risk Level	Summary Description
	CCC/Caa	Low credit quality	Significant credit risk with business weaknesses and vulnerabilities to economic conditions. Low asset quality, limited debt capacity, weak liquidity, or poor management.
	CC/Ca	Very low credit quality	Unacceptable level of credit risk, high volatility of earnings, poor asset quality, limited or non-existing debt capacity, weak liquidity, or ineffective management.
	C/C	Lowest credit quality	Significant weaknesses impacting credit standing that make collection in full difficult or unlikely. Low asset quality, limited or non-existing debt capacity, weak liquidity, or ineffective management.
Default	D	Defaulted or impaired	Obligor defaulted on the debt obligation or repayment seems uncollectible or with little recovery value.

senior *unsecured debt* can have a rating similar to the issuer rating because it benefits from first priority repayments in the event of default. *Secured debt* can have collateral as risk mitigant, which may result in higher recoveries in the event of default relative to unsecured debt. In contrast, *subordinated debt* is subject to a lower priority of claim relative to more senior lenders or investors and has a higher risk. As a result, the credit risk of different debt instruments for the same obligors can vary significantly even when they all share the same intrinsic risk of default. The description of risk assessment becomes more complicated as the obligor could default selectively on different debt instruments.

Rating agencies rely on fundamental analysis and credit assessments using a long-term view of weaknesses and strengths of the obligor, usually referred to as "through-the-cycle" (TTC). In contrast, an assessment based primarily on a short-term view leveraging the most recent information available is referred to as "point-in-time" (PIT).

Rating agencies review and assess their ratings periodically or when there is a significant event that can impact the obligor credit standing. However, because agency ratings are based on a long-term view, they do not change frequently in reaction to short-term changes to market conditions or to changes to the obligor's financial conditions as long as they do not materially affect the obligor's creditworthiness. For example, a change to the regulatory environment for the obligor's industry, a merger or an acquisition, or a change in senior management could lead to a revision of the credit standing of the obligor. This is relevant in the contest of climate risk and ESG as changes in the regulatory environment due to policy changes could result in a significant increase in transition risk to a low-carbon economy for companies in high-carbon emission sectors.

Agency ratings cover a wide range of debt issuers: corporate firms, financial institutions (e.g. banks, insurance companies, broker dealers), sovereign obligors (governments, central banks, government-related entities, sub-sovereign borrowers), multilateral development banks and organizations, real estate entities, and other obligors. The debt could be issued in local currency or foreign currency and could be subject to transfer risk if held locally.

Because rating agencies cover primarily public debt securities, their statistics on default frequency and loss severity may not be applicable to banks and other financial institutions with portfolios with significant number of private borrowers or with small and medium-sized companies with limited access to capital markets. Financial institutions rely on their own internal rating practices to assess the creditworthiness of their clients and counterparties.

Internal Risk Ratings

Internal credit risk ratings are customized to each financial institution's needs, processes, and risk management practices under the appropriate local regulatory guidance governing the activities of the financial institutions (e.g. banks, insurance companies or broker dealers). Regulatory guidance and the design of safe and sound risk management processes drive toward harmonization of credit assessment practices in the industry. Since financial institutions are exposed to a wide range of borrowers and counterparties, the internal risk rating systems need to cover various entities such as corporate and financial companies, sovereign and municipal obligors, real estate entities, and other types of obligors.

A typical internal risk rating system reflects:

1. an assessment of the obligor's risk;
2. an assessment of the support provided by any supporting entity (e.g. a holding company or parent company supporting a subsidiary);

3. an assessment of the risk of the obligation, level of support, seniority or subordination, collateral, and other risk mitigants.

Support plays a key role in the assessment of credit standing and can be formal, in terms of written guarantees and letters of support, or less formal in terms of verbal guarantees and commitments to strategic subsidiaries. For example, if a subsidiary is critical to the business strategy of the parent company, it is likely that the parent company will support it in times of financial stress. In order to determine the overall risk rating of the obligor, the strength of the supporting entity and its ability to support the obligors are evaluated.

For debt obligations, the seniority level, guarantees, and collateral quality are also evaluated to determine the appropriate risk rating. For well-collateralized obligations, risk ratings could be better than those of the obligor, reflecting better recovery in the event of default. These issues are relevant to climate risk and ESG as physical risk could impact the value of collateral directly (e.g. more frequent or more severe floods), while transition risk could impact the long-term viability of the obligor or its supporting entities leading to lower credit standing and, therefore, lower risk ratings (e.g. due to the economic cost of transitioning to low-carbon emissions).

Figure 9.1 illustrates a typical internal risk rating system describing the intrinsic risk of the obligor, the reduction of risk due to third-party support and additional risk mitigation factors (such as collateral or guarantees).

Similarly to rating agencies, financial institutions assign risk ratings to obligors and their debt obligations and transactions with credit risk using their own internally defined risk ratings. However, unlike rating agencies, their portfolios may include a significant portion of entities that are usually not rated by rating agencies because they are private companies or subsidiaries of larger entities, due to their size or their lack of access to capital markets to issue debt.

Additional complexities arise in situations where companies in different jurisdictions can be impacted by sovereign risk, either due to a deterioration of the local economy environment or due to transfer or cross-border risk, when it is no longer feasible to recover from local obligors due to government restrictions on fund transfers. Although at first blush, the risk rating of the obligor could be capped[1] at the sovereign risk rating to reflect the sovereign risk, this approach has several conceptual shortcomings. In principle, the obligor could still be in a good financial position even if the country's credit quality declines. Of course, the obligor's credit quality may be adversely affected if the local economic environment declines, but there could be situations where the impact could be small if the obligor has other foreign sources of revenue, third-party support, or additional risk mitigation strategies.

Similar situations may impact financial institutions (e.g. banks), which are driven by government policies and supervisory processes. When the supervisory process is loose or weak, financial institutions could adopt lax

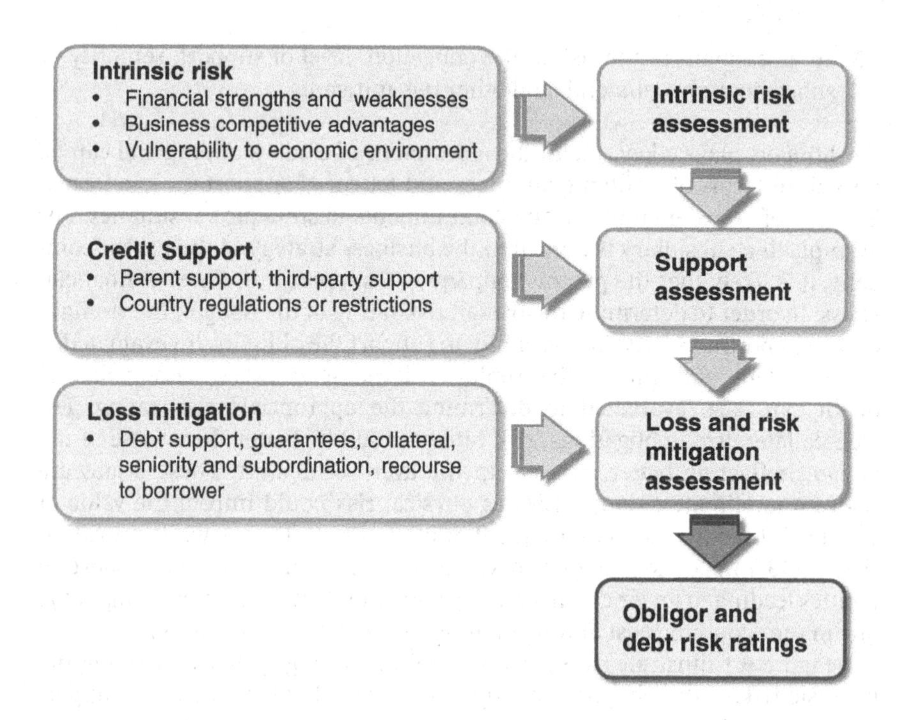

FIGURE 9.1 Illustrative risk rating system.

processes leading to risky strategies and lower credit standing. This highlights the need for more comprehensive risk rating tools that assess the government and supervisory process and that of the financial institutions.

The issues discussed above can impact the modeling of credit quality and its sensitivity to climate risk drivers.

Credit Risk Rating Criteria

In order to assess the credit risk of a company, the risk rating criteria should include quantitative and qualitative assessments of the company's creditworthiness. While the quantitative assessment usually incorporates financial variables and economic indicators, the qualitative assessments can have a richer set of criteria that reflect the many factors that can influence a company's performance and credit quality: quality of management and company strategies, competitive environment and barriers to entry in the target market, regulatory environment, geopolitical and economic situation, and many other factors that are difficult to quantify but play a critical role in determining the business health of the company.

The typical approach to a company's credit assessment focuses on the identification of vulnerabilities or challenges on four basic areas across the obligor's profitability, liquidity, debt capacity, quality of assets, business strategies and management effectiveness (a SWOT analysis):

1. strengths
2. weaknesses
3. opportunities
4. threats.

Vulnerabilities and challenges in these areas are driven by the company's industry or sector, market share and company size, level of competition, diversification of products and services, technology and production issues, financial and economic conditions, regulatory environment, management vision and strategies, and other factors. These issues could also be driven by the cyclical dynamics of sales or business and credit cycles, or by the nature of the industry (e.g. capital-intensive industries, consumer durables, or non-essential services), or by technology and processes that can impact variable and fixed costs of operations.

Appropriate credit risk rating criteria also depend on the nature and type of the obligor or counterparty. For example, while the credit assessment for corporate obligors may focus on the ability of the company to generate revenues and have healthy liquidity and debt capacity; for regulated financial institutions such as banks, ratios measuring capital to assets and allowance for credit losses over total loans, or regulatory capital requirements are very important to understand their credit standing and their ability to meet supervisory expectations.

Risk ratings for debt instruments (e.g. bank loans, facilities, or debt securities) are intended to provide lenders and market participants with a framework for evaluating the credit quality of debt. Risk ratings aggregate a wide range of information into a single symbol, including default probability, loss severity, and financial strength, which refers to the intrinsic creditworthiness of the obligor or debt instrument, removing any potential external support elements, such as a rescue by a third party. This is important since obligors may obtain significant credit strength from external sources of support such as bank regulators, multilateral institutions, and central or local governments.

To illustrate, bonds or loans with the same risk rating may be comparable with respect to overall credit quality but may differ with respect to specific characteristics of these debt instruments. Furthermore, repayment of bonds, loans, or other debt instruments often relies upon third-party support, which could be very likely and predictable, but it cannot be guaranteed completely as the third-party rescue could fail to materialize when it is needed the most.

Within a given industry sector, similar debt instruments with the same rating tend to be quite comparable with respect to credit quality. To create a consistent approach to assessing relative credit quality within a given sector, ratings need to correlate closely with the likelihood of default, that is, with the observed default and loss experience for each rating category. However, the relative contributions of default likelihood and loss severity to credit quality vary across and within industry, geography, and product segments.

Broadly speaking, the analysis of credit quality can be segmented into a few key sectors:

- Commercial and industrial (C&I) corporates
- Utilities
- Financial institutions
- Municipals
- Sovereigns and related entities
- Real estate
- Structured finance
- Specialized products

Debt instruments for these segments can differ across liquidity, price volatility and the timing of expected cash flows. However, the most important distinction across these segments is that the default and loss experiences within a given rating category have historically varied across these segments reflecting different concerns on credit quality and ability to repay obligations.

Table 9.2 illustrates relevant financial factors applicable to the credit assessment of corporate borrowers or counterparties. These variables are usually used in quantitative assessments and in risk rating models. For simplicity of exposition, the example does not cover other obligor segments (e.g. financial institutions or sovereigns) or qualitative factors. A more detailed description is provided in later sections on risk rating modeling.

The use of a common scale for assessing credit quality across segments has its benefits and limitations. To illustrate, different default and loss experience by risk rating category for different segments suggest that the lowest-rated credits in one sector could be more creditworthy than similarly rated credits in other sectors, reducing comparability of ratings across segments.

For debt instruments for C&I corporates and financial institutions, default events tend to follow periods of economic contraction and expansion, but loss severity is more unpredictable, with a wide range of outcomes. However, there is a significant difference in the relative frequency of default events for high-quality (investment grade) obligors, where defaults are rare, relative to lower-quality (non-investment grade) obligors, where defaults are more frequent. For many rating agencies and financial institutions assessing

TABLE 9.2 Illustrative example of financial variables applicable to the credit assessment of corporate firms

Financial Variable	Description
Financial profitability	Return on equity Net income over equity
Operating profitability	Operating returns on assets (e.g. before interests, amortization, and taxes)
Capital structure	Debt to equity structure Financial coverage ratio (EBITDA/interests) Short- and long-term debt, senior vs. subordinated debt
Liquidity	Cash and liquid investments over assets Current assets over total assets and related ratios
Cash flows	Operating cash flows, or free cash flows (available to face debt obligations)
Operating efficiency	Inventory turnover Cash cycle Receivables in days of sale Payables in days of purchases
Operating leverage	Variable and direct costs over fixed costs
Company size	Total assets Total sales Total revenues EBITDA (Earnings before interests, taxes, depreciation, and amortization)
Market value	Market value of equity Equity to book value
Volatility of earnings	Volatility of earnings, sales, or profits

the riskiness of their portfolios, risk ratings for high credit quality commercial and financial corporates primarily reflect relative default probability, while expected severity of loss in the event of default plays an important secondary role. For low credit quality obligors, risk ratings tend to place more emphasis on expected loss than on relative default risk. This emphasis on default likelihood and loss severity is relevant since a significant portion of the investment grade corporate debt has long been held by institutional investors, who have an overall expected-return orientation and are generally averse to default risk.

Risk ratings for regulated utility companies have historically placed more weight on financial measures than on expected loss when compared to risk ratings for corporate borrowers. Even when regulated utilities had experienced financial distress in past situations, the few companies that have declared bankruptcy may have received enough support from regulators to allow them to continue servicing their debt. Regulatory support in the utility sector provides a means to reduce the impact of adverse financial situations. Differences in risk ratings for utilities may account for differences in default and loss experience between similarly rated debt in the corporate sector and the public sector.

In contrast to ratings for C&I companies and financial institutions, risk ratings assigned to sovereigns, related entities, and the general obligations of US municipalities are intended to differentiate, based upon relative financial strength (as measured by traditional statistical ratios), which is not necessarily reflected in historical measures of default risk or expected loss. Compared to the corporate default experience, municipal and sovereign defaults have been rare and idiosyncratic, and recoveries have been relatively high.

While the meanings of risk ratings across these sectors still contain some basic conceptual differences, they are becoming more closely aligned today as some municipal and sovereign-related entities have developed more corporate-like risk profiles (as reflected in industrial revenue and healthcare bonds, issues backed by multi-family housing units and some tax-exempt project financings). Furthermore, local governments have often received state-level support that averted default. Even in the rare event of default, these obligors have rarely declared bankruptcy, often with little loss of principal or even interest.

Credit ratings of debt obligations for real estate property measure credit risk based on the property's risk-adjusted ability to repay the obligation and its structure. There are numerous attributes that impact a property's financial strength, including price appreciation and fair market value, vacancy, ability to rent the property, operating income, mortgages, maintenance expenses, property tax, property management fee, property insurance and many more. Other relevant attributes span markets, housing, government, community, and location. The inherent risks and volatility of these attributes are assessed to arrive at a risk rating.

In the structured finance sector, where default probabilities and expected loss severity are often estimated through statistical analysis due to the rare nature of default events, ratings have placed greater emphasis on the expected loss concept, which often places roughly equal weight on default probability and loss severity (Table 9.3). The characteristics of the asset pools that are typically securitized lend themselves to statistical analyses since structured securities are usually divided into multiple tranches, whose ratings place a heavy emphasis on expected investor loss. Structured finance ratings, however, are not based only on the expected loss concept. For example, investment-grade tranches with relatively high probabilities of default but

TABLE 9.3 Financial variables applicable to the credit assessment for different industry segments

Category	Financial variable		
Business Profile	Product or Service Offered	Target Market	Product Distribution Channels
Competitive position	Market position and brand	Diversity of product categories	Product breadth and diversity
Management	Experience and skills	Business strategy, objectives	Quality, transparency
Profitability	EBIT or EBITDA margin	EBIT/Assets	EBIT/assets
Liquidity	Liquid assets/Liabilities	Access to capital	Current assets/assets
Leverage	Debt/Capitalization	(Cash + marketable securities) / debt	Assets/liabilities
Capital structure	Net debt/Net capitalization		
Capital adequacy	Shareholders' equity/assets		
Market position, size	Revenues or EBIT	Assets or market equity	Net property, plans, and equipment, etc.

extremely low expected loss severity could have ratings that are somewhat lower than would be implied by a simple expected loss calculation.

DEFAULT AND LOSS CONCEPTS AND RISK RATINGS

While credit risk ratings provide an ordinal assessment of risk, measuring and monitoring credit risk required the ability to quantify the likelihood and severity of the losses associated to different ratings. Default occurs when a debt obligor (e.g. bond issuer, bank borrower or counterparty) fails to honor its payment obligations. A loss occurs if the payment is not received in full as agreed, causing an economic loss to the lenders or investors (e.g. because the principal and/or interests are not paid in full, there are collection and legal costs or the amount is recovered later than agreed, resulting in investment opportunity losses (time value of money)).

The likelihood of loss is determined by the Probability of Default (PD) over a fixed time horizon (e.g. one year), while the severity of the loss (or loss rate (LR)) is determined by PD and the fraction of the debt amount lost in the default event, or Loss Given Default (LGD).

$$LR = PD \times LGD \tag{9.1}$$

Note that the loss rate represents the expected loss per unit of exposure. That is, the expected credit loss (EL) is simply $EL = PD \times LGD \times EAD = LR \times EAD$, where EAD is the exposure at default. The fraction of the recovered amount of the loan (or recovery rate) is simply $R = 100\% - LGD$. For corporate obligors, a simple approach is to determine the frequency of default events (or default rate) observed for each rating category over a given time horizon (i.e. 1 year) as a proxy for the probability of default. To illustrate, if out of 1000 companies rated BB/Ba, 10 companies defaulted within one year, the default rate for BB/Ba-rated companies would be 10/1000 = 1%. For debt obligations, the calculation of loss severity requires the estimation of PD and LGD, which is driven by guarantees, seniority, and collateral characteristics and can result in significant variability of the ultimate loss severity. For example, if the same BB-rated company had two loans: an unsecured loan with hypothetical LGD = 50% and another collateralized loan with LGD = 10%, the expected loss rates would be 1% × 50% = 0.5% and 1% × 10% = 0.1% respectively. These different loss rates would translate into different risk ratings for these two loans, even when they share the same intrinsic PD for the obligor. Note that the second loan has significantly lower expected loss and, therefore, a better loan risk rating because of its higher recovery rate.

Default and loss rate statistics can be estimated from historical data on default events and observed losses. Rating agencies publish default and loss statistics regularly to reflect the performance of their issue and issuer ratings at assessing credit risk. Default frequencies tend to increase much more than proportionally when moving across ratings from the highest to the lowest credit quality. Banks and other financial institutions also collect and use their internal historical data to determine the performance of their own risk ratings and to meet regulatory expectations on measuring and monitoring risk. Both rating agencies and financial institutions also collect historical data on rating migration between different risk categories, which reflect both changes in credit quality of the obligors and characteristics of the rating practices. For example, a rating downgrade could result from a deterioration of the credit quality of the obligor but could also be due to a change in credit assessment practices, models, or algorithms used to determine risk ratings.

DEFAULT RATE STATISTICS BY RATING CATEGORY

Default rate statistics for a given rating category reflect the observed frequency of default events over a fixed period T for a population of companies with the same risk rating. The calculation of default rates involves the following basic steps:

1. Define a pool (or cohort) of companies at the beginning of the calculation period T.
2. For each risk rating R:
   a. Identify the number of companies $N_0(R)$ in the cohort at the beginning of period T.
   b. Identify all the defaults $D_T(R)$ that occur during period T.
   c. Calculate the default rate (or default frequency) $f_T(R)$ as the ratio of the total number of defaults $D_T(R)$, in period T, over the number of companies $N_0(R)$ in the rating segment R at the beginning of the period:

$$f_T(R) = D_T(R)/N_0(R) \tag{9.2}$$

In practice, there are some technical adjustments needed to provide better default rate estimates. For example, some companies may exit the population during the observation period T and, therefore, they may not be able to default during the selected period. In this case, keeping them in the population will overstate the number of companies that can default, reducing the default rate. Following earlier discussion, here we remove from the population of companies $N_0(R)$ the number of companies exiting the cohort (or withdrawals) $W_T(R)$ with risk rating R during period T. This result in the adjusted default rates:

$$f_T(R) = D_T(R)/(N_0(R) - W_T(R)) \tag{9.3}$$

The adjusted default rate defined above has some shortcomings: (1) it assumes that the companies that exited the population can default during the period T at the same rate as the remaining companies, and (2) it does not consider the marginal pattern of exits that can occur at different points through period T. Without an adjustment for the marginal pattern of exits or withdrawals, this can result in inconsistent default rate estimates for intermediate time horizons within period T. These issues are exacerbated for longer periods T, as the number of possible companies exiting the population accumulates for longer time periods.

A more general approach is to estimate the marginal contribution of companies exiting the population in each intermediate period $0 \leq t \leq T$ and estimate their cumulative effects across multiple time periods.

$$f_{t(t+1)}(R) = D_{t(t+1)}(R)/(N_t(R) - W_{t(t+1)}(R)) \tag{9.4}$$

Here $f_{t(t+1)}(R)$ is the marginal default rate in the period $[t, t+1]$ derived from the marginal defaults $D_{t(t+1)}(R)$ in the population of companies $N_t(R) = N_0(R) - W_t(R)$ and the marginal number of companies exiting the cohort $W_{t(t+1)}(R)$. The withdrawal-adjusted marginal default rates can be combined with the marginal rating migration rates to calculate the cumulative default rates, including rating changes and the impact of companies exiting the population at different points in time.

These issues are very relevant in the context of climate risk modeling as default rates (more precisely, the probability of default) should cover extensive periods of time (20–30 years) for different transition risk scenarios.

Note that historical default rates reported by the rating agencies for a 1-year period averaged over multiple credit cycles are close to zero for the high-quality risk ratings (AAA/Aaa, AA/Aa, A), increase to about 0.2–0.3% for the last investment grade rating category (BBB/Baa), are around 1–10% for low credit quality (BB/Ba and B). Default rates increase to about 20–40% for the lowest quality risk ratings (CCC/Caa, CC/Ca, and C). Figure 9.2 shows the characteristic pattern of 1-year default rates by rating category $f_1(R)$ averaged over multiple credit cycles, which increase exponentially across the rating scale for decreasing credit quality.

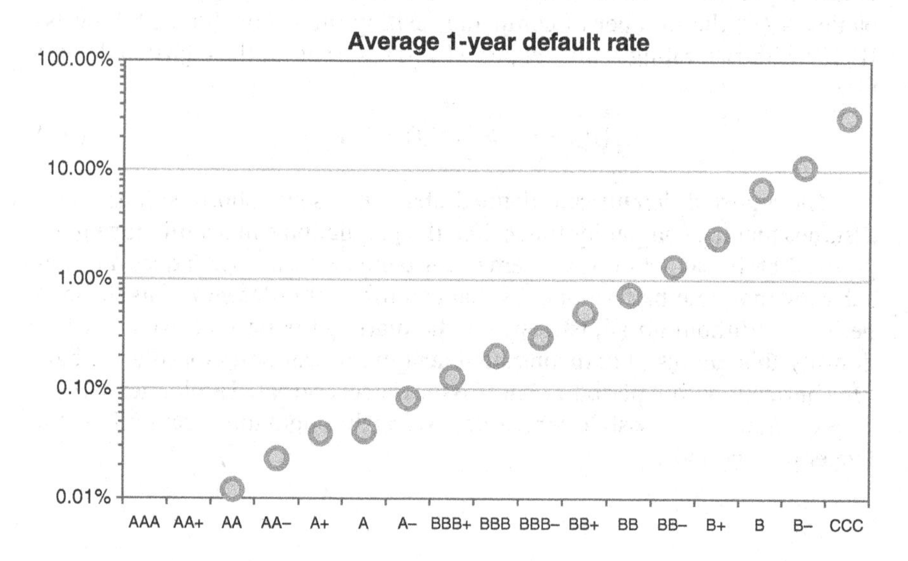

FIGURE 9.2 Stylized average 1-year default rates for different risk rating categories.

Default rates also increase with the observation period T for default events. The longer the period, the higher the chances of observing a default event. The increase of the default rates with the period is not proportional across the rating scale. For high-quality ratings, default rates are low, and the cumulative effects of incremental defaults are somewhat additive. For example, for a BBB-rated company, the 1-year default rate is about 0.2% while the 5-years default rate is about 1–2%, i.e. about (or even more than) 5 times the yearly rate. In contrast, for low-quality ratings, default rates are high, and the cumulative effects are less than proportional as the fraction of non-defaulting (surviving) population decreases significantly with time. For a CCC-rated company, the 1-year default rate is about 20–40%, and the 5-years default rate is around 50–70%, i.e. only about 2–3 times the 1-year rate. Figure 9.3 shows the average cumulative default rates by risk rating and time horizon expressed in terms of the odds of default: $f_T/(1-f_T)$, where $f_T(R)$ is the default rate (default frequency).

Realized default rates change over time in response to changes in the credit quality of the companies or the economic environment that are not always reflected timely in the risk rating assigned by credit analysts. Also, the rare nature of default events for some rating categories is translated into small default counts, creating significant variability in default rates from year to year. For example, let's assume that the 1-year probability of default for a population

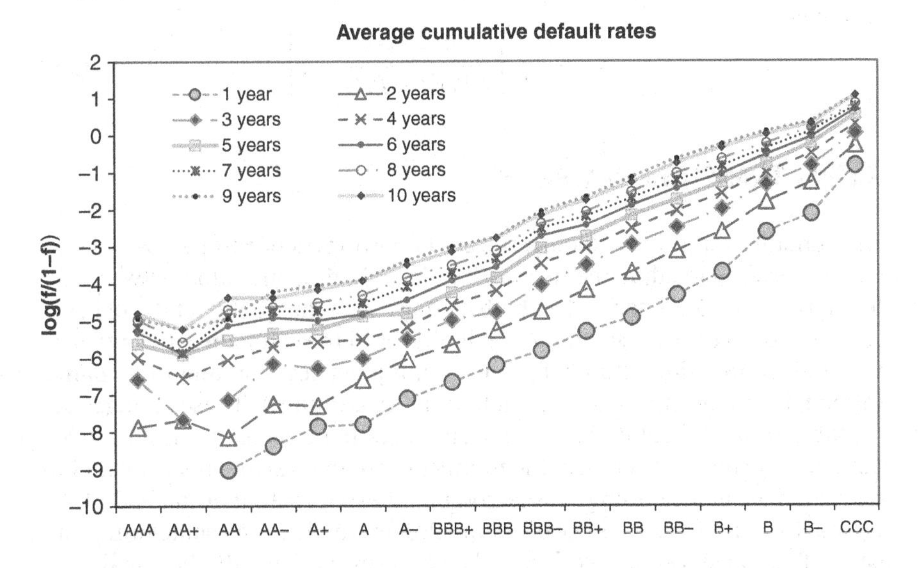

FIGURE 9.3 Stylized average cumulative default rates for different risk rating categories.

of 100 companies with the same risk rating is 1.5%. If we observed two default events this year but in the previous year there was only one default event in the same population, the default rates would have changed from about 1% to 2%. This is a significant change in the observed default rate due only to the rare and discrete nature of defaults. This also highlights the difference between *default rate* (or default frequency), which reflects statistics of realized default events, and is subject to random deviations as default materialize in a pool of companies, and *default probability*, which reflects expected likelihood of default for each individual company in the population. If default events were completely uncorrelated, the volatility of default rates σ for a pool of N companies with default probability PD would be:

$$\sigma^2 = PD(1 - PD)N/(N - 1) \tag{9.5}$$

In practice, historical volatilities of default rates can be calculated from historical data on default events by rating category.

Default rate volatility can also change at different points of the business and credit cycle in response to correlation across obligors or sectors. The intuition is that the higher the correlation ρ, the higher the volatility of default rates because default events trigger additional default events if correlation gets higher. Given a pool of companies with the same probability of default PD and observed volatility σ, the implied correlation of the pool of companies would be:

$$\rho = \frac{1}{N-1}\left[\frac{(N-1)\sigma^2}{NPD(1-PD)} - 1\right] \tag{9.6}$$

RATING TRANSITION MATRICES

As economic conditions change, the creditworthiness of companies can also change, leading to adjustments in risk ratings. Rating migration between different risk categories can reflect both changes in credit quality of the obligors (e.g. due to changes in economic conditions or in the company strategy), new material information and changes in rating practices (e.g. due to a change in models or algorithms used to determine risk ratings). These changes are captured in the transition rates between risk rating categories within a given period. Transition rates reflect the number of companies moving from an initial rating at the beginning of a period to a different rating at the end of the period divided by the number of companies originally in the same rating category. The transition matrix is described in terms of a two-dimensional table where each row represents the frequency of transitioning from an initial rating R to any other rating R', including the default and withdrawal categories.

TABLE 9.4 Stylized average 1-year corporate rating transition matrix over multiple credit cycles (%)

	AAA	AA	A	BBB	BB	B	CCC	D	W
AAA	87.3	8.5	0.5	0.1	0.1	0.0	0.1	0.0	3.4
AA	0.5	86.3	8.3	0.5	0.0	0.1	0.0	0.0	4.1
A	0.0	1.8	87.3	5.3	0.4	0.2	0.0	0.1	4.8
BBB	0.0	0.1	3.6	84.9	3.8	0.7	0.2	0.2	6.5
BB	0.0	0.0	0.2	5.2	76.2	6.9	0.7	0.9	9.9
B	0.0	0.0	0.1	0.2	5.6	73.4	4.4	4.5	11.7
CCC	0.0	0.0	0.2	0.3	0.8	13.8	43.9	26.7	14.4

The sum of the transition rates adds to 100% since each company must either remain in the same rating or change to a different rating, default on its obligations, or withdraw from the population.

Since changes in risk ratings follow primarily changes in credit quality, ratings tend to be concentrated around the initial ratings (transition matrix diagonal) within a few rating subgrades, reflecting the expected changes to the likelihood and severity of losses. Transitions to very different ratings (both updates and downgrades that differ significantly from the initial rating) tend to be rare. Table 9.4 shows stylized average 1-year transition rates aggregated over multiple credit cycles by broad rating category. Note the concentration around the diagonal, indicating that risk ratings tend to be relatively stable, with some upgrades and downgrades around the initial rating reflecting changes in credit quality. Also note that transitions to other rating categories are rare but they do occur with decreasing frequency for larger moves away from the diagonal. The estimated exit and withdrawal rates in Table 9.4 reflect the nature of the population of companies covered by rating agencies. Withdrawals for financial institutions may differ in their nature and number as they reflect bank-client relationships and exit strategies. Similarly to the discussion on default rates above, transition rates could be adjusted to reflect the impact of companies exiting the initial population (rating withdrawals). Different assumptions can be used, which impact the estimation of transition rates and, ultimately, transition probabilities.

RATING TRANSITIONS AND THE TERM STRUCTURE OF DEFAULT RATES

Historical cumulative default rates for different time horizons – segmented by rating category at the beginning of the selected period – reflect the impact of changes in credit quality over the selected horizons. Cumulative default rates are calculated by defining a pool (or *cohort*) of companies with the same risk rating (i.e. same credit quality) at the beginning of the selected observation period and tracking the number of defaults (and exits/withdrawals) that occur in every period in the selected time horizon. Changes in the shape of the cumulative default rates (or term structure of default rates) reflect the impact of different marginal contributions of default events. This effect is observed in cumulative default rates for both rating agencies and financial institutions. The longer the time horizon, the higher the chances of changes in credit quality, and the bigger the potential change in the probability of default in subsequent periods. Furthermore, changes in credit quality through the observation period impact both the rate of change in the number of defaults and the rating of the company. That is, the term structure of default rates and risk rating migration patterns are related.

For simplicity of exposition, let's assume that all companies having the same rating R at the beginning of the observation period will either have a risk rating at the end of the horizon or default. That is, there are no companies exiting the population or rating withdrawals. To simplify notation, we also drop the starting rating R from the notation in the description below. For this cohort of N_0 companies at time $t' = 0$, we observed the number of defaults D_{0t} for every year $0 \leq t \leq T$ during the time horizon T.

The marginal default rate for year t, starting at time t and ending at time $t + 1$, is the ratio of the marginal count of default events $D_{t(t+1)}$ to the number of companies N_t remaining in the cohort at time t as defaults occur. That is, the marginal default rate represents the number of incremental defaults $D_{t(t+1)}$ over the number of surviving companies $N_t = S_{0t}$ in each period t, where S_{0t} is the number of companies surviving at time t from the initial cohort beginning at time 0. Note that the cumulative number of defaults in period t is D_{0t} and the number of remaining companies is $N_t = S_{0t} = N_0 - D_{0t}$. Therefore, dividing the number of cumulative defaults D_{0t} by $N_0 = S_{00}$ provides the cumulative default rate:

$$d_{0t} = D_{0t}/N_0 = D_{0t}/S_{00} = (S_{00} - S_{0t})/S_{00} = 1 - s_{0t} \tag{9.7}$$

Here $s_{0t} = S_{0t}/S_{00}$ is the cumulative survival rate at time t for the initial risk rating R. Similarly, the marginal default rate (which determines the term structure of default events) reduces to the follow expression:

$$d_{t(t+1)} = D_{t(t+1)}/S_{0t} \tag{9.8}$$

Table 9.5 shows the typical term structure of default rates as reflected in the marginal default rates for different risk ratings and time horizons.

TABLE 9.5 Stylized average marginal default rates for selected risk rating category and time horizon (%)

Time Horizon (Years)	AAA	AA	A	BBB	BB	B	CCC
1	0.0	0.0	0.0	0.2	0.7	6.8	30.4
2	0.0	0.0	0.1	0.3	1.7	7.9	18.2
3	0.1	0.1	0.1	0.3	2.6	7.9	14.0
4	0.1	0.2	0.2	0.7	2.7	7.2	12.5
5	0.1	0.2	0.3	0.6	3.1	7.0	14.4
6	0.1	0.2	0.3	0.7	3.1	7.8	8.2
7	0.1	0.2	0.3	0.7	2.7	6.3	5.9
8	0.1	0.2	0.3	0.8	2.8	5.7	5.3
9	0.1	0.2	0.3	0.8	2.7	6.5	7.0
10	0.1	0.2	0.3	0.9	2.2	4.9	7.8

PORTFOLIO RISK RATING

Portfolio risk ratings describe the quality, predictability, and sustainability of the credit portfolio performance and earnings. Table 9.6. shows a three-level rating approach to assessing portfolio risks.

PORTFOLIO TREND RATING

Changes in a portfolio's risk can be described by assessing the trend in the changes of the portfolio's risk characteristics (Table 9.7). A portfolio trend rating can provide a signal for decreasing, stable, or increasing risks to indicate the direction of changes, reflecting the impact of risk factors that could influence portfolio performance (e.g. portfolio growth, strategy and economic or political outlook).

RISK MANAGEMENT RATING

In addition to determining the risk characteristics of individual borrowers and their obligations or the characteristics of the overall portfolio, risk ratings can be applied to the institution's risk management practices themselves. More precisely, the effectiveness of the risk management practices and processes and any existing business concern can be reflected in the risk management rating (Table 9.8). This can help independent functions (such as

TABLE 9.6 Portfolio risk rating characteristics

Rating	Characteristics
Low	Macroeconomic environment is favorable for business activity Portfolio profitability meets or exceeds expectations Portfolio performance is sustainable and predictable Portfolio is characterized by small size, short tenors, high quality, small concentrations, availability of collateral and/or low percentages of past due or criticized credits
Medium	Macroeconomic environment is weaking and business activity is slowing down Portfolio profitability is underperforming expectations Actions are required to achieve predictable and sustainable performance Portfolio characteristics include a mixture of credit quality, significant concentrations, deteriorating collateral, or increases in pass due and criticized credits
High	Macroeconomic environment is negative and business activity is contracting Actions are required to bring portfolio profitability within expectations Actions are required to achieve predictable and sustainable performance Portfolio is characterized by large size concentrations, long tenors, low credit quality, deteriorating collateral, and/or high percentages of past due or criticized credits

TABLE 9.7 Portfolio trend rating characteristics

Rating	Characteristics
Decreasing	Macroeconomic environment is improving, business activity is expanding Portfolio profitability is increasing Portfolio characteristics and quality are improving, reducing overall risk
Stable	Macroeconomic environment is stable, business activity continues to grow Portfolio profitability is stable Portfolio characteristics and quality are stable, overall risk remains constant
Increasing	Macroeconomic environment is deteriorating Portfolio profitability is deteriorating Portfolio characteristics and quality are deteriorating, increasing overall risk

TABLE 9.8 Risk management rating characteristics

Rating	Characteristics
Effective	Risk management processes and practices are effective, proactively identifying, monitoring, and managing risks, implementing controls, and complying with regulations Risk ratings and risk assessments are adequate and well documented The business is staffed appropriately and there are no business concerns
Satisfactory	Risk management processes and practices are generally effective, identifying, monitoring, and managing risks, implementing controls, and complying with regulations consistently Risk ratings are generally adequate and well documented The business is staffed appropriately, and gaps and minor business concerns do not materially impact the ability to manage risks
Deficient	Risk management practices and processes are generally functioning adequately, but some areas may present deficiencies for identifying, monitoring, or managing risks, having effective controls, or complying with regulations There may be gaps in the accuracy of risk ratings, credit analysis may need improvement The business may not be staffed appropriately or there are business concerns that may affect the ability to manage risks
Unsatisfactory	Risk management practices and processes are not sufficient to effectively identify, monitor, or manage risks or to comply with regulations There are material gaps in the accuracy of risk ratings or in the quality of risk assessments There are material gaps in business staffing, or there are significant business concerns

internal audit) to assess the consistency of risk management processes and practices and determine the confidence in the risk mitigation strategies. Although the risk management rating may not have a direct impact on assessing climate risk and ESG effects for obligors and their debt obligations, it can reflect whether existing risk management practices address these risks with the appropriate level of depth to be able to identify vulnerabilities and create risk mitigation strategies. We return to this point in the discussion on disclosures on climate risk and ESG and company vulnerabilities.

RATING STABILITY

Critical to the use of risk ratings to reflect a borrower's or issuer's creditworthiness is the understanding of their quality and timeliness, which may reflect a shortening of rating reviews and a quicker reaction to material events. Related to the timeliness of ratings is the inclusion of market information (which represents the opinions of the market participants reflected in stock prices and credit spreads) into the risk rating process. The use of market opinions may help capture credit information at a faster pace. Several commercial vendors offer market-based credit assessments relying on equity or debt prices. It is certainly possible to rely on market opinions without necessarily allowing them to be a substitute for fundamental credit analysis. Market opinions are not usually a significant component of the analysis of an obligor's fundamental creditworthiness, except in situations where an obligor's loss of market access has liquidity implications. In such cases, market opinion can be an important element in the credit assessment process.

The drawback of this approach is that market opinions are volatile and could produce a procyclical feedback process leading to even greater volatility, which may result in further disruption of the capital markets. Another drawback is that market opinions may reflect changes in market preferences and uncertainty that are not related to the credit quality of the borrowers.

In principle, risk ratings should emphasize medium- to long-term credit fundamentals, as opposed to volatile market conditions. Market participants and institutional investors are usually strongly opposed to volatile ratings, since ratings are deeply engrained in investment guidelines and bond indices. Unexpected rating changes may force asset managers to rebalance their portfolios at inopportune times or under disadvantageous conditions. However, during periods of credit stress, ratings are expected to reflect evolving economic conditions that may affect creditworthiness.

A particular issue affecting the use of agency ratings is that rating agencies may place excessive emphasis on the risk of losing market access. As a result, rating downgrades may become self-fulfilling prophecies as market participants react to downgrades by reducing their investments, limiting the issuers' ability to access capital markets. Another key issue for investors and users of agency ratings is to have more timely rating actions including shorter periods for completing the credit assessment. However, investors also appreciate a lengthier rating outlook signaling process, which gives issuers an opportunity to correct conditions that could otherwise lead to credit deterioration. Furthermore, investors prefer greater transparency in the rating process, with more aggressive views on issues of accounting quality, risk management practices, corporate governance, conduct risk, and disclosures. While rating agencies often request nonpublic information to supplement their credit assessment activities, unlike accounting firms, they have no authority to demand such data and companies are under no obligation to cooperate with them at all. Therefore, agency ratings usually reflect information that has been disclosed or that management has elected to provide.

Although risk ratings should be understood as expert credit opinions that look beyond volatile market events, market information can still play a relevant role in credit assessments. More precisely, market-based models can identify material and systematic gaps between a fundamental credit assessment and the ratings implied by market data. However, the benefits and limitations of market-based approaches need to be clearly articulated to the market participants and users of these tools. In later sections we discuss different quantitative approaches and their possible use in the context of climate risk.

QUANTIFYING ANALYSTS' PERCEPTION OF CREDIT RISK: A BEHAVIORAL MODEL

Additional clarification on the risk assessment process performed by risk analysts could provide some transparency and insight into the notion of risk. Beyond providing transparency of meaning for investors and financial institutions, a more detailed analysis of risk ratings could also help provide a solid basis on which to evaluate the relative performance of ratings against risk scores based purely on quantitative models whose meaning is precise by construction (Cantor et al., 1999; Murray et al., 2000; Krahnen and Weber, 2001; Keenan and Sobehart, 2002; Lyon, 2002; Giesecke et al., 2009). Here we focus on the analysts' perception of credit risk using a quantitative framework based on behavioral models.

The Value of Subjective Expertise and Expert Knowledge in Credit Risk Analysis

Risk is an abstract concept that can be measured on two types of scales: *relative* and *absolute*. Relative scales do not have standards of comparison. They are means of ranking in relation to other values on the same scale. Since relative values are not based on an absolute standard, they can differ in meanings depending on the situation or context. Absolute measures scales quantify different aspects of risk on a common scale, providing a framework for comparison under different situations.

Whereas relative and absolute comparisons are different ways of expressing risk analysis inputs and results, the risk assessment process can be of a *qualitative* or *quantitative* nature. A qualitative risk assessment is a top-down method. It starts with a definition of a standard for comparison. The standard is typically a set of best practices, suggested or mandated, defined by expert judgment or derived from previous experience. A qualitative assessment of a borrower's vulnerability to changing economic conditions can be defined by what the issuer lacks as compared to a peer or benchmark. For example, "How good is the company's management relative to its peers?"

In contrast, a quantitative risk assessment is an objective representation of the borrower, based on reliable financial and market information describing the borrower, its business environment, and the probability of occurrence of credit events that could contribute to borrower's failure (Altman, 1968; Caouette, Altman, and Narayanan, 1998; Duffie and Singleton, 1999). In practice, historical observations are often sparse, limited, or nonexistent and, therefore, credit risk models depend on a significant degree of subjectivity, both in determining relevant financial input variables and determining the relationships between those variables. That is, even credit assessments based on statistical regressions or structural models often depend on significant subjective expertise and judgment of analysts and modelers. Not only is this factor not treated directly in most conventional quantitative analyses, but also its role is rarely indicated to the recipient of the analysis – investors and lenders.

Errors of Anticipation and Habituation

Risk ratings are expert *opinions* of credit quality representing different degrees of belief across many dimensions, such as the probability of default or severe financial distress and loss severity. Analysts' judgment can err in one of two ways: (1) it can indicate low risk when, in fact, the risk is high (type I), or (2) it can indicate high risk when, in fact, the risk is low (type II). Type I errors correspond to highly rated borrowers who nevertheless default on their financial obligations (e.g. an A-rated issuer who defaults on publicly-held bonds). In contrast, type II errors correspond to low-rated borrowers that should be rated higher (e.g. a financially sound company misidentified as a high-risk borrower).

Despite the fact there might be some unavoidable differences of opinion between analysts, long-term obligor default studies for debt obligations show a clear relation between credit rating quality, the severity of losses, and the likelihood of default, with lower ratings generally linked to higher default rates and more severe losses (Keenan et al., 2000). Here we discuss an approach to risk perception oriented toward the behavioral aspects of the rating process using traditional psychological ideas of stimulus and perception (Keenan and Sobehart, 2002; Sobehart and Mogili, 2007).

Let us define a set of potential negative credit events for a given company and the potential outcomes associated with each one of them as its "risk exposure." Because traditional ratings describe different perceptions of risk, preference relations among different risk exposures can be obtained where absolute risk quantification is difficult or impossible. For example, two different start-up companies who have never experienced bankruptcy both have some probability that bankruptcy will occur in the future, but this probability cannot be objectively quantified based on their business histories. Nevertheless, even while there may be no historical or statistical information available on which to assess this probability, a seasoned credit analyst may be able to form an *opinion* on relative riskiness based on discussions with management, on-site visits, or peer comparisons.

Participating as an analyst in the rating process, with its demand for the making of difficult judgments, is a complex behavioral process where other aspects of the situation, besides the evidence being presented, may affect the analyst's response. Any judgmental analysis is particularly susceptible to two classes of errors:

1. error of *anticipation*
2. error of *habituation.*

Both errors may stem from repeated responses that an analyst must make as a series of judgments on a particular borrower or obligation. For illustration, suppose a junior analyst monitors an obligor whose credit quality is continuously deteriorating. As more evidence showing riskier characteristics is accumulated, the analyst repeatedly determines a lower credit quality compared with the initial value but not low enough to elicit a downgrade. After a while, the analyst knows that the credit quality of the issuer is approaching the threshold for a downgrade to a lower rating, and that it is becoming increasingly difficult to distinguish the actual credit quality of the obligor from the appropriate downgrade trigger point. At this point the analyst will be unable to distinguish the two risks after a few additional comparisons. This knowledge, together with her weariness at assigning the same rating on every comparison, may cause her to anticipate when the obligor's risk and the threshold for downgrade become indistinguishable. The analyst may therefore begin assigning a lower rating when in fact she could still sense the difference in riskiness, albeit at a higher cost in terms of time and informational inputs if she made the necessary effort. Similarly, for an issuer whose credit quality improves, the error of anticipation will make the estimate of the upper risk threshold for a rating upgrade somewhat larger than if the error were avoided successfully. In practice, these errors are hard to avoid because it is difficult to instruct and train analysts to assess risk only based on the evidence presented to them at one point in time, ignoring their responses to earlier evidence of credit quality, the obligor's history, and related information.

Some other analysts may persevere in making one response as if by habit, so that this response may persist even beyond the point where their judgment would change if they were responding to new information with identical content. That is, a borrower who is well known to the analyst and has held a given rating for a long time may be able to retain a rating even when the fundamentals deteriorate past the normal downgrade threshold based on expert judgment alone. The error of habituation has opposite effects from the error of anticipation in creating overestimates or underestimates of risk.

Although both types of errors can be reduced substantially through a rating review process involving experienced analysts, it would be unlikely that analysts would avoid making mistakes so consistently as to yield a risk assessment completely free from any behavioral distortion.

The discrete nature of most rating scales for rating agencies and financial institutions and the fact that default and financial distress events are rare pose serious constraints to the statistical analysis of credit risk perception. Due to the differences among industries and financial sectors, analysts often emphasize the likelihood of default as the main descriptor of credit risk as opposed to a more general definition. A more general definition could also include expected losses[2] and/or transition risk. The benefit of focusing on default risk is that it provides a simple and objective framework for estimating rating transitions and default rates.

Quantifying Credit Risk Perception

A qualitative comparison between two risk exposures can be described within a probabilistic framework, since few of the risk components that define the true exposure can be observed with precision. Nor can the dynamic interaction of risk components for any individual obligor be known with certainty. Thus, each qualitative risk assessment is subject to both observational error and model error (analysts' judgment error). On the other hand, since analysts who assign ratings are often well trained and, in many cases, operate through committees endowed with high levels of experience and expertise, a sensible model of risk perception should not imply unrealistically that ratings are wrong most of the time. A realistic model of risk perception may reflect the fact that analysts sometimes make mistakes when comparing different risk exposures, adding an additional level of uncertainty to the assessment of credit risk.

Let us introduce a formal description of the credit assessment process. Let us define the "risk exposure" X of an obligor as an information set that is the observable component of the complete set of risk factors affecting the obligor. That is, X represents the evidence available to the analyst, such as financial and other firm-specific information, industry, and market data. A set of risk exposures is denoted as $\{X\}$ with individual exposure values denoted as X. We treat $\{X\}$ as a portion of the real line; that is, X is a real number. In general, the risk exposure X will be a function $X(E)$ of the evidence at hand $E = \{E_1, \ldots, E_m\}$ representing financial information, industry and market factors and a variety of additional information available to analysts. The risk exposures being considered here may be associated with composite outcomes of the financial distress of a borrower or the default of a particular debt instrument or may be associated with more narrowly specified outcomes like fire in a particular warehouse, or flood damage to a factory.

A qualitative comparison between two risk exposures is clearly probabilistic in nature since analysts could sometimes report the highest risk exposure as smaller. Presumably, two different risk exposures X_1 and X_2 will have a certain degree of psychological similarity that can be operationally defined in terms of the frequency with which they are confused in a pair comparison task. In the following we follow closely the arguments in Thompson (1969) and Burton (1972).

Let $P(X_1 \geq X_2)$ be the frequency that an analyst reports a risk exposure X_1 as greater than X_2. We now ask whether $P(X_1 \geq X_2)$ could define an additive scale on which psychological risk similarity can be measured. That is, is there a monotonic function $G(p)$ satisfying Equation (9.9)?

$$G(P(X_1 \geq X_2)) + G(P(X_2 \geq X_3)) = G(P(X_1 \geq X_3)) \tag{9.9}$$

Then we have

$$G(P(X_1 \geq X_2)) = G(P(X_1 \geq X_3)) - G(P(X_2 \geq X_3)) = H(X_1) - H(X_2) \tag{9.10}$$

Here X_3 is selected as a benchmark risk exposure (e.g. the risk exposure of a typical BBB/Baa rated company) and $H(X) = G(P(X \geq X_3))$. Introducing the function K, the inverse function of H, we have:

$$P(X_1 \geq X_2) = K(H(X_1) - H(X_2)) = K(Y_1 - Y_2) \tag{9.11}$$

Models of this type are called linear (Thompson, 1969). Analysts' response to the risk exposure X is the risk perception $Y = H(X)$, which is a construct introduced to explain the behavior observed in selecting a rating. Intuitively $Y(X)$ is the *utility* analysts assign to an amount of risk X and is reflected as a risk rating.

Now, if $P(X_1 \geq X_2) = 1/2$, that is, if an analyst is unable to determine which risk is larger, then X_1 and X_2 are psychologically similar risk exposures. In some cases, it may be possible to measure the psychological similarity of X_1 and X_2 by the nearness of $P(X_1 \geq X_2)$ to 1/2. If we succeed in fitting this type of *psychological distance* between risk exposures, the qualitative process of credit assessment could be partially quantified.

Under different assumptions, various expressions can be derived to describe the relationship between risk exposure X and risk perception $Y(X)$.

Here we assume that such relation[3] holds approximately and can be derived from simple arguments. For simplicity of exposition, we express the risk perception Y as the number of rating subgrades within a minimum rating Y_m and a maximum rating Y_M. Here, AAA/Aaa ratings are represented by $Y_m = 1$ subgrades and aggregated CCC-C/Caa-C ratings by $Y_M = 17$ subgrades using a typical rating scale.

The basic argument is that, at the fundamental level, the risk assessment process is based on a relative comparison between perceived risk for pairs of risk exposures. If the absolute risk perception of two risk exposures differs by a just noticeable amount when separated by a given absolute exposure increment, then, when the risk exposures are increased, the increment must be proportionally increased for the difference in perception to remain just noticeable. More precisely, let X be the risk exposure, Y be the resultant risk perception, and Δ be an increment of either, then the simplest relation between X and Y is:

$$\Delta Y = \alpha \Delta X \tag{9.12}$$

Here α is the *sensitivity* to variations of the risk exposure for an absolute comparison. Taking the limit $\Delta X \to 0$ and integrating the differential equation we have

$$Y = \alpha(X - X_0) + Y_0 \tag{9.13}$$

Here X_0 and Y_0 are constant to be determined. If $X_0 = X_m$ is the minimum threshold value below which there is a negligible perception of risk, then $Y_0 = Y_m$ is the minimum risk perception (rating).

To keep the risk perception bounded, Equation (9.13) needs to be adjusted:

$$Y = \min(Y_M, \max(Y_m, Y_m + \alpha(X - X_m))) \tag{9.14}$$

Equation (9.14) provides a simple relationship between risk exposure (e.g. one or more financial and market ratios) and risk perception (e.g. risk rating). However, other relationships are possible. If the absolute perception of two risk exposures differs by a just noticeable amount when separated by a given relative risk exposure increment, then, when the relative risk exposures are increased, the increment must be proportionally increased for the difference in perception to remain just noticeable. More precisely:

$$\Delta Y = \beta \frac{\Delta X}{X} \tag{9.15}$$

Here β is the *sensitivity* to relative variations of the risk exposure. Taking the limit $\Delta X \to 0$ and integrating the differential equation we have

$$Y = \beta \log(X/X_0) + Y_0 \tag{9.16}$$

Here X_0 and Y_0 are constant to be determined. Equation (9.16) is known as the Weber-Fechner law in psychology. Note, however, that the logarithmic assignment of perception described in Equation (9.16) is not to be taken literally since risk perception is not well behaved for risk exposures below the threshold value X_0. If, instead of the absolute perception, the relative perception of two risk exposures differs by a just noticeable amount when separated by a given relative increment in risk exposure, then, when the risk exposures are increased, the increment must be proportionally increased for the difference in perception to remain just noticeable. More precisely, the relative relation between X and Y becomes:

$$\frac{\Delta Y}{Y} = \beta\frac{\Delta X}{X} \tag{9.17}$$

Again, taking the limit $\Delta X \to 0$ and integrating the differential equation we have

$$log(Y/Y_0) = \beta \log(X/X_0) \Rightarrow Y/Y_0 = (X/X_0)^{\beta} \tag{9.18}$$

Equation (9.18) exhibits an exponential (power) growth, which holds approximately within the middle ranges of intensity for different risk exposures and perceptions.

We can refine further the exponential model in Equation (9.18) by recognizing that, as the risk exposure X increases or decreases significantly, its contribution to the risk perception Y has diminishing marginal impact ($\Delta Y/Y \to 0$), leading to saturation effects around the minimum Y_m or maximum Y_M. When saturation effects are considered, the relative relation between X and Y is:

$$\frac{\Delta Y}{Y} = \beta\frac{\Delta X}{X}\frac{1}{\left(1-\frac{Y_m}{Y_M}\right)}\left(1-\frac{Y_m}{Y}\right)\left(1-\frac{Y}{Y_M}\right) \tag{9.19}$$

Here the terms $(1 - Y_m/Y)$ and $(1 - Y/Y_M)$ reflect the saturation effects of the sensitivity to the risk exposure X, and $(1 - Y_m/Y_M)$ is a normalization factor for calculation convenience. Taking the limit $\Delta X \to 0$ and integrating the differential equation we have

$$Y = \frac{1}{1+\left(\frac{X}{X_{mp}}\right)^{-\beta}}(Y_M - Y_m) + Y_m \tag{9.20}$$

Here X_{mp} is the risk exposure for the midpoint of the risk perception range $Y_{mp} = (Y_M + Y_m)/2$. The model in Equation (9.20) provides a simple model of risk perception (risk rating) Y as a function of the risk exposure X (e.g. a combination of financial and market ratios).

An alternative model can be derived when the change in relative risk perception is sensitive to the *absolute* change in risk exposure instead of the *relative* risk exposure. The relation between X and Y in Equation (9.19) becomes:

$$\frac{\Delta Y}{Y} = \alpha \Delta X \frac{1}{\left(1 - \frac{Y_m}{Y_M}\right)} \left(1 - \frac{Y_m}{Y}\right)\left(1 - \frac{Y}{Y_M}\right) \tag{9.21}$$

Taking the limit $\Delta X \to 0$ and integrating the differential equation we have

$$Y = \frac{1}{1 + e^{-\alpha(X - X_{mp})}}(Y_M - Y_m) + Y_m \tag{9.22}$$

Equation (9.22) also provides a plausible model of risk perception (risk rating) Y as a function of the risk exposure X (e.g. a combination of financial and market ratios) that saturates as the risk exposure increases or decreases significantly. Note that the logistic transformation in Equation (9.22) is derived naturally from simple assumptions about the sensitivity of risk perception to changes in risk exposure, as described in Equation (9.21).

Risk Perception and Probability of Default

Let's assume one credit analyst assigns a rating to a company with risk exposure X by creating a subjective mapping y, which is only an approximation to the expected risk perception $Y(X)$ for similar risk exposures based on her experience and judgment. We can model the discrepancy of individual risk perceptions $y - Y(X)$ by closely following early seminal work on the subject (Stigler, 1986).

Let's also assume the analyst idealized the risk perception as a line segment AD where A corresponds to the lowest possible risk and D to the highest possible risk, and where an intermediate point C corresponds to the benchmark value $Y(X)$ as shown in Figure 9.4. The risk perception of a borrower's

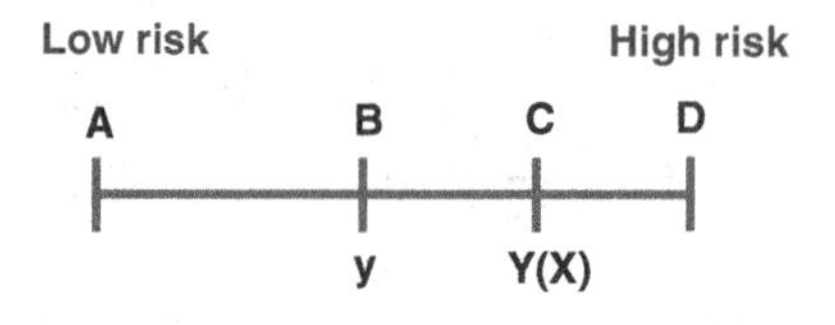

FIGURE 9.4 An analyst or credit officer judges whether the perceived risk AB exceeds the benchmark AC.

credit quality, y, is given by an intermediate point B. The analyst's goal is to judge whether the distance AB should be longer (riskier) than AC. Because the risk perception AC is the analyst's idealized benchmark, we may assume that the analyst forms a mental estimate of the distance AB from the point B to the benchmark point C, and this estimate errs from the expected value by an amount that has a distribution $g(BC, X)$ resulting primarily from two sources: (1) discrepancies in information on the company's credit quality (not contained in the risk exposure X), and (2) errors of habituation and anticipation to changes in credit quality committed by analysts (judgment errors). These errors may stem from repeated responses that an analyst must make as a series of judgments on a particular issuer or debt issue, or they may reflect a misperception of the interrelationship among correctly perceived components of X. In practice, the distribution of errors can be different for each analyst and can depend on a variety of factors such as the analyst's previous experience with troubled issuers, knowledge of the industry, and the business environment. For simplicity of exposition, we assume that analysts' errors are normally distributed and have similar deviation σ_c over a fixed time horizon T_c for the credit assessment process (e.g. $T_c = 1$ yr between credit reviews).

$$g(BC, X) = \frac{1}{\sqrt{2\pi\sigma_c}} e^{-(y-Y(X))^2/2\sigma_c^2} \tag{9.23}$$

Here $\sigma_c(T_f)$ reflects the uncertainty on the rating estimates that can depend, among other things, on the time horizon T_f used by analysts for forecasting credit quality and comparing risk exposures (usually several years or a credit cycle for a through-the-cycle (TTC) view, and roughly one or two years for a point-in-time (PIT) view). Notice that the assumption of a normal distribution in Equation (9.23) is made for analytical convenience only since the difference $y - Y(X)$ cannot be strictly normally distributed due to the discrete and finite nature of risk ratings.

Although judgment errors are unavoidable and usually happen more often than credit analysts tend to admit, Equation (9.23) indicates that it is unlikely to observe significant discrepancies between the assigned risk perception y and the average perception for similar risk exposure $Y(X)$.

Now let us turn our attention to the expected frequency of defaults associated with a given risk perception. After assigning a group of obligors a similar risk perception $Y(X)$, the observed default rate can be calculated as the ratio $d/(n + d)$ of the number of defaults d over the period T_c to the total number of companies $(n + d)$ at the beginning of the period, where n is the number of non-defaulting companies. The expected default rate $E(d/(n + d))$ for a risk perception $Y(X)$ is given by the probability that a type I error has occurred for troubled obligors whose risk perception should be $Y(X_d)$ but were incorrectly perceived at a lower value $Y(X)$. Here X_d is the typical risk exposure of defaulting obligors and $Y(X_d)$ is the risk perception based on the analyst's experience with troubled obligors over the time horizon T_c.

Using Equation (9.23), the probability of default over the time horizon T_c is the expected default rate, which reduces to a standard probit model as a function of the risk perception Y:

$$p_{T_c}(Y(X)) = E(d/(n+d)) = \int_{-\infty}^{(Y-Y_d)/\sigma_c} \frac{1}{\sqrt{2\pi}} e^{-y^2/2} dy = N\left(\frac{Y(X) - Y(X_d)}{\sigma_c}\right) \tag{9.24}$$

Equation (9.24) indicates that the long-term expected default rate will be around 50% for borrowers whose ratings are close to the ideal benchmark for defaulting issuers $Y(X_d)$. Higher default rates are observed when analysts rate obligors who are riskier than their idealized benchmark for distressed firms.

Figure 9.5 shows the empirical relation between the risk perception $Y(X)$ (given by the risk rating) and the average default rate for corporate borrowers for different time horizons T over multiple credit cycles. Figure 9.6 shows the same data transformed using the inverse of the cumulative error function scaled with the time horizon for cumulative default rates: $Z = N^{-1}(\mathrm{E}(d/(n+d)))$. This approximate transformation yields a simple

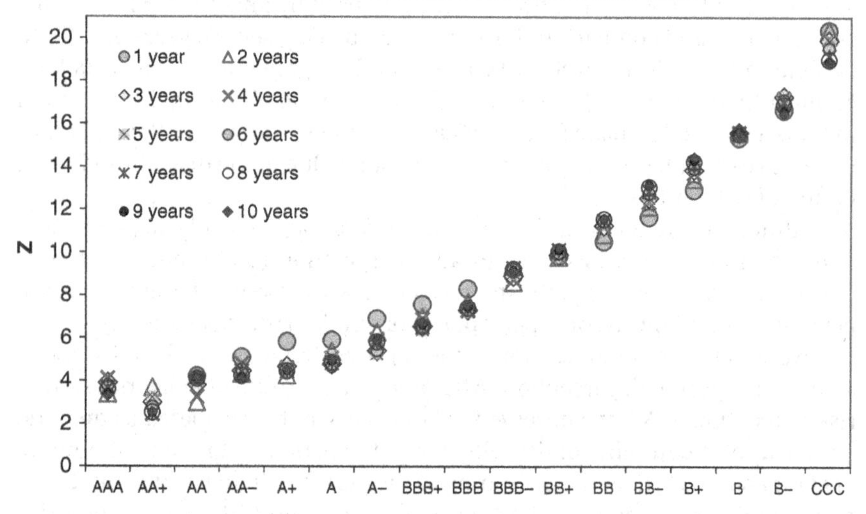

FIGURE 9.5 Transformed default rates for Equation (9.24) for different ratings (risk perception) and time horizons obtained from Figure 9.3.

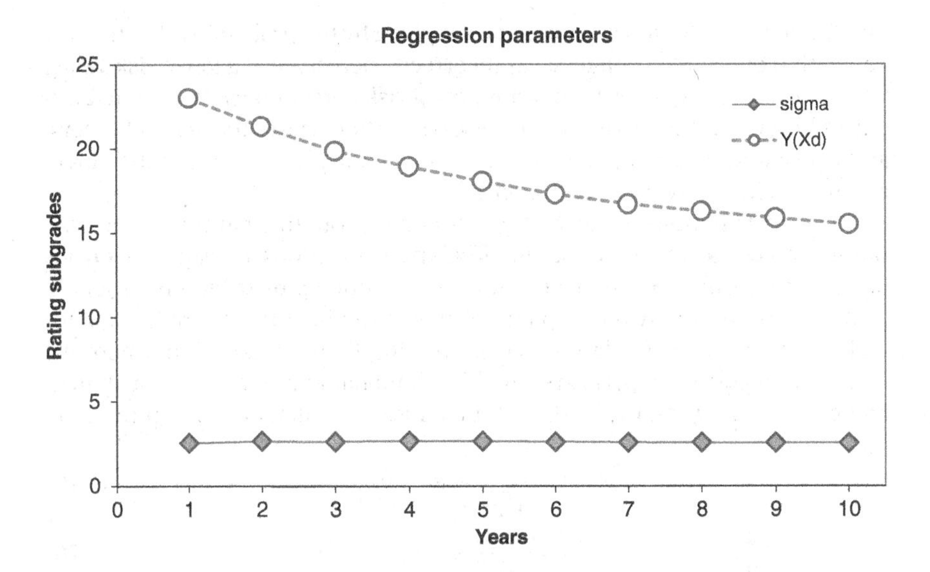

FIGURE 9.6 Regression parameters for different ratings and time horizons obtained from Figure 9.3.

linear relation between the adjusted default rate and analysts' ratings for all time horizons: $Z = (Y(X) - Y(X_d))/\sigma_c$. Note that the general linear trend between analysts' ratings and the transformed default rate Z agrees with the model described in Equation (9.24). The small divergence of the transformed data for different time horizons suggests some additional effects for increasing time horizon T for calculating default rates. Finally, note that the behavioral model introduced in Equation (9.24) provides a simple means for estimating probabilities of default (expected default rates) for different risk ratings and time horizons when historical data are sparse or unreliable. This parsimonious model of probability of default for different risk ratings and time horizons can be modified to include the effects of climate risk drivers (both transition and physical) over extended periods of time.

Risk Perception Implied from Historical Default Rates

Here we discussed an alternative approach based on the expected default rates associated with a given risk perception, assuming that they are unbiased estimates of the (ex-ante) analysts' perception of default likelihood. After assigning the same risk perception to a group of obligors with similar characteristics, the default rate of the group can be calculated as the ratio $d/(n + d)$ of defaulting issuers d to total number of rated companies[4] $(n + d)$. The expected default rate $E(d/(n+d))$ for a risk perception y is given by the probability that the risk perception of troubled obligors was incorrectly perceived at a lower risk value.

Let $p_d(y)$ be the observed distribution of risk perception scores for defaulters and $p_n(y)$ be the distribution of risk perception scores for non-defaulters, and $q_d(T)$ and $q_n(T)$ be the (unconditional) prior probabilities of defaulters and non-defaulters for time horizon T.

The distributions $p_d(y)$ and $p_n(y)$ depend on the nature of the risk rating processes and practices at different institutions or rating agencies but usually exhibit bell-shaped unimodal or bimodal distributions (roughly centered around the two mid-point ratings of the investment grade segment and non-investment grade segment). For simplicity of exposition, here we assume that $p_d(y)$ and $p_n(y)$ are roughly unimodal and normally distributed, and centered at y_d and y_n with width σ_d and σ_n, which depend on the time horizon T:

$$p_d(y) = \frac{1}{\sqrt{2\pi}\sigma_d} e^{-(y-y_d)^2/2\sigma_d^2} \tag{9.25}$$

$$p_n(y) = \frac{1}{\sqrt{2\pi}\sigma_n} e^{-(y-y_n)^2/2\sigma_n^2} \tag{9.26}$$

Using a Bayesian estimate based on the populations of defaulters $p_d(y)$ and surviving companies $p_n(y)$, the probability of default $p(y, T)$ as a function of the rating y and time horizon T reduces to a modified logistic model:

$$p(y,T) = E\left(\frac{d}{n+d}\right) = \frac{q_d(T)p_d(y,T)}{q_d(T)p_d(y,T) + q_n(T)p_n(y,T)} = \frac{1}{1 + e^{-(y(1-\delta y)-b)/a}} \tag{9.27}$$

Here

$$a(T) = \left(\frac{y_d}{\sigma_d^2} - \frac{y_n}{\sigma_n^2}\right)^{-1}$$

$$b(T) = a\left[\frac{1}{2}\left(\frac{y_d^2}{\sigma_d^2} - \frac{y_n^2}{\sigma_n^2}\right) - \log\left(\frac{p_d}{p_n}\right)\right]$$

$$\delta(T) = \frac{a}{2}\left(\frac{1}{\sigma_d^2} - \frac{1}{\sigma_n^2}\right) \tag{9.28}$$

Note that if the distributions of risk perception $p_d(y)$ and $p_n(y)$ were the same for all risk perceptions values y, the analyst's risk perception would be completely uninformative leading to the prior probability for default events. In contrast, if the distributions for healthy and troubled companies were significantly different, the posterior probability in Equation (9.27) would be very informative given the risk perception value y. In later sections we provide additional analysis in the context of credit scoring approaches.

Note that as the time horizon T for observing default events increases, more firms default, leading to a decrease in the mean value y_d and, therefore, a decrease in the reference rating $b(T)$ and an increase in the rating dispersion $a(T)$. Because the number of defaulters increases at the expense of the population of surviving companies, the rating distribution of the surviving companies $p_n(y)$ also changes but less dramatically. Thus, the probability of default in Equation (9.27) will increase as T increases, following a pattern in general agreement with empirically observe default rates.

Expressing the probability of default $p(y, T)$ in terms of the odds of a default event $O_p(y, T)$ yields

$$\log(O_p) = \log\left(\frac{p}{1-p}\right) = (y(1 - \delta(T)\, y) - b(T))/a(T) \tag{9.29}$$

The most prominent feature of Equation (9.29) is the quasi-linear trend between perceived risk (risk rating) and the odds of default. Ignoring the higher-order curvature effects leads to a simple and insightful linear relationship between risk perception (risk ratings) and the odds of default for each time horizon T.

$$\log(O_p) \approx \frac{1}{a(T)}(y - b(T)) \tag{9.30}$$

Parameter $b(T)$ provides the reference risk rating used in the comparison of risk exposures, and the parameter $a(T)$ provides the sensitivity of the risk perception to changes in the odds of default.

Equation (9.30) suggests that a change in risk perception (e.g. one rating subgrade) has the same meaning in terms of the odds of default across the risk rating scale. This highlights why analysts may prefer the rating scale for credit assessments as opposed to estimating probabilities of default directly. When probabilities get very close to zero or very close to one, our intuition does not work very well to assess their magnitude. Does the difference between a probability of 0.0001 and 0.0002 mean a great deal to the reader? However, it is clear a difference of one or two rating subgrades does have meaning both for credit analysts and in the markets for credit instruments. Intuitively human sensations tend to be measured in a relative sense, leading to logarithmic functions of the stimulus, which our minds comprehend more naturally. Figure 9.7 shows observed default rates over multiple credit cycles transformed using the logarithm of the odds of default. Note that the trend between analysts' risk perception (risk ratings) and the logarithm of the odds of default for each time horizon is in general agreement with Equation (9.30).

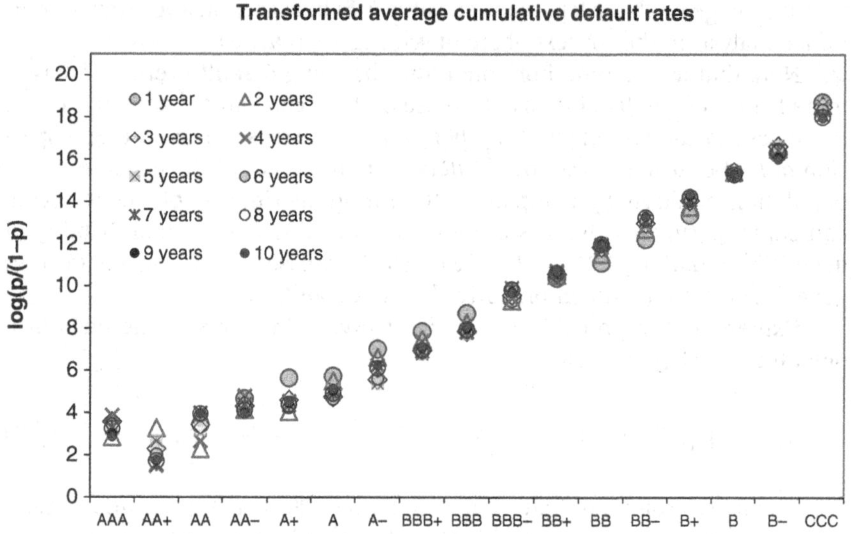

FIGURE 9.7 Observed logarithm of the odds of default for different ratings (risk perception) and time horizons. See also Equation (9.30).

Finally, note that the behavioral model introduced in Equation (9.30) provides a simple means for estimating short- and long-term probabilities of default when historical information is sparse or for the extension to climate risk and ESG by adjusting the rating dispersion $a(T)$ and reference rating $b(T)$ with adequate physical and transition risk drivers that can reflect scenarios with different severity.

Changes in Risk Perception and Rating Transition Rates

Equation (9.30) represents a simple but plausible characterization of the risk perception relationship underlying the assignment of analysts' ratings: a roughly linear relationship between ratings and the log of the odds-ratio for the perceived risk. Note that, in practice, this relationship appears to break down at the higher-rated end of the credit quality range, where analysts could be focusing on other aspects of credit risk when the expectation of default is low.

Common signals that analysts have been inaccurate in their credit assessments include persistent inconsistencies between assigned risk ratings and apparent credit risk spreads in the marketplace and multiple revisions of published ratings. In some situations, new information is revealed, which requires analysts to revisit their credit quality assessments (i.e. catastrophic events or

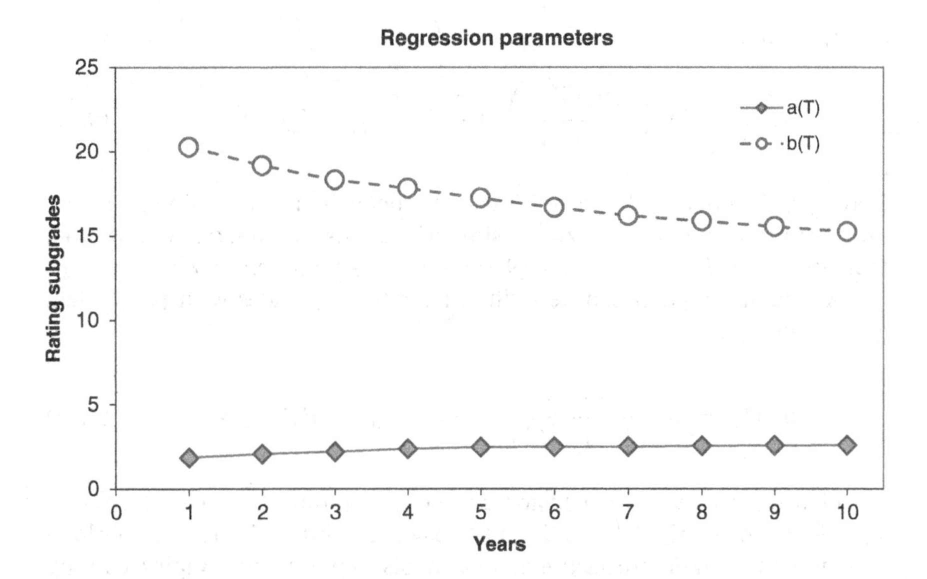

FIGURE 9.8 Regression parameters for Equation (9.30) for different ratings and time horizons obtained from Figure 9.3.

changes in regulations that can lead to unanticipated litigation). In other situations, the value of new information is subject to interpretation and rating revisions could be interpreted as evidence that the previous risk rating was incorrect. Risk rating revisions of multiple subgrades or multiple rating revisions over a short period of time are usually evidence of analysts' difficulty to determine the borrower's true riskiness.

Since large and frequent risk rating revisions are often perceived negatively by investors and financial institutions, credit analysts may consider the likelihood of default and future rating revisions in their credit assessments. Rating agencies and financial institutions usually measure rating volatility over time with empirical rating transition matrices that calculate the frequency of transitions f_{jk} from an initial risk rating y_j to a final risk rating y_k over a fixed time horizon T. The observed rating migration frequencies can be used to estimate the probability of rating transitions p_{jk}.

We can extend the relationship between risk perception (risk ratings) and the odds of default in Equation (9.30) to model the odds of risk rating migration. For each initial risk rating y_j and time horizon T, we calculate the odds-ratio of transitioning to a lower rating y_k by a given change $|y_j - y_k|$ (in terms of rating subgrades) using the empirical transition frequencies as proxies for the transition probabilities p_{jk}. Next, we approximate the logarithm of the odds-ratio of rating downgrades as a function of the magnitude of rating revisions:

$$\log(O_d) = \log\left(\frac{p_{jk}(T)}{1 - p_{jk}(T)}\right) \approx \frac{1}{c_d}((y_j - y_k) - d_d) \quad j < k \tag{9.31}$$

Here $c_d(j, T)$ and $d_d(j, T)$ are empirical parameters to be determined for each initial rating y_j and time horizon T, similarly to those in Equation (9.30). Note that when $y_k + d_d \approx b$, Equation (9.31) resembles Equation (9.30).

A similar equation can be written for rating upgrades with parameters $c_u(j, T)$ and $d_u(j, T)$.

$$\log(O_u) = \log\left(\frac{p_{jk}(T)}{1 - p_{jk}(T)}\right) \approx \frac{1}{c_u}((y_j - y_k) - d_d) \quad j > k \tag{9.32}$$

Figure 9.9 shows 1-year historical risk rating transition rates[5] for rating agencies from an initial risk rating BBB/Baa to any other risk rating over long periods of time. In Figure 9.9 the small symbols[6] represent the empirical rating transition frequencies over multiple credit cycles, the big symbols represent the historical average, and the solid lines are the model in Equations (9.31)

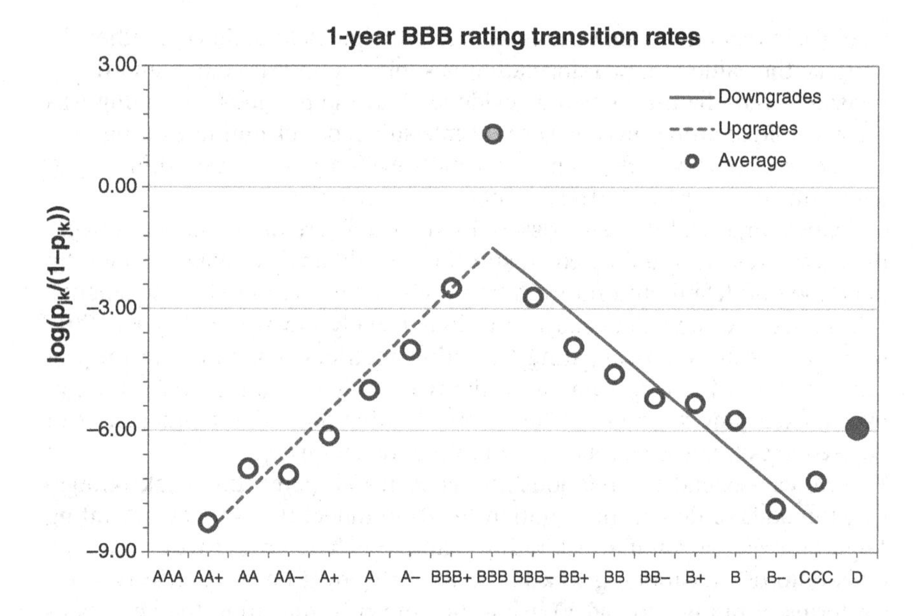

FIGURE 9.9 1-year average transition rates from Baa3/BBB- to any other rating over multiple credit cycles (open symbols), and Equations (9.31) and (9.32) (solid lines).

and (9.32) for rating upgrades and downgrades. Figure 9.9 also shows the transition element p_{kk}, reflecting the frequency of remaining in the same rating BBB-/Baa3, and element p_{kD}, reflecting the default rate. The overall impression one gets from Figure 9.9 is that the linear approximations (9.31) and (9.32) seem in general agreement with the observations, given the significant variability of transition rates for different years.

Note that if $c_u = c_d$ and $d_u = d_d$ for all initial ratings, the risk perception would be *symmetric* with respect to upgrades and downgrades for each time horizon. This would indicate that the separation between rating grades has a consistent meaning in term of the relative change in the likelihood of default. For example, a one rating subgrade downgrade would roughly double the likelihood of default from the previous rating, and a one rating subgrade upgrade would roughly reduce the likelihood of default by half. In practice, the situation is different. For investment grade companies and the high end of non-investment ratings, credit analysts seem to be more reluctant to upgrade ratings than to downgrade them ($c_u \leq c_d$), creating asymmetries in the risk rating migration rates. In contrast, for very low credit quality, analysts seem roughly unbiased to either rating downgrades or upgrades ($c_u \approx c_d$). Furthermore, for each initial risk rating, Equations (9.31) and (9.32) are asymmetric V-shaped functions.

One noticeable feature of Figure 9.9 is the misalignment of the default frequency to the line implied by the behavioral model in Equation (9.31). The empirical default rates lie above the line and represent one of the largest deviations from the model. That is, the expected default rate based on the credit migration pattern is lower than the observed default rate. Of the many possible explanations for this, we would like to emphasize three:

1. Analysts may argue that their primary focus is not the obligor's likelihood of default but the severity of losses.
2. For lower-rated credits, where default frequencies are significant and rating volatility for downgrades seems lower than expected, analysts may need to either make revisions more often or revise their loss expectation in anticipation of default events.
3. Analysts attempting to arrive at a default probability estimate cannot assess the obligor's risk precisely, focusing on transition risk where perception errors are likely to be smaller and occur more frequently.

Finally, the behavioral models described above provide a simple means of constructing rating migration probabilities that preserves the typical relationships observed in the historical transition rates and default rates. This is important for modeling rating migration under different economic and climate risk conditions and for performing what-if scenario analysis beyond the limitation of historical data.

Modeling Credit Quality Migration

The structural regularities observed in risk ratings can be used to characterize credit migration for different scenarios, which could include the effects of credit cycles, economic activity, and climate risk drivers beyond the limitations of historical data. The ability to create probabilities of default and credit rating migration is important for estimating potential losses in credit portfolios as observed default and rating migration rates are often driven by the irregularities and small sample size effects found in published transition data. Furthermore, credit risk transition matrices provide the frequency of migration (or transition rates) between any pair of risk ratings for different time horizons due to changes in the credit quality of the company, changes in the economic environment, the credit assessment process, models, or algorithms used to determine risk ratings.

Given the large number of rating migration transitions for the typical rating scale used by rating agencies or financial institutions, industry- and geography-specific models may require a large volume of historical rating data. Some rating transitions are rare, which makes granular models sensitive to sparse observations and data outliers. To overcome these limitations, we describe rating migration models that leverage simple phenomenological patterns that can be linked to different economic and climate risk indicators to reflect industry or geographic specific behavior.

Rating migration patterns can be divided into two basic contributions:

1. migration to default status reflecting a default event
2. migration from an initial rating to a different rating.

More precisely, for any given period t and time horizon T for observing default events, and risk rating R_j, an obligor can either:

1. default during the observation period T with a probability of default $PD_j(t, t+T)$ (and its status is flagged as default or D); or
2. if it does not default (with probability $(1 - PD_j)$), it can migrate to another rating R_k, during the observation period T. There are four possible ways in which an obligor with initial risk rating can change:
   a. The obligor remains in the same risk rating R_j.
   b. The obligor is perceived as less creditworthy and migrates to a worse risk rating R_k $(j < k)$ (rating downgrade).

c. The obligor is perceived as more creditworthy and migrates to a better risk rating R_k ($j > k$) (rating upgrade).
d. The obligor exits the population during the observation period T and is no longer rated (withdrawn rating, not rated, or population exit). In the following we ignore population exits and withdrawals.

Transition to the default state (reflected in the probability of default) depends on the obligor's ability to repay obligations, the economic environment, and credit and market conditions. In contrast, transition to other risk ratings is determined by two different but related processes:

1. *Rating reviews*, where ratings are reviewed periodically by rating analysts, credit officers, risk managers or derived from credit models with a review period that can be idiosyncratic (in response to changing credit or economic conditions or to new material information on the obligor) or regular (in response to internal rating policy requirements such as annual or semi-annual reviews). The probability of a rating review event during the observation period is $F_j(t)$ at time t.
2. *Rating revisions.* The rating review process evaluates the credit quality of the obligor, which can trigger a rating revision with a probability Q_{jk} that the obligor will remain in the same rating R_j or will migrate to a different rating R_k as result of the rating review.

The probability of a rating review $F_j(t)$ and the probability of a rating revision Q_{jk} can depend on specific characteristics of the obligor, the economic environment and credit and market conditions but can also depend on the characteristics of the credit assessment processes and practices and subjectivity of the analysts' reviews. Figure 9.10 sketches the framework described above.

This model provides a more realistic approach to understanding rating migration since it separates observed rating changes into two distinct steps:

1. rating review (driven by F_j), where information is received and evaluated following specific processes and practices
2. rating revision (driven by Q_{jk}), where information is reviewed and acted upon.

Note that, in this approach, the probability of migration to a different rating is $Q_{jk}F_j$ (conditional no default occurs during period T), while the probability of remaining in the same rating is $[Q_{jj}F_j + (1 - F_j)]$. The total probability is adjusted by the fraction of surviving companies $(1 - PD_j)$.

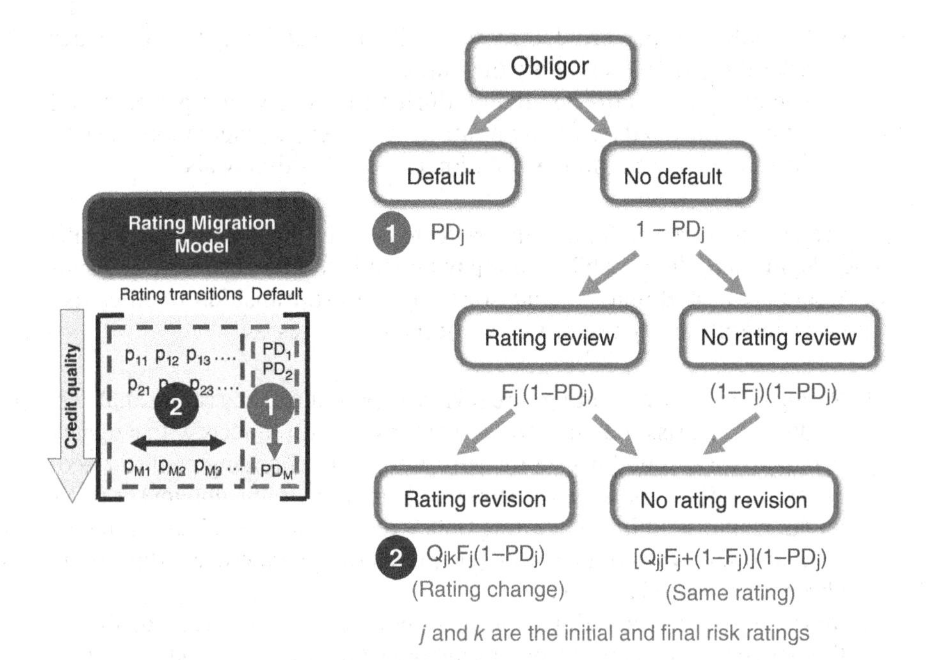

FIGURE 9.10 Credit migration framework: (1) default model; (2) rating migration model.

The model also allows us to understand basic differences in rating migration patterns observed for a wide range of rating processes from those done by rating agencies to market-implied structural models (e.g. Merton-style models), which in principle could produce similar conditional rating migration rates Q_{jk} based on the obligor's credit quality but whose rating review processes can have very different frequencies and, therefore, different probabilities of review F_j, leading to different rating changes and rating migration patterns.

To illustrate, for a 1-year observation period T, a biannual review of risk ratings could have $F_j \sim 0.5$, while a daily equity-based rating model would have $F_j \sim 1$. If rating transition rates Q_{jk} (which represents the actual change in the obligor's rating after the rating is reviewed) are of the same order of magnitude for both rating processes to reflect similar changes in credit quality for the same obligor, then the 1-year rating migration matrix for the biannual rating process would appear less volatile (fewer rating changes) than the 1-year rating migration matrix obtained for the daily equity-based model. In this case, the observed differences in rating migration reflect differences in the rating processes but not necessarily on the underlying credit quality of the obligor.

The description above can explain some of the disparities observed between rating migration derived from market-based models (driven by high-frequency information) and traditional agency ratings, where analysts review ratings less frequently and use forecasts of credit quality over longer periods of time.

In the discussion below, we derive a model for PD_j, F_j and Q_{jk} based on basic assumptions aligned to empirically observed relationships for rating migration patterns. Note that the rating migration probabilities Q_{jk} represent different rating migration patterns that can be derived from three basic assumptions:

1. The number of obligors N_j with an initial rating R_j is conserved during the observation period. That is, companies either default or, if they do not default, they remain in the same rating R_j or migrate to a different rating R_k during the observation period.
2. Regardless of the migration pattern across ratings, the average number of rating subgrade upgrades from an initial rating R_j to a rating R_k is determined by (conditional on) economic conditions.
3. Regardless of the migration pattern across ratings, the average number of rating subgrade downgrades from an initial rating is R_j to a rating R_k is also determined by (conditional on) economic conditions.

Technically, the three assumptions above can be expressed as follows:

$$N_j = \sum_{k=1}^{M} n_{jk} \tag{9.33}$$

$$U_j = \sum_{k=1}^{j} n_{jk}(R_j - R_k) \tag{9.34}$$

$$D_j = \sum_{k=j}^{M} n_{jk}(R_j - R_k) \tag{9.35}$$

The most likely distribution of the number n_{jk} of obligors with rating R_j and final rating R_k can be determined from *maximum entropy* considerations with the fixed constraints of Equations (9.33)–(9.35). More precisely, we maximize the entropy $\log(w)$ for the number of possible configurations w of the rating migration counts n_{jk} with Lagrange multipliers α, β, γ representing Equations (9.33)–(9.35):

$$\begin{aligned} E = \log(\mathrm{w}) &+ \alpha\left(N_j - \sum_{k=1}^{M} n_{jk}\right) + \beta\left(U_j - \sum_{k=1}^{j} n_{jk}(R_j - R_k)\right) \\ &+ \gamma\left(D_j - \sum_{k=j}^{M} n_{jk}(R_j - R_k)\right) \end{aligned} \tag{9.36}$$

Maximum entropy considerations (max(E)) result in the following number n_{jk} of firms with rating R_j and final rating R_k

$$n_{jk} = \begin{cases} N_j e^{\alpha+\beta(R_j-R_k)} & 1 \leq k \leq j \\ N_j e^{\alpha+\gamma(R_j-R_k)} & j < k \leq M \end{cases} \tag{9.37}$$

Here α and β are model parameters that determine the sensitivity of the number of transitions n_{jk} to the change in risk ratings $(R_j - R_k)$. From Equation (9.37) it follows that the most probable distribution of rating upgrades and downgrades is determined by the rating migration frequencies:

$$Q_{jk} = \frac{n_{jk}}{N_j} = \begin{cases} e^{\alpha+\beta(R_j-R_k)} & 1 \leq k \leq j \text{ (upgrades)} \\ e^{\alpha+\gamma(R_j-R_k)} & j < k \leq M \text{ (downgrades)} \end{cases} \tag{9.38}$$

Note that the approach above assumes that default and rating migration rates are completely determined by the initial risk rating R_j or the difference in risk ratings $(R_j - R_k)$, which reflect difference in credit quality for the firm. In principle, there could be other external events leading to default or changes in risk ratings that may not necessarily be reflected in the risk ratings. For example, a sudden change in the market or regulatory environment, unanticipated lawsuits, or reputation impact due to climate risk or ESG issues can make companies susceptible to rating downgrades or default.

More precisely, let the obligor's risk exposure be measured in terms of the likelihood of default PD_j for time horizon T, and let R_j be the resultant risk perception (risk rating). If the absolute perception ΔR_j of two risk exposures differs by a just noticeable amount when separated by a given relative increment of risk exposure $\Delta PD_j / PD_j$, then, when the risk exposures are increased, the perceived risk increment must be proportionally increased for the difference in perception to remain just noticeable. This translates into the following mathematical relation between PD_j and R_j to first-order approximation:

$$\log\left(\frac{PD_j}{1 - PD_j}\right) = \alpha_d R_j + \beta_d \tag{9.39}$$

Here α_d and β_d are model parameters that determine the sensitivity of the default probability PD_j to risk rating R_j.

The probabilities of default and rating migration described above can be modified to accommodate events that are not reflected in the risk ratings. Let p_e be the probability that an independent external event can occur and let q_e be the probability that the event results in the obligor's default regardless of its risk rating (credit quality). For example, due to physical risks such as severe floods or hurricanes, or due to transition risk caused by new regulation or reputation damage related to the obligor's practices. The probability of default adjusted for independent external events is:

$$\widetilde{PD}_j = PD_j(1 - p_e) + p_e q_e \tag{9.40}$$

Similarly, the rating migration frequencies described previously can be adjusted to include unanticipated events unrelated to the credit quality information reflected in the risk ratings, with probabilities p_u and q_u for upgrades, and p_d and q_d for downgrades:

$$\tilde{Q}_{jk} = \frac{n_{jk}}{N_j} = \begin{cases} e^{\alpha+\beta(R_j-R_k)}\,(1-p_u) + p_u q_u & 1 \le k \le j \text{ (upgrades)} \\ e^{\alpha+\gamma(R_j-R_k)}\,(1-p_d) + p_d q_d & j < k \le M \text{ (downgrades)} \end{cases} \tag{9.41}$$

The model parameters in Equation (9.41) can be calibrated for different industry and geography segments using maximum likelihood estimates and leveraging available rating migration histories over multiple business cycles and macroeconomic and market factors driving business and credit cycles (e.g. GDP, unemployment rates, or excess credit demand).

Observed changes in default rates and rating migration rates seem aligned to credit downturns as described in previous sections. For illustrative purposes, Figure 9.11 shows Equations (9.39) and (9.40) for PDs and realized

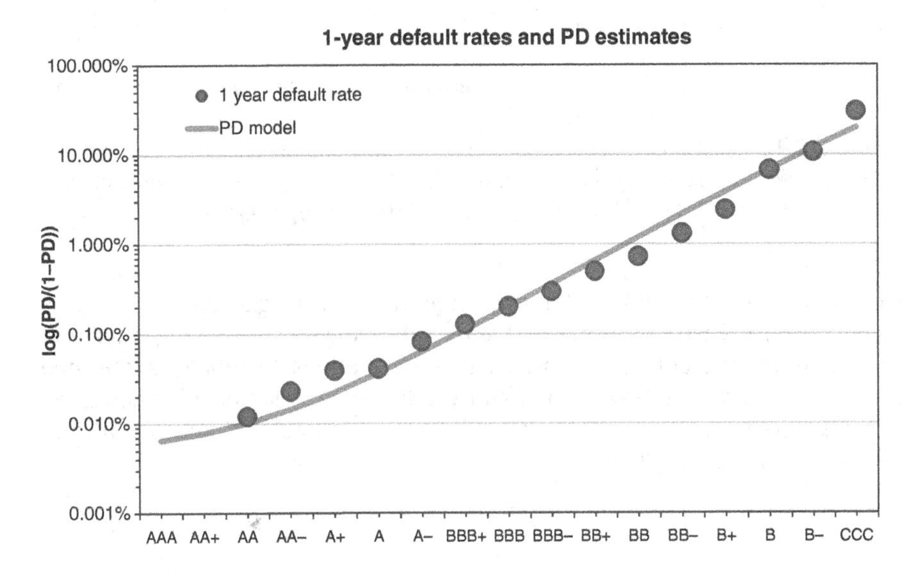

FIGURE 9.11 Observed 1-year agency default rates by rating category (symbols) aggregated across industries and Equations (9.39) and (9.40) (solid line) with $p_e = 0.005\%$ and $q_e = 1$.

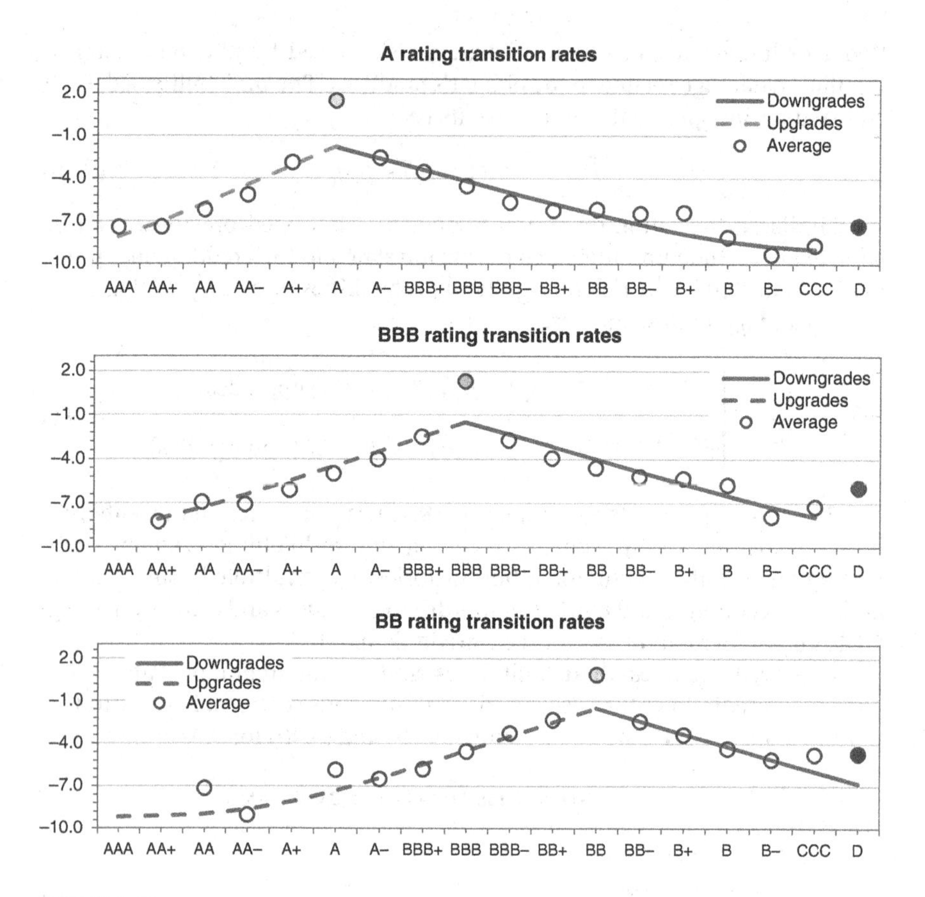

FIGURE 9.12 Observed 1-year corporate rating transition rates (logarithmic scale) from different initial ratings to other ratings (symbols) aggregated across industries and Equation (9.41) (solid line) with $p_u = p_d = 0.01\%$ and $q_u = q_d = 1$.

corporate default rates and rating migration rates aggregated over time and across industries. Similarly, Figure 9.12 shows 1-year rating transition rates from different initial ratings to other ratings (symbols) aggregated across industries and over time, and the log-linear relationships described in Equations (9.39) and (9.41) (solid lines). Additional technical details are provided in later sections.

NOTES

1. That is, the obligor risk rating cannot be better than the sovereign risk rating.
2. For example, rating agency studies of default risk have attempted to control for the severity of losses by considering only senior unsecured or estimated senior unsecured ratings; however, this adjustment is only statistical in nature.
3. The assigned risk rating is $Y(X) = R + \varepsilon$, where ε is an error term that ideally has mean zero and minimum variance. This error is the result of lack of complete information.
4. The total number of companies can be adjusted for the marginal number of rating withdrawals and companies exiting the population in each period in proportion to the defaulting and non-defaulting populations. This provides a consistent default rate estimate with a neutral view on rating withdrawals and company exits.
5. This covers the history of ratings over the numerically modified scale over multiple credit cycles.
6. Years with zero transition rates are not represented in the figures.

REFERENCES

Altman, E.I. (1968). *Financial Ratios, Discriminant Analysis, and the Prediction of Corporate Bankruptcy. Journal of Finance*, September: 589–609.

Burton, G.A. (1972). *Experimental Psychology*. John Wiley, New York, pp. 137–144.

Cantor, R. (2001). Moody's Investors Service Response to the Consultative Paper Issued by the Basel Committee on Bank Supervision 'A New Capital Adequacy Framework'. *Journal of Banking & Finance* 25: 171–185.

Cantor, R., Fons, J.S., Mahoney, C.T., and Pinkes, K.J.H. (1999). *The Evolving Meaning of Moody's Bond Ratings*. Moody's Rating Methodology, August.

Caouette, J.B., Altman, E.I., and Narayanan, P. (1998). *Managing Credit Risk*. John Wiley and Sons, New York.

Charan, R., and Useem, J. (2002). Why Companies Fail. *Fortune*, May 27.

Crouhy, M., Galai, D., and Mark, R. (2001a). Prototype Risk Rating System. *Journal of Banking & Finance* 25: 47–95.

Crouhy, M., Galai, D., and Mark, R. (2001b). Credit Rating Systems. In: *Risk Management*, McGraw-Hill, New York, pp. 259–313.

Duffie, D., and Singleton, K. (1999). *Credit Risk for Financial Institutions: Management and Pricing*. Graduate School of Business, Stanford University.

Giesecke, K., Longstaff, F.A., Schaefer, S., and Strebulaev, I. (2009). *Corporate Bond Default Risk: A 150-Year Perspective*. Stanford University, Working Paper, pp. 1–34.

Griep, C., and De Stefano, M. (2001). Standard & Poor's Official Response to the Basel Committee's Proposal. *Journal of Banking & Finance* 25: 140–169.

Keenan, S.C., Hamilton, D.T., and Berthault, A. (2000). *Historical Default Rates of Corporate Bond Issuers, 1920–1999*, Moody's Special Comment, January.

Keenan, S.C., and Sobehart, J.R. (2002). A Behavioral Model of Rating Assignment. In: M. Ong (Ed.), *Credit Ratings: Methodology, Rationale and Default Risk*. Risk Books, London.

Krahnen, J.P., and Weber, M. (2001). Generally Accepted Rating Principles: A Primer. *Journal of Banking & Finance* 25: 3–23.

Lyon, P. (2002). *Changes Afoot at Rating Agencies*. Risk, March: S10–S11.

Mahoney, C. (2002). "*The Bond Rating Process: A Progress Report*. Moody's Investors Service, Global Credit Research, February.

Murray, C., Cantor, R., Collins, T., Hu, C.M., Keenan, S.C., Nayar, S., Ray, R., Ruttan, E., and Zarin, F. (2000). *Promoting Global Consistency for Moody's Ratings*. Moody's Rating Methodology, May.

Sobehart, J.R., and Mogili, D. (2007). *Modeling Credit Migration Citigroup Global Markets. Quantitative Credit Analyst*, April, pp. 34–41.

Stigler, S.M. (1986). *The History of Statistics: The Measurement of Uncertainty before 1900*. Belknap Press, Harvard University Press, Cambridge, MA, pp. 236–261.

Stumpp, P, Arner, R, Mahoney C., Perry, D., Hilderman, M., and Madelain, M. (2001). *The Unintended Consequences of Rating Triggers*. Moody's Investors Service, Special Comment, December.

Thompson, W.A. (1969). *Applied Probability*. Holt, Rinehart and Winston, New York, pp. 85–89.

Zarin, F. (2000). *Promoting Global Consistency for Moody's Ratings*. Moody's Rating Methodology, May.

CHAPTER 10

Pillar 2: Demand for Credit: The Value of Financial Information

THE VALUE OF FINANCIAL INFORMATION: BALANCE SHEET, INCOME STATEMENT, AND STATEMENT OF CASH FLOWS

The assessment of credit risk for borrowers and debt issuers relies on risk rating and credit scoring methodologies and the estimation of the loss likelihood and loss severity, which often combine quantitative and qualitative approaches. While a range of factors can affect a company's credit assessment, consideration of the fundamental information of the company and its business environment, together with a detailed examination of a company's balance sheet, income statement, statement of cash flows, and market information, remain critical components of any analytical risk assessment framework (Helfert, 1967; Altman, 1968; Foster, 1986). This is important for assessing the impact of climate risk on credit risk as companies, their competitive environment and market reaction can be affected by physical and transitions risks.

Some of the credit risk models described in the literature and in commercial vendor documentation have been advertised as capable of offering wider, more timely and more cost-effective coverage than teams of credit analysts. However, reliance on models to trade and manage credit risk carries significant risk itself. Models are susceptible to errors: from incorrect assumptions about the behavior of companies and markets to inaccurate estimation of volatility and asset correlation or reliance on inputs that are not directly observable and must be estimated or implied. For example, market-based models of default risk are often derived under the idealized assumption of perfect capital markets but, in practice, market imperfections and model simplifications lead to substantial and persistent discrepancies between model predictions and the way markets behave (Mackay, 1841; Cootner, 1964; Malkiel, 1973; Kindleberger, 1978; Grossman and Stiglitz, 1980; Kahneman et al. 1982; Ingersoll, 1987; O'Hara, 1995; Kent et al., 1998;

Shleifer, 2000; Lo and MacKinlay, 2002). Furthermore, there is a range of behavioral effects and market imperfections that can limit the use of market information for credit assessment (Samuelson, 1965; Fama, 1970, 1998; Grossman and Stiglitz, 1980; Shleifer and Summers, 1990; Banz, 1981; Shiller, 1981, 2000; DeBondt and Thaler, 1985; Allan, 1986; Aumann, 1987; DeBondt, 1987; DeLong, 1990, 1991; DeLong et al., 1990; White, 1990; Lee et al., 1991; Conrad and Kaul, 1993; Jegadeesh and Titman, 1993; Barberis et al., 1998; Hong and Stein, 1999; Shumway, 1999; Schwartz and Moon, 2000; Sobehart and Farengo, 2002, 2003; Sobehart, 2003, 2010, 2012).

Before we introduce specific technical details for modeling credit risk, we discuss the scope and quality of the information available to determine a company's financial health. Among the most common accounting reports used for determining a company's credit quality are: (1) the balance sheet, (2) the income statement, and (3) the statement of cash flows, which we describe below. Note, however, that our description focuses primarily on their use for building credit risk models as opposed to accounting and financial reporting practices.

THE BALANCE SHEET

A company's balance sheet describes the financial resources that are under a company's control on a specified date. It consists of three major sections:

1. **Assets:** valuable rights owned by the company.
2. **Liabilities:** funds that have been provided by outside creditors in exchange for the company's promise to make payments or to provide services in the future.
3. **Owners' equity:** funds that have been provided by the company's owners.

A simple balance sheet for a hypothetical company "Any-Company" is shown in Table 10.1. The list of assets shows the way the company's resources are lodged, and the lists of liabilities and the owners' equity indicate where these resources have come from.

The company's assets are usually divided into *current assets* and *non-current assets*. Current assets include cash and marketable securities, accounts receivable (money owed to the company), inventories (raw materials, work in process, supplies, and finished goods ready to be sold) and other assets that are expected to be consumed or can be readily converted into cash during the next operating cycle of production, sales, and payment collection.

TABLE 10.1 Any-Company: the balance sheet ($mm)

Balance Sheet			
Assets ($mm)		**Liabilities ($mm)**	
Current assets	1892	*Current liabilities*	632
Cash	76	Accounts payable	198
Marketable securities	400	Short term loans and notes payable	176
Accounts receivable	576	Other current liabilities	258
Inventories	794		
Other current assets	46		
Non-current assets	1210	*Non-current liabilities*	170
Property, plant, and equipment	1056	Long-term debt	170
Other assets	154	Deferred income taxes	0
		Owners' equity	2300
		Paid-in capital	290
		Common stock	280
		Preferred stock	10
		Retained earnings	2010
Total assets	3102	*Total liabilities and equity*	3102

Non-current assets may include non-current receivables, fixed assets (such as land and buildings), and long-term investments.

Similarly, the company's liabilities are divided into *current liabilities* and *non-current liabilities*. Current liabilities include accounts payable (money the company owes to suppliers and others in the normal course of business), to employees (wages payable), or to governments (taxes payable). Short-term obligations such as commercial paper and notes payable to creditors are also included among the current liabilities. Non-current liabilities consist primarily of amounts payable to holders of the company's long-term bonds or loans, and other items such as obligations to employees under company pension plans.

The credit assessment process focuses primarily on the company's ability to meet obligations, determine the true worth of its assets and the nature of its liabilities, its ability to withstand setbacks from internal and external sources, its ability to get funding when needed, its overall financial condition and its operating efficiency to earn a fair return on its investments. Understanding the quality and nature of the company's assets and liabilities can help identify excessive buildups of inventories, redundant cash lying idle, misaligned short-term and long-term liabilities or the ability to convert assets into cash for supporting business operations. Note that a significant portion of a company's assets and liabilities are non-tradable or, at least, very illiquid, and subject to estimation uncertainty. This contrasts with the unrealistic picture of perfectly liquid assets and liabilities that is assumed in structural models of credit risk based on market information. We will discuss these issues in more detail in later chapters.

One important concept related to assets and liabilities is *working capital* (or net current assets), which is simply current assets minus current liabilities and reflects the excess of assets that can be readily converted to cash for supporting the company's operations:

$$WC = CA - CL \tag{10.1}$$

The relationship between current assets and current liabilities is usually expressed in terms of the *current ratio* R_C, which is the ratio of current assets CA to current liabilities CL:

$$R_C = \frac{CA}{CL} \tag{10.2}$$

A current ratio of 2 or higher ($R_C \geq 2$) is usually desirable although lower ratios could still reflect financial health. Note, however, that a large current ratio does not guarantee financial strength to meet current obligations or the ability to quickly turn current assets (e.g. inventories) into cash. Furthermore, an excessively large R_C value could indicate that the company's assets may not be deployed optimally.

Another related ratio is the *acid test ratio* R_{AT} (or liquidity ratio), which is the ratio of cash and cash equivalents (marketable securities) CE to current liabilities CL:

$$R_{AT} = \frac{CE}{CL} \tag{10.3}$$

The acid test reflects the cash and cash equivalent assets that can be used directly to support the company's operations. The acid test ratio also measures a company's ability to meet obligations using current assets. However, it concentrates on assets that are strictly liquid, whose value is fairly known and are readily available for use.

Separately, the value of receivables (money owed to the company) can be assessed by relating the accounts receivables to the sales from which they arose. This assessment is usually expressed in terms of the number of days of sales represented by the receivables, or *collection period*. The collection period T_{CP} for the accounts receivable AR generated by sales S over the financial reporting period T (in days) is:

$$T_{CP} = \frac{AR}{(S/T)} = \left(\frac{AR}{S}\right)T \tag{10.4}$$

The promptness with which receivables are collected is an indicator of the quality of the receivables and the company's collection effectiveness compared to the credit terms granted to customers. A significant deviation of T_{CP} toward a longer collection period (slower collection), could be a warning signal for the management of current assets and the ability to collect payments.

A similar concept can be applied to accounts payable (money the company owes to suppliers and others) relating them to the company's purchases for the reporting period. The assessment is usually expressed in terms of the number of days of purchases represented by the payables, or *payable period*. The payable period T_{PP} for the accounts payable AP generated by purchases P over the financial reporting period T (in days) is:

$$T_{PP} = \frac{AP}{(P/T)} = \left(\frac{AP}{P}\right) T \qquad (10.5)$$

The promptness with which payables are paid compared to the credit terms granted to the company by its suppliers and others is an indicator of the company's effectiveness for managing payments. A significant deviation of T_{PP} toward a longer payment period, could be a warning signal for the management of current liabilities and the ability to pay bills. The cost of goods sold is usually used as a crude approximation for the company's purchases. Note, however, that the cost of goods sold may reflect labor, repairs, services, and other items beyond the cost of the raw materials needed to build inventory. Also, accounts payable may include payments for purposes other than basic purchases. Thus, the payable period is usually less insightful than the collection period for accounts receivable.

For manufacturing companies, the analysis of the inventory account is relevant for assessing the value of the company's inventory and its size relative to the sales volume it supports. Accurate appraisals of the true value of a company's inventory are rarely available short of a detailed independent review and verification. To overcome this issue, judgment is usually made based on the size of the inventory account relative to current assets, total assets, volume of sales or cost of goods sold during the reporting period. The cost of goods to inventory turnover ratio R_{CT} expresses the number of times the inventory account IA is turned over through business operations to cover the cost of goods sold ($COGS$), which measures roughly the size of the total inventory produced during the reporting period:

$$R_{CT} = \frac{COGS}{IA} \qquad (10.6)$$

In general, the higher the turnover ratio, the better the performance for operating with a relatively small commitment of funds, indicating that the inventory must be relatively current. Note, however, that a high turnover ratio could also highlight inventory shortages and limited ability to produce goods to meet demand.

Similarly, the sales to the inventory turnover ratio R_{ST} expresses the number of times the inventory account IA is turned over through business operations to cover the sales (S):

$$R_{ST} = \frac{S}{IA} \tag{10.7}$$

This ratio represents the level of inventory investment required to support the level of sales and current capital needs. Additional financial ratios are usually defined for assessing different items of current and non-current assets and liabilities.

In addition to the company's assets and liabilities, Table 10.1 shows the breakdown of the *owners' equity* (or *net worth*), which represents the stockholders' interest in the company. The owners' equity is the amount of money that has been invested directly into the company and all earnings reinvested back into the company up to date of the balance sheet. In the United States, the owners' equity of a company is usually divided between (1) paid-in capital and (2) retained earnings. Paid-in capital represents the amounts paid to the company in exchange for shares of the company's preferred and common stock. The amount of retained earnings is the difference between the amounts earned by the company in the past and the dividends that have been distributed to the stockholders. In countries that follow different accounting practices, the breakdown of the owners' equity may distinguish between those amounts that cannot be distributed except as part of a formal liquidation of all or part of the company (capital and legal reserves) and those amounts that are not restricted in this way (free reserves and undistributed profits).

Several financial ratios can be used to express the balance between ownership funds and borrowed funds that provide the funding resources for the company (or *capital structure*). These debt ratios provide basic insight into the size and quality of the cushion provided by ownership funds, which lenders and investors rely upon to weather periods of economic stress, meet short-term and long-term obligations, and absorb losses from the company's operations, reductions in asset values or poor business performance. While a small cushion of ownership funds reveals potential risks for creditors, an excessively large cushion may not always be in the best interest of the company and its owners (stockholders) since the company may have favorable risk characteristics that could benefit from the use of low-cost debt to improve profits.

The capital structure of the company reflects the total money invested in the company, including short-term debt STD, long-term debt LTD and owners' equity E (or net worth). Four frequently used debt ratios are: (1) total debt to total assets, (2) long term debt to total capitalization, (3) net worth to capitalization, and (4) net worth to total debt.

The ratio of total debt to total assets compares the company's total debt *TD* to its total assets *TA*, and provides the fraction of assets funded by all the company's creditors.

$$R_{TD} = \frac{TD}{TA} \tag{10.8}$$

Here $TD = STD + LTD + DTX$ is the total debt, which includes short term debt *STD*, long-term debt *LTD* and deferred income taxes *DTX* that are taxes owed but are not due to be paid until a future date. A more selective measure of the company's debt structure is the ratio of long-term debt *LTD* to the company's capitalization *C*, which does not include short-term debt.

$$R_{LTD} = \frac{LTD}{C} \tag{10.9}$$

Here $C = LTD + E$ is the company's capitalization, *LTD* is the long-term debt and *E* is the owners' equity (net worth), which includes common and preferred stock, retained earnings and other surplus. Similarly, the ratio of equity (net worth) to capitalization is:

$$R_E = \frac{E}{C} \tag{10.10}$$

Finally, the ratio of net worth *E* to total debt *TD* provides with the relative size of equity and debt.

$$R_{ED} = \frac{E}{TD} = \frac{E}{(STD + LTD + DTX)} \tag{10.11}$$

Note that these ratios measure the company's reliance on equity from the point of view of the creditors who provide long-term financing.

From Table 10.1, the ratios $E/C = 93\%$ and $LTD/C = 7\%$ for AnyCompany show that the company is managed in a conservative way, given its small amount of debt relative to the owners' equity. This provides the company's management with greater flexibility during periods of difficult economic conditions but limits the company's ability to leverage external sources of funds for funding business operations.

Both market participants and creditors may adopt common principles to assess the company's credit quality, for example, measuring assets at their value to the company's owners. This amount depends on what the company expects to be able to do with the assets, and it ordinarily reflects forecasts of the cash inflows the company will receive in the future. When cash inflows are expected to be delayed, the value of the assets is less than the anticipated cash flows. For example, if the company must pay lenders interest at the rate of 10% per year for the purchase of a new asset needed for production (which is expected to last for only one year), an investment of $100 in the asset today will not be worthwhile unless it will return at least $110 over one year ($100 plus $10 of interest). In this example, $100 is the present value of the right to receive $110 for the asset one year later. The present value is the maximum amount the company would be willing to pay for a future inflow of cash after deducting interest on the investment at a specified rate for the time the company has to wait before it receives its cash.

In simpler terms, asset values depend on at least three key factors:

1. the amount of the anticipated future cash flows
2. their timing
3. the level of interest rates.

The higher the expectation of cash inflows, the shorter the timing or the lower the interest rate, the more valuable the asset will be. This is the meaning of the "value of the company's assets" required in structural models such as the Merton model and its extensions.

This (unobservable) "market value of assets" could be very different from the book value in reported financial statements as accountants are traditionally reluctant to accept value as the basis of asset measurement in going concern assessments. Although monetary assets such as cash or accounts receivable are usually measured by their value, most other assets are measured at their cost. In general, accountants, investors and lenders could find it difficult to verify market value estimates if they were based on a different valuation approach. The balance sheet does not show how much the company's assets are worth but how much the company has invested in them.

Furthermore, the book value of the assets can provide a rough benchmark for their liquidation or replacement value. Note that the historical cost of an asset is the sum of all expenditures the company made to acquire it, although this amount is not always easily measurable. For example, if a company introduces a new process for manufacturing products, which requires support from other units within the company, then judgment must be used in determining how much of the costs of the supporting activities should be placed on the balance sheet as part of the "capitalized" cost of the new process.

The issues above highlight that there is always uncertainty in the interpretation of a company's balance sheet for assessing its credit quality. This uncertainty affects both institutional lenders who extend credit to the company and investors, who value the company's assets, sustainability of revenues and growth prospect.

THE INCOME STATEMENT

A company's success depends primarily on the way it uses its assets to produce goods and services and is usually measured by the amount of profit (measured by the company's net income) it earns in each period, where net income is revenues minus expenses and taxes. The company's income statement shows how the net income for that period was derived. For example, Table 10.2 shows the company's net sales revenues for the period: the assets obtained from customers in exchange for the goods and services that constitute the company's stock-in-trade. Table 10.2 summarizes the company's revenues and expenses for the selected period: the assets that were consumed while the revenues were being created. The costs and expenses are usually broken down into several categories indicating how the assets were used, starting with the cost of the merchandise that was sold during the period and continuing down through the payment of interest expenses and, finally, the provision for income taxes.

The recognition of income is reflected through two different estimates:

1. *Revenues*, representing the value of the cash that the company expects to receive from customers (either as actual cash payments or as promised cash payments at a future date).
2. *Costs and expenses*, representing the resources that have been consumed in the creation of the revenues.

TABLE 10.2 Any-Company: income statement ($mm)

Income Statement	
Revenues	4504
Sales to customers	4450
Interest income	24
Other income	30
Costs and expenses	3890
Costs of goods sold	2200
Selling, general and administrative expenses	1510
Depreciation	138
Interest expenses	35
Other expenses	7
Non-recurring items	6
Proceeds from the sale of property	6
Earnings after taxes	368
Earnings before taxes	620
Less federal, state, and foreign taxes	252
Net income	368

Revenue estimation requires good judgment and a deep understanding of the industry, business, and regulatory environment. One of the key problems with revenue estimation is the determination of the level of competition and acceptance of the company's products and services, and the fraction of gross sales impacted by unpaid bills ("bad debt") or because customers returned defective products or declined services. In contrast, cost and expense estimates are generally based on the historical costs of the resources consumed and the costs of doing business -such as cost of goods sold, administrative and general costs or interest expenses. As a result, net income can be viewed as the difference between the value received from the use of resources and the cost of the resources that were consumed in the process. The estimation of revenues and expenses (and net income) is particularly important for assessing the impact of climate risk and ESG issues as changes to the regulatory environment and policies, incremental carbon taxes, or restrictions to high carbon-emission activities can result in higher company expenses and changes in consumer behavior, impacting the demand for products and services and, ultimately, the company's revenues.

Another key issue for income estimation is determining the portion of the cost of an asset that has been consumed during the reported period. Some assets give up their services gradually rather than all at once. The cost of the portion of these assets that the company uses to produce revenues in any period is that period's depreciation expense. The amount reported for these assets on the balance sheet is their historical cost less an allowance for depreciation, representing the cost of the portion of the asset's anticipated lifetime services that has already been used. To estimate depreciation, the company must estimate both how long the asset will continue to provide useful services and how much will be used up in each period.

There are two popular methods for depreciation:

1. straight-line depreciation
2. declining-charge depreciation.

In the straight-line approach, the same amount of depreciation is recognized each period, while in the declining-charge approach, more depreciation is recognized during the early years of life than during the later years. This reflects the assumption that the value of the asset's service declines as it ages and eventually becomes obsolete.

For companies whose source of revenue is primarily the sale of products, a similar situation occurs for the estimation of the cost of goods sold, which is the sum of the cost of the beginning inventory and the cost of goods purchased in the reported period. However, this sum must be divided between the cost of goods sold and the cost of the ending inventory. When inventory purchase prices are rising (e.g. in an inflationary economy), last-in-first-out (LIFO) inventory costing keeps many gains from the holding of inventories out of net income. If purchases of goods equal the quantity sold, the cost of goods sold will be measured at the higher current prices, and the ending inventory will be measured at the lower prices shown for the beginning-of-period inventory. The difference between the LIFO inventory cost and the replacement cost at the end of the reporting period is an unrealized gain.

The bottom portion of the income statement in Table 10.2 also reports the effects of events that are outside the usual flow of operating activities, or non-recurring items.[1] A non-recurring item refers to a financial entry that is considered infrequent or unusual and is unlikely to happen again. The illustrative example in Table 10.2 shows the sale of property for more than the original purchase price. Because this revenue was not part of the company's normal operations, the sale price and costs of the sale were reported separately from regular operating revenues and expenses. Furthermore, the income statement shows only a single entry for non-recurring items representing the net gain on the asset sale.

Understanding the relationship between the company's earnings and the funding and investment commitments required to attain these earnings is one of the most fundamental issues for credit assessments. Below we present several financial ratios frequently used for describing the company's profitability.

The ratio of earnings before interest and taxes (*EBIT*) to total assets (*TA*) measures the company's ability to generate raw earnings relative to all its assets.

$$R_{EBIT} = \frac{EBIT}{TA} \tag{10.12}$$

A related ratio (*return on assets*) replaces *EBIT* with net income *NI* to reflect the remaining income available to the company's owners after paying interest on funds provided by creditors and paying owed taxes.

$$R_{ROA} = \frac{NI}{TA} \tag{10.13}$$

A similar ratio (*return on equity*) focuses on the earnings power of the owners' investments that is calculated as the ratio of net income to the owners' equity (net worth).

$$R_{ROE} = \frac{NI}{E} \tag{10.14}$$

By adjusting the owners' equity to reflect only tangible net assets (that is, subtracting intangible assets such as goodwill, patents, and similar assets from net worth) leads to the profitability ratio on tangible equity (*return on tangible net worth*).

$$R_{TNW} = \frac{NI}{TNW} \tag{10.15}$$

The return on tangible net worth measures the ability of net productive assets to generate earnings. Similarly, the company's efficiency of operations can be measured by the relationship between company's earnings before income and taxes and sales:

$$R_{TNW} = \frac{EBIT}{S} \tag{10.16}$$

Additional ratios can be constructed from the balance sheet and income statement for measuring different aspects of a company's operations.

The income statement is usually supplemented with a statement that shows the change to the company's retained earnings during the reported period (Table 10.3). Net income increases retained earnings, while net operating loss or the distribution of cash dividends reduces it. The hypothetical company in the example above had positive net income (Table 10.2) and distributed a portion of it as dividends to the shareholders during the year. The remaining income increased the retained earnings reported on the end-of-period balance sheet (Table 10.1).

While the income statement provides valuable information on the company's financial health, accounting income does not reflect all the company's gains or losses or increases or decreases in the market value of its assets. Furthermore, financial statements reflect a company's past and present but what concerns lenders and investors making investment decisions is the company's *future*. To illustrate, the launch of a successful new product or service may increase the company's future cash flows. The anticipated increase in future income makes the company more valuable today. However, this anticipated future income is not reflected in the income statement or the balance sheet,

TABLE 10.3 Any-Company: change in retained earnings ($mm)

Change in Retained Earnings	
Net income	368
Less cash dividends	98
Preferred stock dividends	4
Common stock dividends	94
Change in retained earnings	270

as they have not been realized yet. Accountants and banking supervisors have been reluctant to accept this type of asset valuation (key to most market-based credit models) since it would rely heavily on assumptions of what could happen in the future, which may never materialize in practice and may not be readily susceptible to independent verification. Furthermore, financial reporting is transactional in nature and recognizes income that can be substantiated from actual transactions that take place when services are provided to customers, goods are delivered to customers or when customers are billed.

THE STATEMENT OF CASH FLOWS

The statement of cash flows shows how changes in the company's balance sheet and income statement accounts affect the company's cash and cash equivalents and its ability to pay bills, breaking the analysis down to basic business activities carried out by the company. More precisely, the statement of cash flows is concerned with the inflows and outflows of cash for three major business activities:

1. operating activities
2. investing activities
3. financing activities.

Cash flows from *operating activities* measure the cash generated by a company's business operations to determine whether a company can produce sufficient cash flow to cover expenses and pay debts. Cash flows from *investing activities* measure the amount of cash generated from activities such as the purchase or sale of physical assets and investment securities. Cash flows from *financing activities* measure the net flows of cash used for funding the company, including paying dividends and issuing or retiring equity and debt.

The cash flow information helps investors and lenders to evaluate the company's liquidity and ability to pay bills when they come due. Furthermore, the statement of cash flows reveals the quality of a company's earnings and whether there is sufficient cash to operate (that is, separating actual cash flows received by the company from promised payments and accounting treatment of asset depreciation). Despite its fundamental nature for identifying a company's liquidity problems to service its debt, detailed structural models based on cash flows analysis have been sparse or limited. This contrasts with the proliferation of market-based models of credit risk (such as the Merton model and its extensions), which rely primarily on a market view of the company's leverage (assets and liabilities). In fact, many of the "cash flow" structural models

TABLE 10.4 Any-Company: sources and uses of funds ($mm)

Source and Uses of Funds	
Source of funds – Cash inflows	554
Funds from operating activities	516
Net income	368
Depreciation	138
Other sources	14
Less non-recurring items (Sale of property – net of paid taxes)	4
Funds from other sources	38
Increase in long-term debt	26
Proceeds from employees' stock options	8
Proceeds from sale of property (net of paid taxes)	4
Uses of funds – Cash outflows	410
Additions to property, plant and equipment	272
Cash dividends	98
Decrease in long-term debt	20
Other uses	20
Change in working capital	144

found in the literature are simply asset-based models with a default boundary that reflects a cash outflow such as interest payments (Leland, 1999). We return to this issue in later chapters.

Table 10.4 illustrates the sources and uses of funds for Any-Company from Tables 10.1 and 10.2. Cash was received from different sources, including payments from customers on goods and services (through net income), cash from new long-term debt, proceeds from stock options and the sale of property (non-recurring item). Note that the income statement included only the net gain on the sale of property (the difference between the sale proceeds and the amount at which the investment had been shown in the balance sheet before it was sold). Since the net income in Table 10.2 included the net gain amount of the non-recurring item, the company couldn't include the cash proceeds (net of taxes) from the sale of the property in the funds from regular operating activities. Instead, Any-Company subtracted the non-recurring item from net income and reported the amount (net of taxes) under cash from other sources. Cash was paid for capital expenditures (property, plant, and equipment), for distributing dividends and for paying down earlier long-term debt.

Note that the sources and uses of funds in Table 10.4 describe cash inflows and outflows combining multiple activities of the company. The cash inflows

TABLE 10.5 Any-Company: statement of cash flows for operating, investment, and financing activities (indirect method) ($mm)

Cash Flows	
Net cash flow from operating activities	516
Net income	368
■ Depreciation and amortization	138
■ Adjustments to net income (non-recurring items)	−4
■ Decrease (increase) in accounts receivable	
■ Increase (decrease) in liabilities (accounts payable, taxes payable)	
■ Decrease (increase) in inventories	
■ Increase (decrease) in other operating activities	14
Net cash flows from investing activities	−268
Capital expenditures	−272
Investments	
Other cash flows from investing activities	4
Net cash flows from financing activities	−104
Dividends paid	−98
Sale (repurchase) of stock	
Increase (decrease) in debt	6
Increase in long-term debt	26
Decrease in long-term debt	−20
Other cash flows from financing activities	−12
Proceeds from employees' stock options	8
Other uses of funds	−20
Effect of exchange rate changes	
Net increase (decrease) in cash and cash equivalents (Cash flow)	144

and outflows can be separated explicitly for each one of the operating, investing, and financing activities to determine their contributions to the company's liquidity. Table 10.5 illustrates the statement of cash flows including the operating, investment, and financing cash flows for the hypothetical company, Any-Company.

The statement of cash flows differs from the income statement in many respects since it excludes non-cash items such as depreciation, write-offs on bad debts, deferred income taxes and sales on credit (promised payments) where the actual receivables have not been collected. For illustration, in Table 10.5, cash was received from new long-term debt while cash was paid to stockholders as dividends, neither of those figured in the income statement. The payment of dividends can be viewed as a transfer of wealth from debtholders to stockholders. In practice, debtholders attempt to protect themselves by including protective covenants in the indenture agreement of debt instruments. These provisions may limit the amount of dividends paid to stockholders and, to some extent, control investment and production decisions of management by specifying the purpose of the proceeds of the debt.

Cash was also paid to purchase equipment (capital expenditures). This amount was added to the *plant and equipment assets* but was not subtracted from current revenues because it would be used for many years, not just during the current reporting period.

Cash from operations is not the same as net income because not all revenues are collected in cash. Revenue is usually recorded when a customer receives products or services and either pays for them or *promises to pay* the company in the future (in which case the revenue is recorded in accounts receivable). On the other hand, cash from operating activities reflects the actual *cash collected*, not the inflow of accounts receivable, which are not actual payments to the company but promises of payments at a future date. Similarly, expenses may be recorded without an actual cash payment. For example, accounts payable reflect future promised payments as opposed to realized payments made by the company.

Table 10.5 adds items not requiring immediate cash payment to income (e.g. depreciation) and subtracts items that appear in the income statement but are not part of regular business operations (e.g. the gain on the sale of property). Cash flow analysis can provide helpful insight into the company's profits and ability to finance projects, and it is particularly useful when comparing companies since not all of them depreciate their assets at the same rate or using the same methodologies. Depreciation is a source of capital because it is a non-cash expense in the income statement. Table 10.5 also shows that the company's cash and marketable securities increased during the year. This provides the firm with liquidity to cover its short-term operational needs.

The calculation of cash flows in Table 10.5 started with net income, making adjustments that recognize transactions resulting in actual cash collected or paid. This approach is an "indirect method" for cash flow estimation derived from the income statement. An alternative calculation approach is the "direct method," which calculates the cash flows directly from the assessment of cash inflows and outflows as illustrated in Table 10.6. The benefit of direct calculation is that cash inflows and outflows reflect the sources and uses more clearly, which could be used for building refined credit risk models that can determine whether a company can produce sufficient cash flow to cover expenses and pay debts (*cash shortfall* models). This approach is relevant for assessing the impact of climate risk and ESG issues as changes to the regulatory environment and policies or restrictions to high

TABLE 10.6 Any-Company: cash flows statement of operating, investing, and financing activities (direct method) ($mm)

Cash Flows	
Net cash flow from operating activities	516
Cash generated from operations	787
Cash receipts from customers	4450
Interest income	24
Other income	30
Cash paid to suppliers and employees	−3710
Costs of goods sold	−2200
Selling, general and administrative expenses	−1510
Other expenses	−7
Interest paid	−35
Income taxes paid (removing $2 million for the sale of property)	−250
Increase (decrease) in other operating activities	14
Net cash flows from investing activities	−268
Capital expenditures	−272
Investments	
Other cash flows from investing activities (net of taxes)	4
Net cash flows from financing activities	−104
Dividends paid	−98
Sale (repurchase) of stock	
Increase (decrease) in debt	6
Increase in long-term debt	26
Decrease in long-term debt	−20
Other cash flows from financing activities	−12
Proceeds from employees' stock options	8
Other uses of funds	−20
Effect of exchange rate changes	
Net increase (decrease) in cash and cash equivalents (Cash flow)	144

carbon-emission activities can affect cash flows from operating, investing, and financing activities differently.

FINANCIAL INFORMATION AND UNCERTAINTY

As discussed above, in preparing financial statements companies have several methods for estimating assets, income, and cash flows. Assets may be measured at their past cost or at what they could be sold for now. Also note for large companies that own other companies, financial statements could be consolidated statements, reflecting the total assets, liabilities, owners' equity, net income, and cash flows of all the corporations in the financial group. In cases where companies are not wholly owned by their parent companies, the equity of these minority shareholders in the subsidiary companies is shown separately on the balance sheet.

In addition, many companies do business in multiple markets and geographies with different macroeconomic conditions and, therefore, are exposed to different changes in the prices of goods and services, which have two effects. First, net monetary assets (basically cash and receivables minus liabilities calling for fixed monetary payments) lose purchasing power as the general price level rises (e.g. due to inflation). These losses are not reflected in accounting statements. Second, holding gains measured in nominal currency units may merely result from changes in the general price level and, therefore, they may not represent true improvements in the company's financial condition. The issues discussed above reflect additional uncertainty in the interpretation of the reported financial figures and, therefore, the forecast of the company's financial health for different geographies and industries.

CASH LIQUIDITY AND DEBT CAPACITY

In view of both the many potential distortions and the lack of timeliness in financial reporting (BW, 2001, 2002; WSJ, 2002a, 2002b), academics and practitioners have looked to the company's assets implied from market equity or debt to help measure its creditworthiness (Black and Scholes, 1973; Merton, 1973, 1974). The standard approach to modeling the company's assets is to assume that their value follows a random walk. That is, the value of the company's assets drift randomly over time reflecting information on the present and future capacity to generate revenues. In this idealized framework, a company defaults when the value of its assets falls below a "reorganization" or "default boundary" (usually a function of the company's liabilities). In this framework, market participants evaluate the value of the company's assets and liabilities to determine the likelihood of default, resulting in the observed market equity and debt prices (which are equivalent to financial options on the value of the company's assets). Under additional technical assumptions, the market value of the company's equity and debt can be linked directly to the value of the company's assets. Therefore, the (often unobservable) "value of the company's assets" can be implied from observable equity or bond prices. The implied value of the company's assets is then compared to the default boundary to determine an effective market-implied leverage measure (or distance to default) to determine the company's financial health (Black and Scholes, 1973; Merton, 1973, 1974; Lang and Stultz 1992; Longstaff and Schwartz, 1995; Anderson et al., 1996; Fridson and Jonsson, 1997; Jarrow and van Deventer, 1998; Nandi, 1998; Kealhofer, 1999; Leland, 1999; Sobehart and Keenan, 1999, 2002; Jarrow and Turnbull, 2000; Kao, 2000; Keenan and Sobehart, 2000; Sobehart et al., 2000; Duffie and Lando, 2001; Janossi et al., 2001; Hillegeist et al., 2002; Duffie and Singleton, 2003).

Although assets-based (or equity-based) structural models of credit risk provide powerful insight into the valuation of risky debt, they often introduce unrealistic assumptions to make the problem analytically tractable. Uncertainty as to the default point lowers the predictive power of these models, an effect that gets worse as the default point is approached. Even more important, models based on the company's assets fail to capture one crucial aspect of actual default events: companies do not default on their obligations because their assets are below their liabilities (or, more generally, below a default boundary). In the real world, companies default on their obligations because they exhaust their capacity for borrowing and run out of cash and sources of funds to service their obligations. As the credit quality of the company deteriorates, its liquidity and capacity for borrowing or refinancing can be the determining factor as to whether it defaults or not. There are two commonly used idealized situations for modeling the company's borrowing capacity:

1. **Negligible borrowing capacity:** represented by single balloon payment models such as the European-style option model proposed by Merton (1973; 1974) and the American-style (first passage time) model of Longstaff and Schwartz (1995).
2. **Infinite (costless) borrowing capacity:** represented by models with continuous rollover of debt, such as the perpetual options model proposed by Leland (1999) and Kealhofer (1999) among others.

The negligible borrowing capacity assumption is a worst-case scenario where the borrowing capacity of the company is completely exhausted, and the company must pay the total amount of its debt outstanding at maturity or as contractually agreed. In contrast, the infinite borrowing capacity is an unrealistic benign scenario where the company's debt never matures or can be rolled over at no cost, regardless of the fact its borrowing capacity may deteriorate, and refinancing costs may increase as the value of its productive assets decrease.

Companies neither finance themselves with perpetual debt nor liquidate their assets to repay simple bullet bonds and other obligations if they have other financing options available. They can use debt obligations with finite horizons either with coupons (e.g. corporate bonds) or without coupons (e.g. commercial paper), and roll them over, paying them down or borrowing additional capital as needed. Some of these debt obligations may include sinking funds, call features, convertible features, and early amortization schedules. Companies that borrow to undertake investment projects will tailor the debt payments to the expected cash flows from the projects. Capital-intensive companies that construct long-term plants and equipment will prefer to issue long-term debt rather than rolling over short-term debt. Less capital-intensive companies will prefer borrowing short term. Companies will accommodate changes in their expected cash flows by restructuring debt obligations to support their needs, leveraging their available debt capacity. To illustrate, let's assume a company has a committed line of credit from a bank available for a partial rollover of debt. This type of contingent claim may not appear on the company's balance sheet unless the committed line is used. Under the terms of the contract, the bank will guarantee the supply of funds at an agreed rate plus a fee on the unused balance. A committed line gives the company the ability to fund its needs when it is cheaper to draw on the committed line than to borrow in the open market. This provides the company with a means for managing its capital structure and the likelihood of default on its debt obligations.

The management of the capital structure refers not only to the ratio of debt to equity and the different types of debt but also to the maturity structure of the debt. What portion of total debt should be short-term and what portion long-term? These issues have been addressed extensively in academic literature (see Copeland and Weston, 1992; Ravid, 1996, and references therein). The situation gets more complicated when we include additional variables such as accounts payable, taxes and wages, and the company's option to change the amount of short- and long-term debt. Another important factor is the relationship between the obligor and its creditors, who can extend additional credit based on the company's borrowing capacity. This indicates that the variability of the company's liabilities – due to the uncertainty of the business environment, its financing strategy, and its capacity to borrow and renegotiate obligations – should play a more relevant role in credit risk models.

Also note that the inclusion of industry and geography effects in advanced credit risk models is usually limited, partially due to restrictive modeling assumptions and partially due to the limited number of bankruptcies available to calibrate models. Yet common sense suggests that industry effects should be an important component in default prediction. Different industries face different levels of competition and uncertainty in their business environment and, therefore, the probability of default can differ widely for companies in different industries with otherwise identical balance sheets and income statements. For example, companies that require heavy investment in plants and equipment, such as in the utility and steel industries, may have high financial leverage compared to companies in the service sector, such as accounting companies and financial consulting companies. Therefore, similar reported numbers can imply very different probabilities of default in different industries and geographies.

Furthermore, when a company in financial distress in a concentrated industry needs to dispose of assets, its industry peers may be experiencing similar problems themselves, leading to stranded assets or asset sales at prices below the fair value in best use. Material bankruptcy announcements can also lead to contagion effects involving competitors and suppliers. This issue is extremely important in the context of climate risk and ESG as high-carbon emission sectors, such as thermal carbon extraction and thermal carbon and oil power generation, may result in a significant number of stranded assets and asset fire sales with cascading effects for other sectors of the economy.

The issues described above can be addressed in credit risk models at the expense of losing the simplicity and insight provided by the standard structural models. More realistic approaches would require an intermediate point between the extremes of a fully depleted borrowing capacity and an infinite and costless borrowing capacity, considering the dynamics of borrowing and lending. The company's borrowing capacity is the result of the borrower's ability to generate revenues for servicing its debt and the value of specific assets available for use as collateral, and the lending policy of investors and/or the institution extending credit. Both issues are driven by key accounting information such as the borrower's profitability, liquidity, and capital structure as well as information on the business environment and the borrower's competitiveness. These factors can be affected directly by climate risk and ESG issues, from the impact of low-carbon and net-zero strategies to reputation issues related to the company's environmental and social policies and practices. A generalized framework for debt capacity and a behavioral approach to risk perception and reputation damage is presented in later sections.

CASH SHORTFALL, BUSINESS UNCERTAINTY, AND FINANCIAL DISTRESS

Building credit models that can leverage financial and market information to better identify the default point as a function of the residual debt capacity of the company can provide a more realistic view of companies experiencing financial distress. It seems natural that financial and market information in the balance sheet, income statement and statement of cash flows widely used by credit analysts and loan officers to make credit assessments and credit decisions should play a more significant role in models of credit risk.

Here we discuss a cash liquidity model based on the cash required by the company to support its operations. From Tables 10.1–10.6 we construct the cash surplus/shortfall required to service debt and operating expenses for the next business cycle.

We define the cash shortfall C_{SF} in terms of the operating, investing, and financing activities of the company, and its ability to refinance debt based on its debt capacity:

$$C_{SF} = X + Y + Z + W - I \tag{10.17}$$

Here X is the cash flow from operating activities, Y is the cash flow from investment activities, Z is the cash flow from financing activities, W is the amount of forthcoming debt that could be refinanced, and I is the incremental interest expense due to debt refinancing. The cash shortfall reflects the company's cash flow surplus or deficit to cover existing obligations, adjusted for the company's debt capacity to restructure its debt or take new debt to reduce its short-term default risk. If the company has not exhausted its debt capacity (i.e. its ability to take more debt by leveraging assets as collateral, pledging expected future cash flows, or managing working capital), the existing cash flow deficit could be reduced or rolled forward by refinancing or restructuring its debt, reducing forthcoming liabilities.

A company that refinances its obligations will tailor the debt payments to its expected cash flows. Capital-intensive firms may prefer long-term debt rather than rolling over short-term debt. Less capital-intensive firms may prefer short-term debt. For illustration, consider the simplified case where a company refinances a fraction $0 \leq F \leq 1$ of its short-term liabilities D_{ST} (due at time T_{ST}) to be paid at future time T_F. In this example, the company's long-term debt D_{LT} remains unchanged. The refinanced amount $W = FD_{ST}$ requires an additional interest charge $I = I_F(D_{ST}, F)$ that will be paid at maturity T_{ST} of the original obligation as an up-front refinancing fee. The net effect of refinancing is to reduce the level of the company's short-term liabilities from D_{ST} to $D_F = (1 - F)D_{ST} + I_F$, rolling over an amount FD_{ST} to a future period $T_F \geq T_{ST}$. This is illustrated in Figure 10.1. How much the company can refinance and how much needs to be charged in interest depend on the lender's practices and the expectation of the borrower's ability to repay at future date T_F, which is determined by its remaining debt capacity (Hubbard, Kuttner, and Palia, 1999).

Note that, if we only consider assets and liabilities for estimating the company's debt capacity, the default risk adjusted refinancing expense would be $I_F = FD_{ST} - FV(FD_{ST}, A(T_{ST}) - D_F; T_F - T_{ST})$. Here FV is the default risk adjusted fair value of the refinanced amount FD_{ST}, and $A(T_{ST})$ is the value of the company's assets at time T_{ST} before paying the debt amount D_F. In this example, the default risk adjusted fair value can be derived approximately as the fraction $FD_{ST}/(FD_{ST} + G_{LT})$ of the fair value of a bullet loan of face value $(FD_{ST} + G_{LT})$ and maturity $T_F - T_{ST}$, given the residual value of the

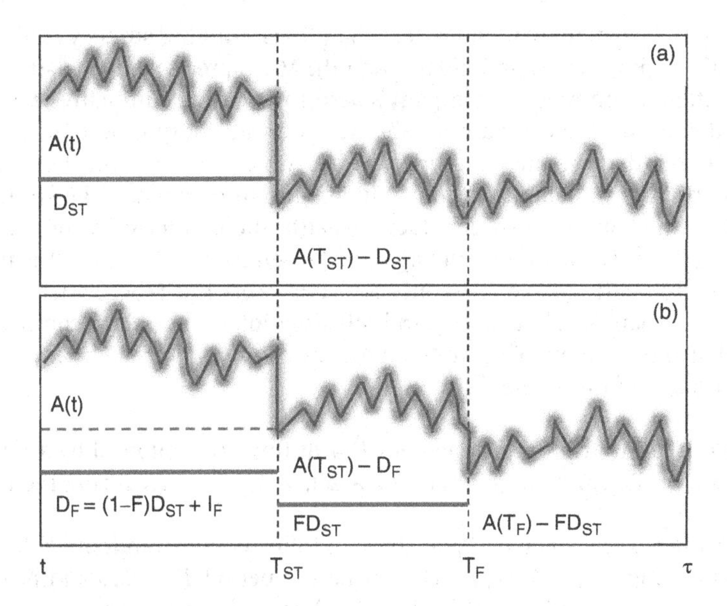

FIGURE 10.1 Debt refinancing to reduce short-term liabilities: (a) single payment; (b) two-period rollover.

company's assets $A(T_{ST}) - D_F$ at time T_{ST}. Here G_{LT} represents any amount of long-term liabilities D_{LT} and interest payments I_{LT} due at T_F. Of course, this may not be the actual interest expense I_F paid by the company since there are other contributions to the company's debt capacity, and the lender can have a different assessment of interest charges from the one described above. Note, however, that this example highlights that companies have alternative options for managing their capital structure and likelihood of default on obligations.

The management of the company's capital structure refers not only to the ratio of debt to equity and the different types of debt but also to the maturity structure of the debt. Deciding what portion of total debt should be short-term and what portion long-term can help manage credit risk and the company's ability to avoid business failure. Furthermore, the company's management has the option of choosing the timing of payments and the best fraction F of the existing debt that needs to be refinanced based on the company's expected cash flows and the impact on shareholders' wealth (Shleifer and Vishny, 1990, 1992; Copeland and Weston, 1992; Ravid, 1996; Hubbard et al., 1999). The adoption of different business strategies impacts refinancing decisions, leading to alternative valuation models. For example, one possible refinancing strategy could be to maintain or improve the performance of the company as a going concern after the full payment of its short-term liabilities. Another strategy could be to maintain or improve shareholders' wealth. More precisely, if the strategy is to maintain or improve the company's performance, the cumulative probability of default on the company's liabilities at an arbitrary time $\tau \geq T_F \geq T_{ST}$ with intermediate refinancing must be *the same or lower than* the cumulative probability of default without refinancing. In contrast, if the strategy is to maintain or improve shareholders' wealth, the market value of equity at time $\tau \geq T_F \geq T_{ST}$ with refinancing must be *equal or greater than* the market value without refinancing. If we only consider assets and liabilities for estimating the company's debt capacity, in both situations, we need to compare the cumulative probability of default and the distribution of the company's assets at an arbitrary future time τ:

1. **no refinancing:** single payment D_{ST} at time T_{ST} followed by a random walk of the company's remaining assets $A(T_{ST}) - D_{ST}$ during the period $\tau - T_{ST}$;
2. **refinancing:** payment D_F at time T_F followed by a random walk of the remaining assets $A(T_{ST}) - D_F$ during the period $T_F - T_{ST}$ and payment $(FD_{ST} + G_{LT})$ at T_F, with a final random walk of the remaining assets $A(T_F) - (FD_{ST} + G_{LT})$ during the period $\tau - T_F$ (Figures 10.1(a) and 10.1(b)).

The cumulative probability of default over period τ for any of the two strategies above, combined with the condition that the company must meet its obligations ($C_{SF} \geq 0$), can be used to determine the refinancing strategy, leading to the refinanced amount $W = FD_{ST}$, interest expense $I = I_F$ and refinancing maturity T_F.

This example illustrates that debt refinancing is a compound-option problem where the first strike-value (D_F) depends on the second strike-value $(FD_{ST} + G_{LT})$ and vice versa. Note that in practice the problem is significantly more complex because management has the option to refinance or retire the company's debt or create new debt with different payment schedules, which generates uncertainty in the assessment of future liabilities (volatility of the company's liabilities). Although at first blush, the number of debt refinancing strategies may seem unlimited, the number of practical alternatives may be limited by access to capital markets or to lenders, refinancing costs, and lengthy credit approval processes, and other relevant factors affecting the company's ability to borrow. Furthermore, the refinanced fraction is a function of the debt capacity, which can be described in terms of key financial ratios $F(P, L, C, \ldots)$, where $P, L, C, \ldots$ represent the measures of profitability, liquidity, capital structure, business efficiency and other financial ratios defined in previous sections on the balance sheet, income statement and statement of cash flows. These risk drivers can vary by industry, business environment, and level of competition. Note the significant contrast with the standard asset-based structural models of credit risk where financial variables and industry and geography effects are claimed to be completely redundant due to unrealistic descriptions of the company's borrowing capacity and the nature of the information reflected in equity and debt prices.

NOTE

1. There are some accounting distinctions between non-recurring items and extraordinary items, which are usually reported separately from operating income, net of taxes.

REFERENCES

Allan, W. (1986). Variance Bounds Tests and Stock Price Valuation Models. *Journal of Political Economy* 94(5): 953–1001.

Altman, E.I. (1968). *Financial Ratios, Discriminant Analysis and the Prediction of Corporate Bankruptcy. Journal of Finance,* September: 589–609.

Anderson, R.W., Sundaresan S., and P. Tychon P (1996). Strategic Analysis of Contingent Claims. *European Economic Review* 40: 871–881.

Aumann, R. (1987). Correlated Equilibrium as an Expression of Bayesian Rationality. *Econometrica* 55: 1–18.

Banz, R. (1981). The Relationship Between Return and Market Value of Common Stocks. *Journal of Financial Economics* 9(1): 3–18.

Barberis, N., Shleifer, A., and Vishny, R. (1998). A Model of Investor Sentiment. *Journal of Financial Economics* 49: 307-343.

Black, F., and Scholes, M. (1973). The Pricing of Options and Corporate Liabilities. *Journal of Political Economy* 81: 637–659.

BW (*BusinessWeek*) (2001). The Fall of Enron. *BusinessWeek,* December 17, 30–38.

BW (*BusinessWeek)* (2002). Restoring Trust in Corporate America. *BusinessWeek, June* 24, 30–35.

Conrad, J., and Kaul ,G. (1993). Long-Term Market Overreaction or Biases in Computed Returns? *Journal of Finance* 48(1): 39–64.

Cootner, P. (1964). *The Random Character of Stock Market Prices.* MIT Press, Cambridge, MA, pp. 17–78.

Copeland, T.E., and Weston, J.F. (1992). *Financial Theory and Corporate Policy.* Addison-Wesley, Reading, MA, pp. 471–472, 480–481.

De Bondt, W.F.M. (1987). Further Evidence on Investor Overreaction and Stock Market Seasonality. *Journal of Finance* 42(3): 557–581.

DeBondt, W.F.M., and Thaler, R. (1985). Does the Stock Market Overreact? *Journal of Finance* 40(3): 793–805.

DeLong, J.B. (1990). Positive Feedback Investment Strategies and Destabilizing Rational Speculation, *Journal of Finance* 45(2): 375–395.

DeLong, J.B. (1991). The Survival of Noise Traders in Financial Markets. *Journal of Business* 64: 1–19.

DeLong, J.B., Shleifer, A., Summers, L., and Waldeman, R. (1990). Noise Trader Risk in Financial Markets. *Journal of Political Economy* 98: 703–738.

Duffie, D., and Lando, D. (2001). Term Structures of Credit Spreads with Incomplete Accounting Information. *Econometrica* 69: 633–564.

Duffie, D., and Singleton, K. (2003). *Credit Risk: Pricing, Measurement, and Management*. Princeton University Press, Princeton, NJ.

Fama, E.F. (1970). Efficient Capital Markets: A Review of Theory and Empirical Work. *Journal of Finance* 25(2): 383–417.

Fama, E.F. (1998). Market Efficiency, Long-term Returns, and Behavioral Finance. *Journal of Financial Economics* 49: 283–306.

Foster, G. (1986). *Financial Statement Analysis*. Prentice-Hall, Englewood Cliffs, NJ, pp. 533–571.

Fridson, M., and Jonsson, J.G. (1997). Contingent Claims Analysis. *Journal of Portfolio Management*, Winter: 31–43.

Grossman, S., and Stiglitz, J.E. (1980). On the Impossibility of Informationally Efficient Markets. *American Economic Review* 70(3): 393–408.

Helfert, E.A. (1967). *Techniques of Financial Analysis*. Richard D. Irwin, Homewood, IL, pp. 37–82.

Hillegeist, S.A., Keating, E.K., Cram, D.P., and Lundstedt, K.G. (2002). *Assessing the Probability of Bankruptcy*. Kellogg School of Management, Northwestern University, Working Paper.

Hong, H., and Stein, J. (1999). A Unified Theory of Underreaction, Momentum Trading, and Overreaction in Asset Markets. *Journal of Finance* 54(6): 2143–2184.

Hubbard, R.G., Kuttner, K.N., and Palia, D.N. (1999). *Are There Bank Effects in Borrowers' Cost of Funds? Evidence from a Matched Sample of Borrowers and Banks*. Federal Reserve Bank of New York, Working Paper.

Ingersoll, J.E. (1987). *Theory of Financial Decision Making*. Rowman and Littlefield Press, Savage, MA.

Janossi, T., Jarrow, R.A., and Yildirim, Y. (2001). *Estimating Default Probabilities Implicit in Equity Prices*. Cornell University, Working Paper.

Jarrow, R.A., and Turnbull, S.M. (2000). The Intersection of Market and Credit Risk. *Journal of Banking and Finance* 24: 271–299.

Jarrow, R.A., and van Deventer, D.R. (1998). *Integrating Interest Rate Risk and Credit Risk in Asset and Liability Management*. Asset and Liability Management: *The Synthesis of New Methodologies*. Risk Books, London, UK.

Jegadeesh, N., and Titman, S. (1993). Returns to Buying Winners and Selling Losers: Implications for Stock Market Efficiency. *Journal of Finance* 48: 65–91.

Kahneman, D., Slovic, P., and Tversky, A. (1982). *Judgment Under Uncertainty: Heuristics and Biases*. Cambridge University Press, Cambridge.

Kao, D. L. (2000). Estimating and Pricing Credit Risk: An Overview. *Financial Analysts Journal* 56(4): 50–66.

Kealhofer, S. (1999). *Credit Risk and Risk Management.* AIMR, Charlottesville, VA, pp. 80–94.

Keenan, S.C., and Sobehart J.R. (2000). *Credit Risk Catwalk. Risk,* July: 84–88.

Kent, D., Hirshleifer, D., and Subrahmanyam, A. (1998). Investor Psychology and Security Market Under- and Overreactions. *Journal of Finance* 53(6): 1839–1885.

Kindleberger, C. (1978). *Manias, Panics, and Crashes.* Basic Books, New York.

Lang, L., and Stultz, R. (1992). Contagion and Competitive Intra-industry Effects of Bankruptcy Announcements: An Empirical Analysis. *Journal of Financial Economics* 32: 45–60.

Lee, C.M., Shleifer, A., and Thaler, R. (1991). Investor Sentiment and the Closed-end Fund Puzzle. *Journal of Finance* 46: 75–110.

Leland, H.E. (1999). *The Structural Approach to Credit Risk. Frontiers in Credit-Risk Analysis:* AIMR Conference Proceedings, 36–46.

Lo, A. and MacKinlay, A.C. (2002). *A Non-Random Walk Down Wall Street.* Princeton University Press, Princeton, NJ.

Longstaff, F.A., and Schwartz, E.S. (1995). A Simple Approach to Valuing Risky Fixed and Floating Rate Debt. *Journal of Finance* 50: 789–819.

Mackay, C. (1841). *Extraordinary Popular Delusions and the Madness of Crowds* Vol. I. Richard Bentley, London.

Malkiel, B. (1973). *A Random Walk Down Wall Street.* W.W. Norton & Company, New York, pp. 15–98.

Merton, R. (1973). The Theory of Rational Options Pricing. *Bell Journal of Economics and Management Science* 4: 141–183.

Merton, R. (1974). On the Pricing of Corporate Debt: The Risk Structure of Interest Rates. *Journal of Finance* 29: 449–470.

Nandi, S. (1998). *Valuation Models for Default-Risky Securities: An Overview.* Federal Reserve Bank of Atlanta, Economic Review, Fourth Quarter: 22–35.

O'Hara, M. (1995). *Market Microstructure Theory.* Blackwell Publisher, Malden, MA, pp. 1-12, 53–75.

Ravid, S.A. (1996). Debt Maturity: A Survey. *Financial Markets, Institutions, and Instruments* 5(3): 1–69.

Samuelson, P.A. (1965). Proof That Properly Anticipated Prices Fluctuate Randomly. *Industrial Management Review* 6: 41–49.

Schwartz, E.S., and Moon, M. (2000). Rational Pricing of Internet Companies. *Financial Analyst Journal* (May–June): 62–75.

Shiller, R. (1981). Do Stock Prices Move Too Much to be Justified by Subsequent Changes in Dividends. *American Economic Review* 71: 421–436.

Shiller, R. (2000). *Irrational Exuberance.* Princeton University Press: Princeton NJ.

Shleifer, A. (2000). *Inefficient Markets – An Introduction to Behavioral Finance.* Oxford University Press, New York, pp. 1–53.

Shleifer, A., and Summers, L. (1990). The Noise Trader Approach to Finance. *Journal of Economic Perspectives* 4(2): 19–33.

Shleifer, A., and Vishny, R. (1990). The Limits of Arbitrage. *Journal of Finance* 52: 35–55.

Shleifer, A., and Vishny, R. (1992). Liquidation Values and Debt Capacity: A Market Equilibrium Approach. *Journal of Finance* 47: 1343–1366.

Shumway, T. (1999). The Delisting Bias in CRSP Returns. *Journal of Finance* 52(1): 327–340.

Sobehart, J.R. (2003). *A Mathematical Model of Irrational Exuberance and Market Gloom. GARP Risk Review*, July/August: 22–26.

Sobehart, J.R. (2010). Follow the Money from Boom to Bust. *Model Risk -Identification, Measurement and Management*. Ed. D. Roesch and H. Scheule, Risk Books, London, 19–43.

Sobehart, J.R. (2012). *Market Reaction to Price Changes and Fat Tailed Returns*. Risk, June: 76–81.

Sobehart, J.R., and Farengo R. (2002). *Fat Tailed Bulls and Bears. Risk,* December: S20–S24.

Sobehart, J.R., and Farengo R. (2003). A Dynamical Model of Market Under- and Over-reaction. *Journal of Risk* 4(5): 91–116.

Sobehart, J.R., and Keenan, S.C. (1999). *Equity Market Value and its Importance for Credit Analysis: Facts and Fiction*. Working Paper.

Sobehart, J.R., and Keenan, S.C. (2002). *Hybrid Default Models*. Risk, February: 73–77.

Sobehart, J.R., Stein, R., Mikityanskaya, V., and Li, L. (2000). *Moody's Public Firm Risk Model: A Hybrid Approach to Modelling Short Term Default Risk*. Moody's Investor Service, Rating Methodology.

WFJ (*Wall Street Journal*) (2002a). Nasdaq Drops as Accounting Worries Hit More Stocks. *The Wall Street Journal,* February 20: C1–C12.

WSJ (*Wall Street Journal*) (2002b). Creative Accounting, How to Buff a Company. *The Wall Street Journal,* February 21: C1–C18.

White, E. (1990). The Stock Market Boom and Crash of 1929 Revisited. *Journal of Economic Perspectives* 4(2): 67–83.

CHAPTER 11

Pillar 2: Demand for Credit: Models of Business Failure

CREDIT RISK MODELS OF BUSINESS FAILURE

Here we discuss different approaches to credit risk modeling to support applications for climate risk and ESG. Our discussion covers standard statistical and econometric models, the structural and reduced form models. We also provide a generalization to broader approaches to modeling companies in businesses with uncertainty, covering machine learning approaches and other nonlinear models. Finally, we introduce a mathematical derivation based on a forward-looking, multiple time scale, singular perturbation analysis of the joint distribution of the state variables.

Business, legal, and regulatory pressures drive financial institutions toward more comprehensive approaches to measure, monitor, and manage their financial risks. Developing models that can describe a broad range of stressful conditions leading to default or credit deterioration is key for performing the monitoring and managing functions required for assessing climate risk and ESG for active portfolio management. With the increased focus on the impact of climate change on different economies and industries, the estimation of credit risk and risk drivers has taken on a new importance.

Credit risk can be defined as the potential that a debt issuer, borrower, or counterparty to a transaction will fail to meet the obligations in accordance with the terms of a debt agreement, contract, or indenture. For both individual and institutional investors, debt securities are their main source of credit risk, while for banks, loans and banking products are often the primary source of credit risk. Since banks and other financial institutions often lend to companies that are not rated by rating agencies, they need supplemental credit assessments. Ordinarily it is uneconomical for these institutions to devote extensive internal resources to the analysis of each individual borrower's credit quality. Not surprisingly, these economic factors have caused banking institutions to be among the earliest adopters of advanced quantitative models for credit assessments (Caouette et al. 1998; BCBS, 1999, 2000; Crouhy et al., 2001).

Quantitative models lend themselves to what-if scenario analysis and stress testing, providing a valuable and cost-effective means for forward-looking risk assessment and to uncover credit vulnerabilities objectively. There are three basic types of quantitative approaches for assessing an obligor's credit quality:

1. statistical representations of the qualitative judgment process primarily based on patterns found in historical data
2. structural or fundamental assessments based on prescribed relationships between risk drivers and default events
3. reduced form representations of defaults as unexpected/random events driven by hazard rates calibrated to historical observations.

The first type (*statistical approach*) is the most frequently found in the industry literature. It maps a set of financial variables and other relevant information to a risk scale or default flags to mimic the expert judgment process of credit assessment. This approach relies primarily on correlations between financial variables and default events found in historical data. The linear Z-score model developed by Altman (1968) is an example of a statistical representation that has remained relevant for many decades due to its simplicity and insightful view of creditworthiness. The second approach (*structural approach*) describes default events as the direct consequence of financial variables reaching a threshold or tipping point. The contingent claims model of risky debt introduced by Merton (1973, 1974) is an example of a structural credit risk model where default only occurs when the (market) value of the company's assets is below the value of the company's liabilities. In this situation, the company owners (equity stakeholders) have no incentive to pay the company's obligations. In the third approach (*reduced form approach*), default events are not casually modeled but occur unexpectedly with a given (hazard) rate of events given exogenously. The hazard rate is usually described in terms of obligor characteristics and market risk drivers obtained from econometric regressions of accounting ratios and market prices as done for statistical models.

A major challenge in developing models that can effectively assess the credit risk of individual obligors is the limited availability of relevant and reliable high-frequency information to use as inputs for credit models. In cases where historical data is sparse or is not available, both model development and model validation must rely on heuristic methods and domain experts to determine a model's reliability and robustness for practical applications. In contrast, when historical observations are available, model development and validation can proceed in a more objective and rigorous way.

Credit models often estimate creditworthiness over a period of one year or more and are calibrated using several years of historical financial data for each borrower. While reliable and timely financial data can usually be obtained for the largest corporate borrowers, they are difficult to obtain for smaller borrowers and, more importantly, are more difficult to obtain for companies in financial distress or default, which are key to constructing accurate credit risk models. The scarcity of reliable data required for building credit risk models stems from the infrequent nature of default events. There is no shortage of theoretical models to choose from. However, the true performance at default prediction is not a theoretical issue but an empirical one. Credit risk models that perform well over a given historical data set may not perform reliably under the situations found in practice, where data can be limited, sparse or unreliable. Furthermore, this becomes more relevant in the context of climate risk, as credit and default models need to translate the effects of physical risks and transition risks to a low-carbon economy into credit losses over extended periods of time for hypothetical scenarios with no previous historical reference or benchmarks.

Estimated risk ratings or, more generally, probability of default (PD) (i.e. the likelihood of a default event) and the distribution of losses given default (LGD) (i.e. the loss severity) are fundamental inputs to credit risk models. PDs and LGDs can be estimated explicitly from patterns found in observed default frequencies and loss rates and their correlation to accounting variables and market data (as in the statistical approach), or can be implied from equity, debt prices and other relevant information (as in the structural and reduced-form approaches).

The statistical approach determines the empirical relationship between the default event and market information and accounting variables using econometric techniques. The selected variables usually include financial ratios or combinations of items from the balance sheet, income, and cash flow statements. In these models there is no prescribed structural relationship between the dynamics of the firm and its credit quality. The statistical models rely primarily on empirical correlations found in the historical default and financial data.

In contrast, the structural approach makes explicit assumptions about the dynamics of a company's assets and liabilities, or its ability to service debt, and the relation to default events. In his seminal paper on the valuation of risky debt, Merton (1973, 1974) introduced an insightful model based on the relation between the company's assets, its equity and debt. The model describes the company's common equity stock as a call option on the market value of the company's assets as suggested in Black and Scholes (1973). Many extensions of the Merton model were introduced over the years to refine assumptions, including jumps in the value of the company's assets, bond indenture provisions, taxes, bankruptcy costs, and stochastic volatility and interest rates (Vasicek, 1984, 1995; McQuown, 1993; Kim et al., 1993; Levin and van Deventer, 1997; Shumway, 1998; Sobehart and Keenan, 1999; Sobehart et al., 2000, and references therein.) Although these models and their extensions include more refined features, they preserve the fundamental assumption of equity-as-an-option proposed by Merton or Black and Scholes.

While the structural approach is insightful and economically appealing, the predicted default probabilities and credit spreads hardly match observed default frequencies and spreads. Furthermore, most structural models often show poor predictive power for default events in precisely those cases where these models are expected to hold strongest (Fridson and Jonsson, 1997; Sobehart and Keenan, 2001, 2002). There have been cases where solvent companies have sought the protection of bankruptcy courts for previously unanticipated legal or regulatory related liabilities. In other situations, companies that are solvent according to structural models may default on their obligations due to severe liquidity problems, fraud, or mismanagement. Most importantly, these idealized structural models cannot readily incorporate debt refinancing, restructuring, renegotiation of debt contracts, or the distressed exchange of securities.

In the third approach (reduced form approach), default events are not casually modeled in terms of the company's behavior. More precisely, defaults are assumed to occur unexpectedly and randomly with a given rate of events (or hazard rate) (Duffie and Singleton, 2003). The rate of default events is usually obtained from simple econometric regressions of the hazard rates as a function of market prices, market equity and accounting ratios as in the statistical models. More refined reduced form models can reflect the term structure of hazard rates for different time horizons and the correlation structure of defaults across different segments, leading to more complex models. Although reduced form models are usually preferred by academics and credit derivative trading desks, detailed studies on the probability of default inferred from stock and bond prices have been limited (Janossi, Jarrow, and Yildirim, 2001). These studies are important to gain insight into the causes of default since reduced form models do not prescribe structural relationships between the company's risk characteristics and the event of default but rely heavily on regressions of the hazard rates based on historical observations.

The impact on the credibility in accounting practices that usually follows the collapse of large companies highlights the relevance of having timely and accurate information for the assessment of credit risk. These business failures show that market equity and the implied "market value of assets" are usually built on capitalized confidence about the company's prospects and on conceptual rather than physical capital or assets. These conceptual assets can vanish overnight when investors lose confidence or are exposed to deficient information. Certainly, there is a clear need for more realistic models that can reflect credit deterioration reliably.

The reconciliation of structural and reduced form models has been attempted multiple times in the academic literature with limited success due to the technical difficulties for integrating these very different frameworks (Jarrow and Turnbull, 2000; Duffie and Lando, 2001). Despite these challenges, market information can be extremely valuable for credit assessments, and it is most useful when it is combined with a clear understanding of a company's business environment, its balance sheet, income statement and cash flows. These remain core components of any risk assessment approach and are critical for assessing the impact of climate risk on credit risk, for both physical and transition risks.

Here we highlight two fundamental issues that arise in determining the accuracy and reliability of default and loss severity models:

1. *what is measured,* or the metrics by which model "goodness" should be defined;
2. *how it is measured,* or the framework that should be used to ensure that the observed performance can reasonably be expected to represent the behavior of the model in practice.

MODEL SELECTION

Models are simply stylized mathematical abstractions of reality. As such, they represent only a partial and codified view of the problem they describe. This limitation is particularly important for modeling default risk, where models are usually estimated with sparse or limited data on credit events and realized losses. If an unreliable or unsound model is used, then inference based on the model and its underlying data will often be poor. Therefore, it is important to select an appropriate model for the analysis of a specific data set.

There are three basic aspects of valid inference in default risk modeling:

1. model specification and selection
2. estimation of the model parameters
3. estimation of the precision of the parameters.

There are scores of articles in professional journals and books on model fitting and parameter estimation (items (2) and (3) above). In contrast, discussions around model specification (what candidate models to consider) and model selection (what models to use for inference) in the context of climate risk have been sparse. Because model specification and selection are usually believed to be outside the field of mathematical statistics, most of the practical work on this area has been related to hypothesis testing as a means of selecting a model.

Model specification and selection are conceptually more difficult than estimating the model parameters and their precision. Model specification and selection should be the point where economics, finance and transition and physical risk arguments formally enter the research. Because in practice data could be unreliable, sparse, or nonexistent, default risk models often depend on significant subjectivity, in both, determining relevant financial variables and determining the relationships between those variables. Thus, even default risk models based on structural models or statistical regressions often depend on significant judgment of analysts and modelers, which can usually go unnoticed by practitioners and users of these models. Theoretically sound model selection methods are still required and, particularly, methods that are easy to use and widely applicable. "What default risk model to use?" remains the key question in making a valid inference from data on default and credit events. Here we provide readers with practical advice on selecting default risk models.

STATISTICAL AND ECONOMETRIC MODELS

Statistical models link risk characteristics and other attributes of borrowers and debt issuers to risk ratings (risk rating replication models) or to observed default and loss events (probability of default and loss severity models). Note, however, that statistical models do not rely on underlying assumptions about the drivers and causes of default but on statistical relationships observed in historical data. These models are primarily driven by observed *correlations* between selected risk drivers and the dependent variables as opposed to direct *causation* resulting from the evolution of risk drivers (Foster, 1986; Mensah, 1984).

Statistical models have evolved greatly over the years, from the early statistical models linking risk ratings and default events to financial characteristics of companies and market information using standard econometric techniques, up to complex Bayesian and decision trees models, neural networks, and fuzzy logic models in the realm of machine learning (ML) and artificial intelligence (AI).

In the following sections we focus on statistical models, their benefits, and limitations. Later sections and chapters focus on structural models, reduced form models and their generalizations, which are based on the identification of drivers of credit quality, and explicit assumptions about their impact on default likelihood and loss severity.

Statistical models of risk ratings focus on mapping the financial characteristics of companies and risk characteristics of their obligations to observed risk ratings, which are categorical variables that can be ranked order in terms of riskiness. That is, risk ratings can be viewed as ordinal numbers and used as dependent variables in statistical regressions. In contrast, statistical models of default probability or loss severity focus on mapping observed default frequencies or loss severity rates (cardinal number) to company attributes to produce estimates of probability of default and loss severity. In this case, the binary default and non-default events are assigned a probability of occurrence as a function of multiple characteristics or variables. When the distribution of probabilities is related to the normal (or Gaussian) distribution of outcomes, the regression is known as *probit model*. In contrast, a *logit model* refers to the logistic distribution of outcomes. These approaches are discussed in more detail in later sections.

Statistical models of default likelihood (usually known as credit scoring models) use mathematical techniques to separate the population of defaulters from non-defaulters, viewed as either a single category or multiple categories aligned to different risk ratings. The determination of the model's ability to separate between two or more populations can be based on a linear combination of model input variables or more complex nonlinear relationships. A common characteristic of all models is that they usually make errors in predicting default events, risk ratings or credit quality. Therefore, an optimization or error cost function is required to determine the cost of errors when fitting the model to observed data. Error costs depend on the nature of the model and its outputs. For example, in an *ordinary least squares* (OLS) regression, the error cost function is the observed variance of errors (or quadratic errors) around the expected model outcomes. The larger the error in quadratic sense, the larger the cost. Therefore, OLS attempts to minimize quadratic errors around the model by reducing large deviations relative to smaller ones. In contrast, in the *maximum likelihood* approach, usually associated with probit or logit regression models, the error cost function determines the probability that the observed data came from the assumed distribution of errors around the model. Therefore, maximum likelihood methods attempt to maximize the chances that the observed data is aligned to the model. These conceptual differences about the value of errors are usually reflected directly in the estimated values of the model parameters, which reflect different error optimization strategies.

The simplest statistical methods fit observable attributes (independent variables), such as a company's financial variables, to variables to be predicted (dependent variables), such as a default or non-default event, or the company's risk rating using univariate or linear multivariate analysis, which assumes a linear relationship between the observed attributes. When these models are used to separate populations, they are referred to as *linear*

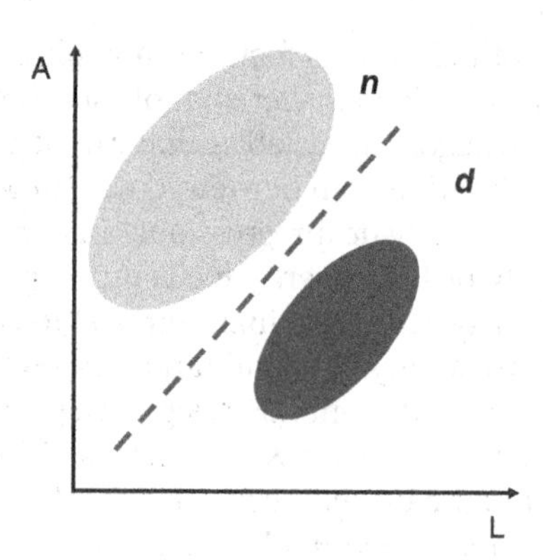

FIGURE 11.1 Linearly separable distributions of defaulters (d) and non-defaulters (n).

discriminant analysis (LDA), while the use of nonlinear methods and nonlinear relationships between observed attributes are referred to as *nonlinear discriminant analysis* (NLDA).

Linear methods such as LDA are based on finding combinations of financial and market variables to produce lines or planes of separation between distinct regions of the space of variables. In the case of credit risk, the objective is to develop models that can partition the data such that companies with similar characteristics would be grouped in the same region of the space of variables, which would be bounded by separating planes. Figure 11.1 shows an example of linear separation. For simplicity, there are two classes: defaulters (d) and non-defaulters (n), and only two variables: assets (A) and liabilities (L). The broken line indicates a linear discriminant function that separates the two regions in the space of variables. In this example, the model effectively uses two variables, A and L, to classify the data into defaulters and non-defaulters.

Linear models have been widely used for credit risk applications with varying degrees of success. Linear methods are extremely useful for linearly separable problems, that is, those problems for which lines and planes are the natural boundaries between different groups. However, linear methods are often inadequate to describe the complex nature of the relationships of financial and market variables, often characterized by nonlinear relationships between variables. For example, the relationship between the company's risk rating and the company size (as measured by assets, sales, or revenues) is usually nonlinear. Large companies tend to have better risk ratings than smaller companies. Note, however, that size may not be as useful to differentiate between smaller companies because their credit quality could depend

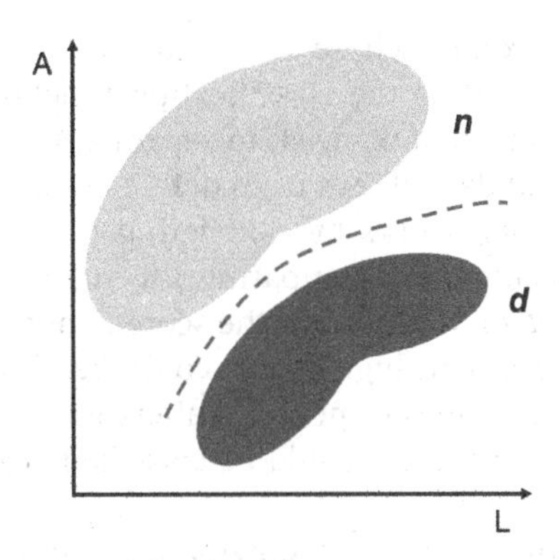

FIGURE 11.2 Nonlinear separable distributions of defaulters (d) and non-defaulters (n).

on other factors such as profitability, liquidity, leverage, business efficiency or capital structure. This would be reflected in a nonlinear relation between risk rating and size.

Nonlinear methods are not subject to many of the limitations of linear methods. This allows nonlinear methods to achieve better results at separating groups than linear methods in cases where relationships between variables are more complex. Figure 11.2 shows a case for which nonlinear methods are more suitable than linear methods. Note that in this case a linear method is less suitable because of the complex boundary separating the group of defaults (d) vs. non-defaults (n). In this case, a nonlinear method can be used to fit a boundary that matches more closely the observed groups.

CREDIT SCORING AND STATISTICAL DISCRIMINANT ANALYSIS

Determining the relationship between a response variable (or score) and key attributes or model drivers is a fundamental problem in statistics and is usually divided into two related by distinct topics: *regression*, where the response variable is continuous and needs to be estimated or projected, and *classification*, where the response variable is categorical or discrete in nature, and the response variable is used to separate classes or groups.

The fundamental goal of credit scoring is to use a metric to differentiate good from bad credits or, more generally, differentiate two or more groups with different credit quality. This type of analysis (known as *discriminant analysis*) attempts to do this by constructing linear or nonlinear combinations of relevant variables to obtain a classification function that can be used for separating the target groups, whose values are the credit scores. Discriminant analysis works by fitting the classification function to data, including observations in all groups that we wish to separate. Fitting the classification function to the observed data allows us to define cutoff values for separating the groups. The approach works by calculating the classification function output (score or decision axis), comparing the score to cutoff values, and assigning a group based on the value of the score relative to the cutoff values. For example, companies with high scores could be classified as high credit quality if related to risk ratings or non-defaulters if related to default events, while companies with low scores could be classified as low credit quality or defaulters.

For clarity of exposition, in the following we discuss the simplest case of two distinct groups of defaulting (d) and non-defaulting (n) companies and a classification function $Z(X_1, X_2, X_3, \ldots X_m)$ based on a linear model of m relevant financial ratios labeled $X_1, X_2, X_3, \ldots X_m$. That is,

$$Z = \sum_{k=1}^{m} w_k X_k + w_0 \tag{11.1}$$

Here the model coefficients w_k are weights determined by fitting the function Z to observed data on the company's attributes $X_1, X_2, X_3, \ldots X_m$ and default and non-default observations over a fixed time horizon. These weights measure the relative contribution of each variable X_k to the classification function Z.

The classification function serves to assign each company as a member of a group for which the value of the classification function is the highest. Where there are only two groups, the discriminant function is aligned to the classification function for defaulter and non-defaulter groups. The cutoff is determined as the score (value of the function Z) that provides the best separation of the two groups. Figure 11.3 illustrates the nature of the classification function Z (represented by the arrow) and the cutoff value (represented by the dashed line separating the groups). For multiple groups or nonlinear relationships, the classification functions are more complex as the classification can occur across multiple directions in the space of variables.

Plotting the number of observations in each group having a score within predefined ranges of values provides the distribution of scores or histogram. The distribution of scores for each group can help us visualize how well the scores separate the groups and to what extent these groups overlap in scores. That is, this allows us to understand to what extent the selected attributes can differentiate the characteristics of each group and determine appropriate criteria (or cutoff points) as shown in Figure 11.4.

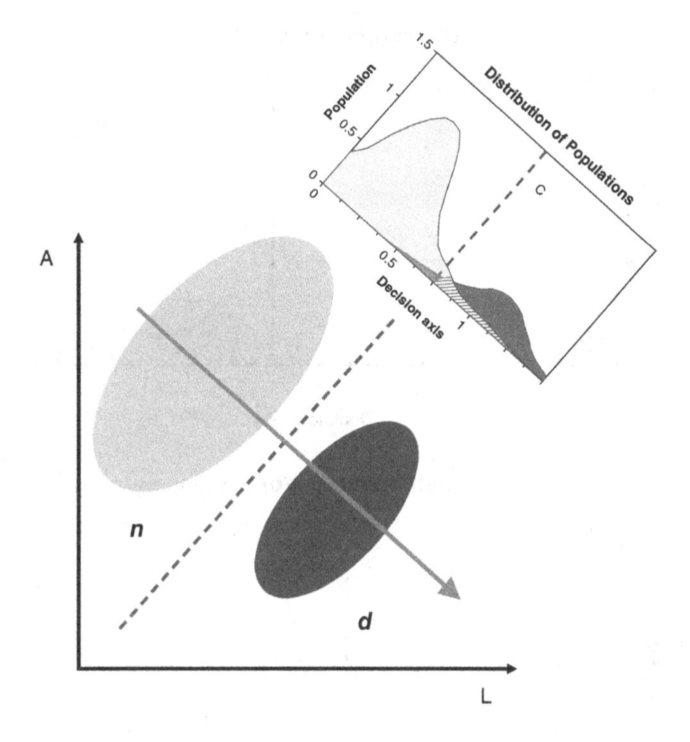

FIGURE 11.3 Linearly separable distributions of defaulters (d) and non-defaulters (n) and classification function $\boldsymbol{Z}$.

The goal of the classification function Z is to distinguish defaulters (right distribution (d)) from non-defaulting borrowers (left distribution (n)), as shown in Figure 11.4. Note that the population of defaulters has been exaggerated for illustration purposes. In practical situations, defaulters are a small fraction of the population of borrowers, and the two populations overlap considerably. Figure 11.4 shows a hypothetical decision axis representing the scores of the classification function Z, which reflect the borrowers' credit quality. In Figure 11.4(a) we assume that the cutoff point C_1 adopted for classifying a borrower as a defaulter has been established at a fairly high value along the decision axis. The selected cutoff point C_1 reflects a lax decision in judging the credit quality of a borrower. Borrowers whose risk scores are above C_1 should be classified as likely defaulters. The cut-off C_1 results in a small fraction of defaulters to be classified correctly, misclassifying a significant fraction of the defaulters as non-defaulters (type I error). In Figure 11.4(b) a lower cutoff C_2 is adopted following a stricter criterion in

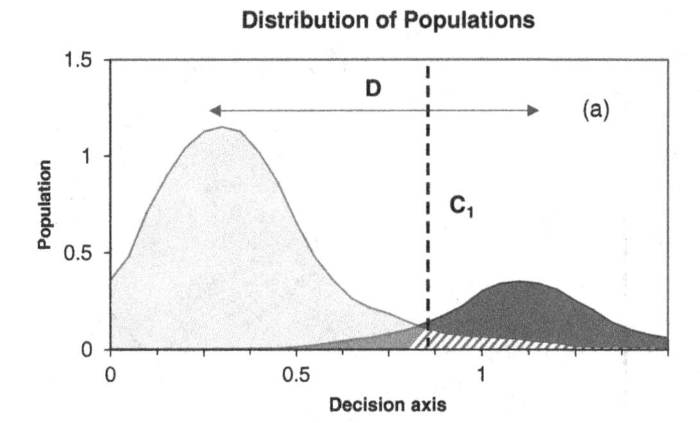

FIGURE 11.4 Distributions of defaulters (*d*) and non-defaulters (*n*) along the decision axis (score values) for different cutoff values. Panels (a) and (b) show the hypothetical decision axis for a lax cutoff C_1 or a stricter cutoff C_2. The population of non-defaulters is on the left, and the population of defaulters is on the right. The parameter *D* reflects the separation between these populations.

judging a borrower's credit quality. Cutoff C_2 results in a large fraction of non-defaulters to be classified incorrectly as defaulters (type II error).

For each borrower, there are four possible outcomes in the identification of defaulters:

1. **Hit:** the model correctly classifies a troubled borrower as a defaulter.
2. **Miss:** the model assigns low risk to a defaulter.
3. **False alarm:** a financially sound borrower is classified as a defaulter.

4. **Correct rejection:** the model assigns low risk to a financially sound borrower.

The relative magnitudes of the outcomes above depend on two issues: (1) the cutoff C used to classify borrowers and (2) the separation D of the populations of defaulters (d) and non-defaulters (n) along the decision axis. The latter depends on the classification function's ability to differentiate these populations of borrowers. That is, the selection of adequate attributes to characterize the borrowers' credit risk.

Figure 11.4 indicates how the cutoff C and the separation D contribute to the proportion of hits, misses, false alarms, and correct rejections. The shaded area under the population of defaulters represents the proportion of correct *hits*. The cross-hatched area represents the proportion of *false alarms* based on the risk characteristics of the borrower. The unshaded area under the population of defaulters represents *misses* occurring when the score assigns low risk to a defaulter. Finally, the unshaded area under the non-defaulter population represents *correct rejections*. In Figures 11.4(a) and 11.4(b), the population separation D is relatively small, reflecting that the populations of defaulters and non-defaulters overlap significantly along the decision axis.

A key question about the classification function is how effectively it separates the populations of defaulters and non-defaulters for different decision cutoffs C. This requires constructing a classification function that maximizes the distance D between the two populations and a cutoff C that minimizes misclassification errors for defaulters and non-defaulters, driven by the selected attributes of the borrowers' credit quality.

A good classification model should result in a small number of misclassifications. More precisely, the probability of misclassification should be as small as possible. When one population has a greater likelihood of occurring than another because one population is much larger than the other, the classification approach should take the prior probabilities of occurrence into consideration explicitly. This situation is particularly relevant for credit and climate risk impact, where there are more financially sound companies than financially troubled or defaulted companies.

Another relevant aspect of classification is the *misclassification error cost*. For example, suppose that misclassifying a troubled company (defaulter d) as a financially sound company (non-defaulter n) represents a more costly error than misclassifying a non-defaulter as a defaulter. Then the classification approach should, whenever possible, reflect this information in the classification criteria.

More precisely, let $p_d(s)$ be the distribution of scores for defaulters and $p_n(s)$ be the distribution of scores for non-defaulters for a score value $s = Z(X_1, X_2, X_3, \ldots)$ as illustrated in Figure 11.4, and let q_d and q_d be the (unconditional) prior probabilities of defaulters and non-defaulters. Let Λ be the space of variables $X = \{X_1, X_2, X_3, \ldots\}$ for all possible scores s, let Ω_d be the collection of all scores s for which companies are identified as defaulters ($s \geq C$), and $\Omega_n = \Lambda - \Omega_d$ be the remaining scores for which companies are identified as non-defaulters ($s < C$), where C is the cutoff value for separating the two defaulter and non-defaulter classes. The sets Ω_d and Ω_n (represented by the conditions ($s \geq C$) and ($s < C$)) are exhaustive and mutually exclusive.

The conditional probability $P(n|d)$ of misclassifying a defaulter as non-defaulter (*miss*) is:

$$P(n|d) = P(X \in \Omega_n|d) = \int_{X \in \Omega_n} p_d(s(X))dX = \int_{s<C} p_d(s)ds \tag{11.2}$$

The conditional probability $P(d|n)$ of misclassifying a non-defaulter as defaulter (*false alarm*) is:

$$P(d|n) = P(X \in \Omega_d|n) = \int_{X \in \Omega_d} p_n(s(X))dX = \int_{s \geq C} p_n(s)ds \tag{11.3}$$

Furthermore, the probabilities of correctly classifying or misclassifying defaulters and non-defaulters are based on the product of the conditional probabilities of classification and the unconditional prior probabilities:

1. *Hit:* the model correctly classifies the borrower as a defaulter with probability:

$$Prob(Hit) = q_d P(d|d) = q_d P(X \in \Omega_d|d) = q_d \int_{s \geq C} p_d(s)ds \tag{11.4}$$

2. *Miss:* the model assigns low risk to a defaulter with probability:

$$Prob(Miss) = q_d P(n|d) = q_d P(X \in \Omega_n|d) = q_d \int_{s<C} p_d(s)ds \tag{11.5}$$

3. *False alarm:* a low-risk borrower is classified as a defaulter with probability:

$$Prob(False\ alarm) = q_n P(d|n) = q_n P(X \in \Omega_d|n) = q_n \int_{s \geq C} p_n(s)ds \tag{11.6}$$

4. *Correct rejection:* the model assigns low risk to a financially sound borrower with probability:

$$Prob(Correct\ rejection) = q_n P(n|n) = q_n P(X \in \Omega_n|n) = q_n \int_{s<C} p_n(s)ds \tag{11.7}$$

Classification approaches evaluated only in terms of the misclassification probabilities described above would ignore misclassification costs, which could lead to significant credit losses or missed investment opportunities. The costs of misclassification errors can be defined using a classification cost matrix for the four cases described above (Table 11.1).

In Table 11.1, the classification costs are assumed: zero for correctly classified defaulters and non-defaulters, $c(d|n)$ when a financially sound company is misclassified as a defaulter and $c(n|d)$ when a troubled company is misclassified as non-defaulter. In practice, information collection and processing costs, opportunity costs and other costs could change the classification cost structure assumed in Table 11.1.

We can introduce a practical cost-based classification criteria by making the expected cost of misclassification (ECM) from Table 11.1 and the probabilities of occurrence of each event as small as possible:

$$\begin{aligned} ECM &= q_d P(n|d)c(n|d) + q_n P(d|n)c(d|n) \\ &= c(n|d)q_d \int_{s<C} p_d(s)ds + c(d|n)q_n \int_{s\geq C} p_n(s)ds \end{aligned} \tag{11.8}$$

In principle, the generalization of the approach discussed above from two classes to multiple classes is conceptually straightforward, although computationally more demanding.

More precisely, let $p_k(s)$ be the distribution of scores $s_k = Z_k(X_1, X_2, X_3, \ldots, X_m)$ and q_k be the (unconditional) prior probability for class $k = 1, \ldots h$. Let the classification costs be $c(j|k)$ when a borrower is assigned to class j when, in fact, it belongs to class k. We assume that correctly classified borrowers have no classification costs: $c(k|k) = 0$ for class $k = 1, \ldots h$.

TABLE 11.1 Classification costs

		Model Classification	
		d	N
Actual population	d	0	$c(n\|d)$
	n	$c(d\|n)$	0

Let Λ be the space of variables $X = \{X_1, X_2, X_3, \ldots\}$ for all possible scores s_k for all classes $k = 1, \ldots h$. Let Ω_k be the collection of all scores s_k for which companies are identified as belonging to class k. The sets Ω_k are selected as exhaustive and mutually exclusive regions of the space of variables that minimize the generalized expected misclassification cost:

$$ECM = \sum_{k=1}^{h} q_k \left(\sum_{\substack{j=1 \\ j \neq k}}^{h} P(j|k)c(j|k) \right) = \sum_{k=1}^{h} q_k \left(\sum_{\substack{j=1 \\ j \neq k}}^{h} \int_{X \in \Omega_k} p_k(X)dX\, c(j|k) \right) \tag{11.9}$$

In later sections we revisit this approach and introduce alternative criteria for model selection.

MODELS OF PROBABILITY OF DEFAULT

Note that the classification function scores provide information on the credit quality of the companies. Therefore, the probability of default given the score would differ from the probability of default without that information. Furthermore, without the additional information on credit quality provided by the classification function score, we would assign the same default probability to all companies given by the simple ratio of defaulters over the total number of defaulters and non-defaulters (unconditional prior probability based on the observed frequency of events). Therefore, the probability of default given the classification function score (posterior probability) transforms the scoring model (classification function Z) into a default risk model.

The probability of default $P_d(s) = Prob(d|s)$ given a score value $s = Z(X_1, X_2, X_3, \ldots)$ can be determined using Bayes' rule of probability for the populations of defaulters (d) and non-defaulters (n):

$$P_d(s) = Prob(d|s) = \frac{p_d(s)q_d}{p_d(s)q_d + p_n(s)q_n} \tag{11.10}$$

Here $p_d(s)$ is the observed distribution of scores for defaulters and $p_n(s)$ is the distribution of scores for non-defaulters as illustrated in Figure 11.4, and q_d and q_n are the (unconditional) prior probabilities of defaulters and non-defaulters. Notice that if the distributions of scores $p_d(s)$ and $p_n(s)$ were the same (i.e. they overlapped completely and $D = 0$), the scores would be completely uninformative, leading to the prior probability for default events. In contrast, if the distributions were significantly different (i.e. they barely overlapped and D was large), the posterior probability would be very informative given the score s. Below we describe common models for the probability of default $P_d(s)$.

Logit Model

One of the simplest probability models is the represented by the logit model. For simplicity of exposition, let's assume that a change Δs in the credit score s represents a relative change in the probability of default $\Delta P_d/P_d(1 - P_d)$, where the additional term $(1 - P_d)$ accounts for saturation effects of the relative changes as the probability approaches the value one. That is,

$$\frac{\Delta P_d}{P_d(1 - P_d)} = \alpha \Delta s \tag{11.11}$$

Here α is the sensitivity to changes in score s. Taking the limit of small changes in the credit score and integrating Equation (11.11) above leads to the standard logit model:

$$log\left[\frac{P_d}{(1 - P_d)}\right] = log\left[\frac{Prob(d|s)}{Prob(n|s)}\right] = \alpha s + \beta \tag{11.12}$$

Here β is an integration constant. Equation (11.12) represents the cumulative logistic probability of distribution for the score s:

$$P(d|s) = \frac{1}{1 + e^{-\alpha s + \beta}} \tag{11.13}$$

Note that the probability distribution above becomes close to zero for large negative values of the score s, and becomes close to one for large positive values of s.

Probit Model

After assigning a group of obligors a similar score s, the default rate can be calculated as the ratio $d/(n + d)$ of defaulting obligors d to total number of defaulter and non-defaulter companies $(n + d)$. The expected default rate $E(d/(n + d))$ for a score s is given by the probability that a type I error has occurred for defaulted companies which were incorrectly assigned a score s.

For simplicity, let's assume that score errors are normally distributed and have similar deviation. The distribution of the normalized errors is simply:

$$f(y) = \frac{1}{\sqrt{2\pi}} e^{-y^2/2} \tag{11.14}$$

Although judgment errors in assigning a credit score are unavoidable, Equation (11.14) indicates that it is unlikely to observe large deviations of credit quality assessments given a score. Then, the probability of default given a company's score s is given by:

$$P(d|s) = E(d|(n+d)) = \int_{-\infty}^{\alpha s+\beta} \frac{1}{\sqrt{2\pi}} e^{-\frac{y^2}{2}} dy = N(\alpha s + \beta) \tag{11.15}$$

Note that the probit and logit models provide similar results, except for extreme values, where the difference of the distributions become significant. Figure 11.5 illustrates the differences between the probit and logit models for the same transformed score $x = \alpha s + \beta$. Furthermore, using a scaled width for the logit distribution to match the variance of the normal distribution (same standard deviation), the alignment between the logit and probit models is remarkably close for moderate values of x. Both the probit and logit models are attractive transformations for modeling credit quality problems where the response variable can take only two values. There are other transformations that offer similar properties, but they are usually limited in their use due to their complexity and the interpretation of the model parameters.

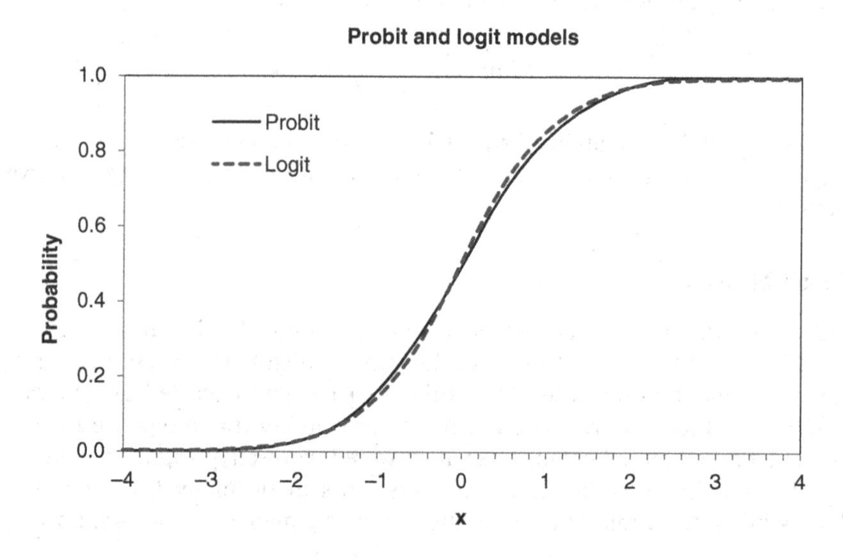

FIGURE 11.5 Comparison of the logit and probit models. Both models are normalized to provide the same mean and standard deviation for variable x.

Ordered Probit Model

The ordered probit model is an extension of the probit model for multiple classes that are ordered in terms of riskiness along a single decision axis (score). More precisely, let's assume there are h classes describe by the score $s = Z(X_1, X_2, X_3, \ldots)$ (for example, h different risk ratings). The decision axis is partitioned into h exhaustive and mutually exclusive segments $\Omega_k\ k = 1, \ldots, h$ that include all scores s for which companies are identified as in class k. The boundaries of the sets $\Omega_1, \Omega_2, \ldots \Omega_h$ are defined by $(h-1)$ cutoff values $C_1, C_2, \ldots C_{h-1}$ selected to minimize the expected misclassification cost or the probabilities of misclassification (equal costs) across all segments. That is, the segments are defined as follows:

$$\begin{cases} \Omega_1\colon \{X \in \Omega_1\} & s < C_1 \\ \Omega_k\colon \{X \in \Omega_k\} & C_{k-1} < s \leq C_k \\ \Omega_h\colon \{X \in \Omega_h\} & s \geq C_{h-1} \end{cases} \tag{11.16}$$

The conditional probability of a company being classified in the ordered segment Ω_k (e.g. getting a risk rating k), given the score s, is:

$$P(\Omega_k|s) = \begin{cases} N(\alpha s + \beta) & s < C_1 \\ N(\alpha s + \beta) - N(\alpha C_{k-1} + \beta) & C_{k-1} < s \leq C_k \\ 1 - N(\alpha s + \beta) & s \geq C_{k-1} \end{cases} \tag{11.17}$$

Note that the probability of being classified in any of the h segments is determined by the value of score s relative to the partition boundaries determined by the cutoffs C_k $(1 \leq k \leq h)$.

NONLINEAR MODELS

In many situations, empirical observations show nonlinear relationships between financial attributes of companies such as company size or revenues and risk ratings, default frequencies, or credit risk. Despite recent advances in machine learning and artificial intelligence techniques promoting their use for financial applications, nonlinear methods such as neural networks, fuzzy logic models, Bayesian methods, and decision trees have been used in the industry extensively for modeling risk ratings and default probabilities for decades (Goonatilake and Treleaven, 1995; Refenes, 1995).

Broadly speaking, a neural network can be seen simply as a higher-order regression technique based on conventional statistical methods. Instead of creating a *single* regression equation, *several* coupled or nested regressions need to be solved simultaneously. The regression coefficients of the model are found by the mutual reinforcement and feedback of the intermediate regressions (or "nodes") based on the statistical distribution of the sample data. Conceptually, a simple neural network with no intermediate nodes (or "hidden layers") and with a logistic (sigmoidal or S-shape) output is equivalent to the standard logistic regression model. Figure 11.6(a) illustrates a simple logistic regression

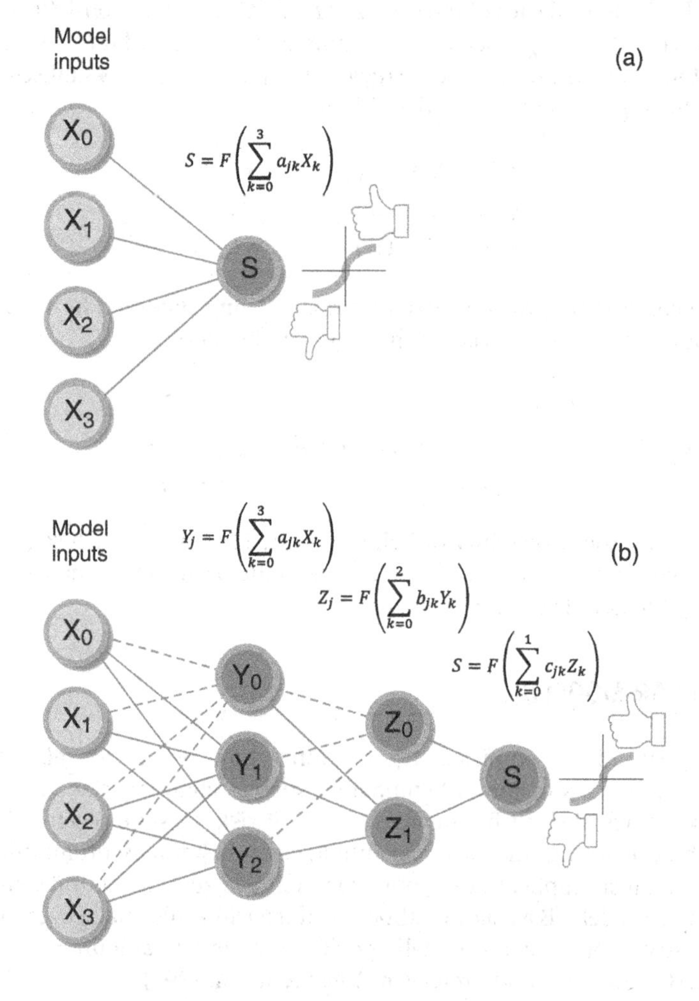

FIGURE 11.6 Illustrative example of (a) a simple logistic regression, and (b) a neural network as nonlinear nested regressions. Subscript "0" refers to regression constants represented by pseudo-variables $\boldsymbol{X_0 = Y_0 = Z_0 = 1}$.

$S = F(x)$. Figure 11.6(b) illustrates a nested regression nonlinear model as a combination of three layers of logistic regressions $F(x)$.

There are different approaches (or "learning processes") to determine the regression coefficients of a neural network. For example, when a network is trained on observations of default and non-default events, a back-propagation algorithm can be used to minimize the network's errors by propagating these errors back through the network, adjusting the connection weights (model coefficients) between the different regressions until errors are minimized. At this point the network is "trained" and ready to provide estimates for different companies. In later sections we provide a more detailed description after introducing key concepts on statistical inference.

STATISTICAL INFERENCE AND BAYESIAN METHODS

Models are approximations to real-world situations and, except for rare exceptions, no single model can completely capture the actual relationship between the response variable and the drivers of the response. We often select the "best" model among alternatives that most closely captures the observed relationships between the response variable and its drivers using specific criteria for weighting regression or classification errors. For example, if our primary criteria included minimizing squared errors between the model estimates and the observed outcomes, we could adopt a model derived from ordinary least squares or similar metrics. If our criteria included maximizing the probability that the model estimates are aligned to the distribution of observed outcomes, a maximum likelihood approach could be used. The model selection criteria provide a means to quantify the fit-for-purpose nature of the model.

In principle, the quality of the models could be improved by incorporating *a priori* knowledge available to us in the form of qualitative or quantitative information. This view leads naturally to the use of Bayesian methods, which allow us to assign prior distributions to the model structure and its parameters and then update the model structure with additional observations or data, resulting in a posterior distribution of the model structure and parameters combining information from observations with prior information:

$$\text{Posterior distribution} \propto \text{Prior distribution} \times \text{Likelihood}$$

Before proceeding to discuss advanced methods, first let's review basic concepts on probability and likelihood estimation. Assume a random sample of N independent observations $x_1, x_2, \ldots, x_N$, each drawn from a distribution function $f(x;\theta)$, where $\theta = \{\theta_1, \theta_2, \ldots, \theta_w\}$ represents a set of w unknown parameters that characterize the distribution. The join density distribution (or likelihood function) of the observations is:

$$L(\theta) = \prod_{k=1}^{N} f(x_k; \theta) \tag{11.18}$$

The likelihood function represents the likelihood that the sample came from the distribution $f(x;\theta)$ given the parameters θ. The *maximum likelihood* (ML) estimation approach is based on the insight that the parameters θ can be estimated by choosing estimates $\hat{\theta}$ that maximize the observed sample likelihood. That is, making the probability that the sample of observations $x_1, x_2, \ldots, x_N$ came from the distribution $f(x;\hat{\theta})$ as high as possible. For computational simplicity, the maximum likelihood approach is usually expressed in terms of the log-likelihood function:

$$\ln(L(\theta)) = \sum_{k=1}^{N} \ln(f(x_k; \theta)) \tag{11.19}$$

For continuous distribution functions with continuous first and second derivatives, the condition for $\ln(L(\theta))$ to have a maximum at $\hat{\theta}$ is that the first derivatives F_j with respect to $\theta = \{\theta_1, \theta_2, \ldots, \theta_w\}$ are zero.

$$F_j = \sum_{k=1}^{N} \frac{\partial}{\partial \theta_j} \ln(f(x_k; \theta)) = 0 \quad j = 1, \ldots, w \tag{11.20}$$

In addition to the condition above, the second-order derivatives of G_{ij} evaluated at $\hat{\theta}$ must be negative semidefinite.

$$G_{ij} = \sum_{k=1}^{N} \frac{\partial^2}{\partial \theta_i \partial \theta_j} \ln(f(x_k; \theta)) \quad i,j = 1, \ldots, w \tag{11.21}$$

Satisfying the conditions for the first and second order derivatives above may not be sufficient to ensure that a single global maximum exists. For complex situations where the parameters θ enter in a nonlinear way into the likelihood function $\ln(L(\theta))$, multiple maxima may exist. Note, however, that, as the observation sample size grows, the distribution of the ML estimators converges to a multivariate normal distribution with mean θ and a variance-covariance matrix $\sum_{ij} = (I(\theta)^{-1})_{ij}$ given by the inverse of the information matrix (or Cramer-Rao lower bound), defined as follows:

$$I(\theta)_{ij} = -E\left[\sum_{k=1}^{N} \frac{\partial^2}{\partial \theta_i \partial \theta_j} \ln(f(x_k; \theta))\right] = -E[G_{ij}] \quad i,j = 1, \ldots, w \tag{11.22}$$

MODEL SELECTION CRITERIA: LEAST SQUARES AND LIKELIHOOD METHODS

A common approach to determine model parameters is to choose those parameters that minimize the sum of the squares of the residuals between the observations and the model predictions. This is known as the *least squares* (LS) approach. That is, selecting model parameters that come as close as possible to all observations as measured by the squares of the residuals. Estimation methods using the LS approach are intuitive and simple to implement. In contrast, methods based on the likelihood approach often require iterative or advanced numerical methods that are computationally demanding. Note, however, that the LS approach shares many similarities with the likelihood approach and yields identical parameter estimators when the residuals are independent and normally distributed. In the following we follow closely the description of Burnham and Anderson (1998).

For illustration purposes, consider a multivariable linear regression model where the response variable y is a function $Z(X_1, X_2, X_3, \ldots, X_m)$ of m explanatory variables $x = \{X_1, X_2, X_3, \ldots, X_m\}$ and one constant (i.e. the regression includes $m+1$ parameters). Residuals ε are assumed independent and normally distributed with zero mean and standard deviation σ. That is,

$$y = Z + \varepsilon = \sum_{k=1}^{m} w\theta_k X_k + w\theta_0 + \varepsilon \quad \varepsilon \sim N(0, \sigma) \tag{11.23}$$

Next let's consider a random sample of N independent multivariable observations $y_1, y_2, \ldots, y_N$, each drawn from the distribution function $f(y; \theta)$, where $\theta = \{\theta_0, \theta_1, \ldots, \theta_m\}$ represents the unknown parameters that characterize the distribution. Because the residuals are independent and normally distributed, the likelihood function is simply the product of the probabilities for all observations:

$$L(\theta) = \prod_{k=1}^{N} f(y_k; \theta) = \prod_{k=1}^{N} \frac{1}{\sqrt{2\pi}\sigma} e^{-\frac{\varepsilon_k^2}{2\sigma^2}} = \left(\frac{1}{\sqrt{2\pi}\sigma}\right)^N e^{-\frac{1}{2}\sum_{k=1}^{N}\frac{\varepsilon_k^2}{\sigma^2}} \tag{11.24}$$

The log-likelihood function is:

$$\ln(L(\theta)) = -\frac{1}{2\sigma^2}\sum_{k=1}^{N}\varepsilon_k^2 - \frac{N}{2}\ln(\sigma^2) - \frac{N}{2}\ln(2\pi) \tag{11.25}$$

Since the standard deviation σ is usually unknown, we can use its ML estimator $\hat{\sigma}^2 = RSS/N$, where $RSS = \sum_{k=1}^{N}\varepsilon_k^2$ is the residual sum of squares of the N observations. In this case, the log-likelihood reduces to the following expression:

$$\ln(L(\theta)) = -\frac{N}{2}\ln(\hat{\sigma}^2) - \frac{N}{2}(1 + \ln(2\pi)) \tag{11.26}$$

Note that the number of estimated parameters is $K = m + 2$, including the m regression parameters, the model constant and the residual variance σ^2. Furthermore, Equation (11.26) indicates that there is a close relationship between the LS and ML approaches for independent and normally distributed residuals: LS minimizes RSS, which maximizes the log-likelihood function for the regression parameters. For more general cases, LS and ML can lead to different estimates of the model parameters, which highlights the need to determine a clear objective function for model selection as part of the model design process.

INFORMATION ENTROPY METHODS AND MODEL SELECTION

As computation tools for pattern recognition and data analysis become more readily available, academics and practitioners have increased their use in the selection of model variables by searching for significance of relationships across a wide range of variables and their combinations, cross-products of variables, powers of variables and other variable transformations. This type of number-crunching analysis results in models that are tailored to data, often diluting the validity of statistical inference made about the data and the model structure. This data-driven approach to modeling often results in model overfitting. Furthermore, the statistical inference tests for models derived from this approach may provide a false sense of confidence as basic assumptions about the data are often violated (e.g. independence of observations, stationarity, etc.). This approach, which includes searching for a range of plausible variables and model structures with no *a priori* target model, can be misaligned with the true nature of the hypotheses to be tested in the presence of multiple plausible models and relationships across variables. This issue can go unnoticed, leading to incorrect inference and model selection. Critical to model selection is to have a conceptually sound model structure, well-defined assumptions to be tested and a clear understanding of the data to support model calibration and hypothesis testing.

Broadly speaking, the more parameters used when building a model, usually the better the fit of the model to the data. However, we must also recognize that an excessive number of parameters can lead to overfitting, which can result in stability and reliability issues with the model and its estimates. The balance between these issues leads to a key question in making valid inferences from data and background information about credit risk assessment:

which model to use? The literature on model specification, model selection, and estimation of parameters is quite broad. Among the range of methodologies, the use of statistical inference and Bayesian methods in the selection of models provides a simple and sound means for evaluating competing models objectively and identifying the best parsimonious model based on the available information and model selection uncertainty. When model selection uncertainty is ignored, model precision and confidence in parameters can be overestimated, and model estimates can be less accurate than expected. This method can also reduce the chance of including spurious variables or parameters in the models that are not supported by the available information. That is, is the model's mathematical complexity adequate to describe the data and information at hand?

The literature on model selection has increased significantly in the last decades, from simple approaches for minimizing the squares of the discrepancies between models and observations (e.g. least squares methods), to increasing the likelihood that the model is aligned to the observed data (e.g. maximum likelihood methods), to advanced information entropy-based concepts (e.g. Kullback-Liebler information, Akaike information criteria, and other generalizations of conditional information entropy). See, for example, Press (1977), Cox and Tiao (1992), and Burnham and Anderson (1998).

The information entropy approach is attractive because it is applicable across a wide range of models and is a powerful way of objectively measuring how much information is gained or lost when the model is introduced. Conceptually, information entropy is a measure of the overall amount of "uncertainty" represented by a probability distribution. The entropy concept has its origin in the fields of *statistical mechanics* in Physics and *communication theory* (Shannon and Weaver, 1949; Jaynes, 1957; Pierce, 1970). In the following we discuss information-based methods in more detail.

First, let's review basic concepts on information entropy. For simplicity of exposition, let's assume a random event with only two possible outcomes:

1. obligor defaults with probability p
2. obligor does not default with probability $1 - p$.

The amount of additional information I we require to determine which outcome occurred is defined as

$$I = -\log_2(p) \tag{11.27}$$

Here $\log_2(p)$ is the logarithm of p in base 2.

If only the first outcome is possible with certainty, then $p = 1$ and the information required is $I = -\log_2(p) = 0$ (bit). In this case, there is no uncertainty about the outcome and, therefore, there is no relevant information that was not previously known. If the two events are equally likely (uninformative case), then $p = 1/2$ and the amount of information required reaches a maximum value of $I = -\log_2(p) = 1$ (bit). Exactly one bit of information (the equivalent to a yes-or-no answer) is the information we need to know which of the two equally likely possibilities have occurred.

The use of logarithm base 2 makes this example easy to understand but any logarithm base can be used. The amount of information depends on the logarithm base, which determines the measure unit of information. Using the natural logarithm, the information entropy of the event is defined as

$$H = -[p\ln(p) + (1-p)\ln(1-p)] \tag{11.28}$$

Figure 11.7 shows the information entropy as a function of the probability p. Note that the entropy reaches its maximum (1 bit $= \log_2(2)$) when the probability is $p = 1/2$. This is a state of absolute ignorance because both possibilities are equally likely. If the assigned probability of an event is lower than 1/2, one outcome is more likely to occur than the other. That is, we have less uncertainty on the possible outcomes, which is reflected in the reduction of entropy.

Now, let's turn our attention to conditional information entropy and model selection. The conditional information entropy methods determine a probability-weighted discrepancy (or "distance") between two arbitrary probability models, say $f(x)$ and $g(x)$. When $f(x)$ is considered the idealized/true model and $g(x)$ is an approximate model, the conditional information entropy

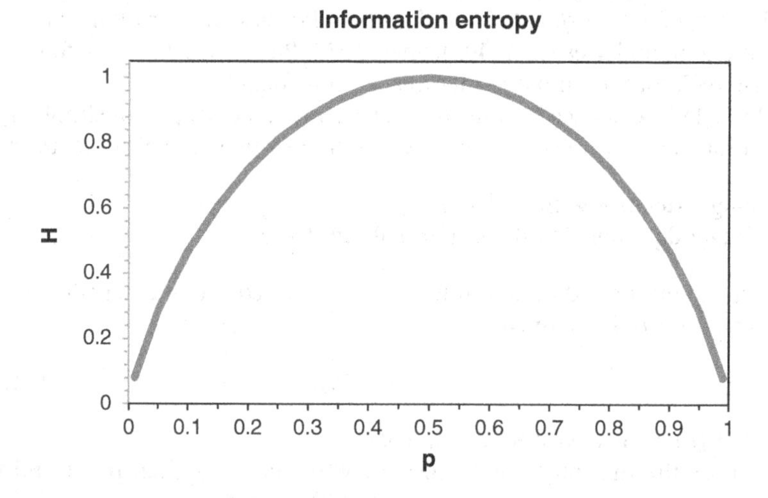

FIGURE 11.7 Information entropy as a function of the probability p.

approach can be used to determine the best approximating model. Since the true model $f(x)$ is rarely known, additional estimates are required to find the best approximating model given all available information. Let x represent the model variables, y represent the available data with the same probability model $f(x)$, θ for the model parameters of the probability model $g(x|\theta)$ and $\hat{\theta}$ for the estimated parameters $g(x|\hat{\theta})$, for example, using LS or ML methods.

When $g(x)$ is used to approximate the true model $f(x)$, there is a loss of information, which can be measured using the Kullback-Leibler information distance:

$$KL(f,g) = \int f(x)\ln\left(\frac{f(x)}{g(x|\theta)}\right)dx = \int f(x)\,[\ln(f(x)) - \ln(g(x|\theta))]dx \quad (11.29)$$

The KL information distance is positive, except when $f(x)$ and $g(x)$ are the same and, therefore, $KL = 0$. Note, however, that the KL information distance is not symmetric for $f(x)$ and $g(x)$ as it is measured relative to a reference model. That is, $KL(f,g) \neq KL(g,f)$.

The KL information distance can be expressed as the difference between two expectations with respect to the true probability model $f(x)$ integrated over all values of x:

$$KL(f,g) = E_x[\ln(f(x))] - E_x[\ln(g(x|\theta))] = C - E_x[\ln(g(x|\theta))] \quad (11.30)$$

Where the first term depends only on the (unknown) true probability model and can be considered a constant reference value C for any approximate function $g(x|\theta)$. Since the model parameters for the approximate model $g(x|\theta)$ are unknown and need to be estimated from the sample y, which has the same probability distribution $f(x)$, we obtain:

$$\begin{aligned} E_y[KL(f,g)] &= E_yE_x[\ln(f(x))] - E_yE_x[\ln(g(x|\hat{\theta}(y)))] \\ &= C - E_yE_x[\ln(g(x|\hat{\theta}(y)))] \end{aligned} \quad (11.31)$$

Thus, the minimization of the KL information distance requires the estimation of $E_yE_x[\ln(g(x|\hat{\theta}(y)))]$, where $\ln(g(x|\hat{\theta}(y))) \sim \ln(L(\hat{\theta}(y)|y))$ resembles the log-likelihood function. Akaike showed that the maximized likelihood is biased upwards as an estimator of the model selection criteria, with a bias approximated equal to the number K of parameters in the approximate model. More precisely,

$$E_yE_x[\ln(g(x|\hat{\theta}(y)))] \approx \ln(L(\hat{\theta}(y)|y)) - K \quad (11.32)$$

The relationship between the KL information distance and the maximized likelihood for the model parameters led to a practical criterion for model selection, known as the Akaike's information criterion (AIC):

$$AIC = -2\ln(L(\hat{\theta}(y)|y)) + 2K \tag{11.33}$$

The factor two in the AIC definition is included for historical reasons. The second-order corrected criterion (AIC_c) includes an adjustment for small samples N in relation to the number of model parameters K:

$$AIC_c = AIC + \frac{2K(K+1)}{N-K-1} \tag{11.34}$$

If the sample size N is large with respect to the number of parameters K, the correction is negligible. The estimation of a more general bias adjustment term to the relationship between the KL information distance and the log-likelihood $-2\ln(L)$ leads to the Takeuchi's information criterion (TIC):

$$TIC = -2\ln(L(\hat{\theta}(y)|y)) + 2\mathrm{tr}(J(\hat{\theta})I(\hat{\theta})) \tag{11.35}$$

Here $J(\hat{\theta})$ and $I(\hat{\theta})$ are matrices based on first and second partial derivatives of the log-likelihood function and "tr(A)" is the trace of a matrix A. TIC reduces to AIC when $\mathrm{tr}(J(\hat{\theta})I(\hat{\theta})) \sim K$. Additional generalizations of the AIC concept can be found in the literature of model selection. Similar concepts can be derived from a Bayesian viewpoint with equal priors for each tested model, leading to the Bayesian information criteria (BIC):

$$BIC = -2\ln(L(\hat{\theta}(y)|y)) + K\ln(N) \tag{11.36}$$

Note, however, that BIC is not an estimator of the KL information distance.

For most practical applications, Akaike's information criteria AIC and AIC_c are sensible approximations to the KL distance that provide a consistent and objective approach for statistical inference. The combination of these criteria with the *principle of model parsimony* (i.e. selecting the simplest model structure, number of variables and parameters consistent with the available information) provides a practical framework for model selection. The principle of model parsimony is aligned to Occam's razor: "prefer simplicity."

To illustrate these ideas, assume we are fitting a simple multivariate model to historical data on default events. Let y_k $k = 1, \ldots, N$ denotes the k-th dependent variable of a sample of N observations. The dependent variable can take only two possible values: $y_k = 1$ when the company risk exceeds certain (random) threshold level leading to default during a time period T, and $y_k = 0$ if the company remains viable. Let ξ_k denotes the threshold response level for company k. Now if $z_k = AX_k + B$ denotes the risk exposure of the company, where $X_k = \{x_{k1}, \ldots, x_{kM}\}$ is a vector of the M regression variables of the model, A is a vector of unknown model coefficients and B is an unknown constant, then $y_k = 1$ if $z_k > \xi_k$ and $y_k = 0$ if $z_k \leq \xi_k$.

We now need to determine how y_k and ξ_k are related in probabilistic terms. This can be done by introducing the probability of the event $P(y_k)$ and the cumulative distribution function $F(z)$ of the random threshold ξ_k given the model inputs X_k:

$$P(y_k = 1) = P(\xi_k < AX_k + B|X_k) = F(AX_k + B) = F(z_k)$$
$$P(y_k = 0) = 1 - F(z_k) \tag{11.37}$$

Once the functional form of $F(z)$ is specified, the model reduces to finding the coefficients A and B such that:

$$E(y_k) = P(y_k = 1) = F(AX_k + B) \tag{11.38}$$

Here the coefficients A and B can be estimated, for example, using the maximum likelihood function described earlier:

$$L = \prod_{k=1}^{N} F(AX_k + B)^{y_k}(1 - F(AX_k + B))^{1-y_k} \tag{11.39}$$

When the model selected for the company's credit quality $z(X)$ is more complex than the simple multivariate linear model $z = AX + B$ described above, other techniques such as error minimization, information entropy, Bayesian methods and cross-validation-based methods can also be used to estimate the parameters of the model. For a technical discussion on alternative methods, see, for example, Press (1977), Cox and Tiao (1992), Burnham and Anderson (1998) and references therein.

Let us abstract from the functional form of the model for the company's credit quality $z(X)$ and focus on the function $F(z)$. When $F(z)$ is the normal cumulative distribution function, the approach reduces to the probit model discussed previously.

$$F(z) = \Phi(z) = \int_{-\infty}^{z} \frac{e^{-x^2/2}}{\sqrt{2\pi}} dx \tag{11.40}$$

For the probit model, estimation using maximum likelihood is straightforward but computationally difficult since $F(z)$ has no simple analytical representation. A more appealing alternative is to select a cumulative distribution function that has a tractable form such as the logistic distribution (logit model):

$$F(z) = \frac{1}{1 + e^{-z}} \tag{11.41}$$

The probit and logit models are numerically close to each other except in their tails. The alignment of these two models over a moderate range of values z can be obtained by simply scaling the mean and width of the logit distribution to match the first two derivatives of both distributions.

Although there is a variety of natural extensions to the logit model, neural networks with sigmoidal transformations of variables represent perhaps one of the most widely used alternatives (Hertz, Krogh, and Palmer, 1991). Neural networks with sigmoid functions are mainly nested logit regressions where the output of each logit regression (or node) serves as the input to other logit regressions. Because the sigmoid (logistic) function saturates for large values of its input, nested logistic regressions allow to model both thin-tail and fat-tail distributions at the expense of additional mathematical complexity. Figure 11.8 shows the difference between a probit model, a logit model, and a simple neural network composed of two layers of nested logit models:

$$F(z) = \frac{1}{1 + e^{-G(z)}}, \quad G(z) = \frac{a}{1 + e^{-\alpha\, z + \beta}} + \frac{b}{1 + e^{-\gamma\, z + \delta}} + c \tag{11.42}$$

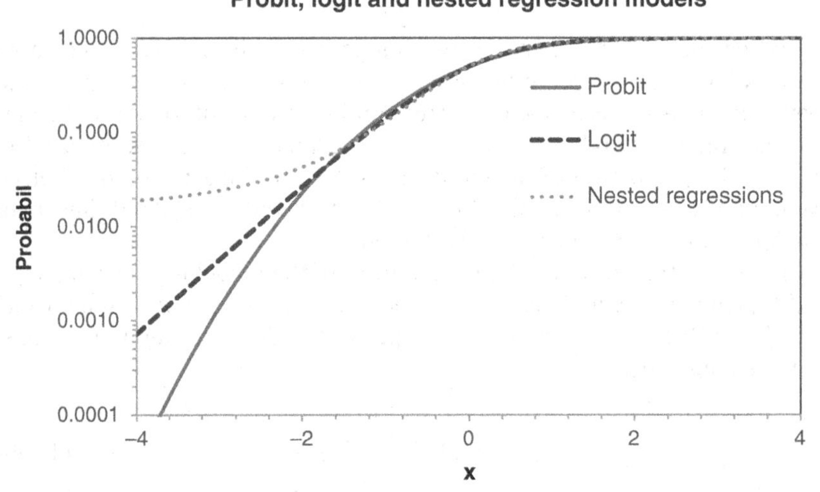

FIGURE 11.8 Probit model, logit model, and nested logit regression with a fat-tail distribution.

TABLE 11.2 Model selection using normalized ***AIC***: (a) logit model, (b) probit model, and (c) nested logit regressions. 150 observations were drawn from a hypothetical univariate logit distribution.

Model	K	AIC	Selection Order
Logit	2	100	1
Probit	2	109	2
Nested logit	7	124	3

The parameters of the nested models 1 and 2 in Figure 11.8 have been selected to highlight that models can generate thin-tail and fat-tail distributions. Note that this type of model provides greater flexibility than the logit and probit models at the expense of additional mathematical complexity in terms of the number of model parameters and relationships between variables. The regression coefficients of the model are found using iterative methods that leverage the mutual feedback of the intermediate regressions (neural network nodes).

While the derivation of the Akaike's information criteria (AIC and AIC_c) for model selection is rooted in mathematical statistics and technical derivations, its application is straightforward. Table 11.2 illustrates the model selection among the 3 models discussed above using AIC for a hypothetical dataset of $N = 150$ observations (default and non-default events) sampled from a logistic probability distribution driven by single risk factor X. Based on the AIC values and number of model parameters K, the logistic model provides the best approximate model to the data, aligned to intuition.

Other advanced computational techniques such as adaptive neural networks, tree structures, fuzzy logic, genetic algorithms, and statistical Bayesian methods are also useful for modeling default risk at the cost of adding additional mathematical complexity. Development in this area has increased substantially in the past decades fueled by work on machine learning and artificial intelligence.

A quick way of assessing whether a model of default probability may be specified incorrectly is to plot a quantile vs. quantile comparison (or Q-Q plot) of the output of the default risk model (x-axis) against the observed frequency of default events (y-axis). If the points fall along a straight line, the model assumptions are acceptable as estimates and observations align. If the points follow a straight line for the central portion of the estimates but they tail off horizontally, the underlying distribution might be fat-tailed relative to the distribution assumed for the model. In contrast, if the points tend to tail-off vertically, the underlying distribution is thin-tailed relative to the distribution assumed for the model. In either case, the assumed distribution is not fully justified by the observations and an adjustment is needed. If the actual underlying distribution can be found from the data, it may be possible to make statistical inference based on it. Transformation of data can also be carried out to reduce the impact on model performance and produce a better fitting. Often, however, this type of problem cannot be eliminated completely.

When the misalignment between the observed and the theoretical default probabilities is small, or when it is not possible to find a more appropriate model specification, a simple approach is to estimate the frequency of default events $W(F, T)$ each period T as a univariate function of the misaligned model estimate $F(z)$. Then the estimated probability of default (PD) for company k during period T is $PD(k, T) = W(F(z_k), T)$. In simple terms, this transformation aligns what is predicted (F) to what is observed (W).

A PRIMER ON NEURAL NETWORKS

One of the most common misconceptions among practitioners is that models based on neural networks are "black boxes" with hidden or unclear variable relationships. Neural networks are higher-order regression techniques based on conventional statistical methods that can be used for performing inductive inference (Hertz et al., 1991; Refenes, 1995; Kubat, 2017; Aggarwal, 2018; Skansi, 2018). Neural networks replace single regression equations with several coupled or nested regressions that need to be solved simultaneously. Depending on the complexity of the network (topology), finding the coefficients on the multiple regressions could be a daunting task usually involving advanced nonlinear techniques. The regression coefficients of the neural network model are found by leveraging the mutual reinforcement and feedback of the intermediate regressions (or "nodes") based on the statistical distribution of the sample data. The overall process of finding the best regression coefficients for the multiple nested regressions is known as "training" the neural network. For a relatively simple neural network with no intermediate nodes (or "hidden layers") and a logistic (S-shape) output, the network structure is equivalent to the standard logistic regression model. The novelty of the neural network approach lies in the ability of the network to model nonlinear relationships across model variables and response variables with little or no *a priori* information on those relationships, which are found by analyzing patterns in the data as the network is trained.

To have a better understanding of neural networks, we start by aligning the terminology for machine learning, statistics, and econometrics. For example, neural network *inputs* are model *independent variables*, neural network *outputs* are *dependent (or response) variables*, *convergence* refers

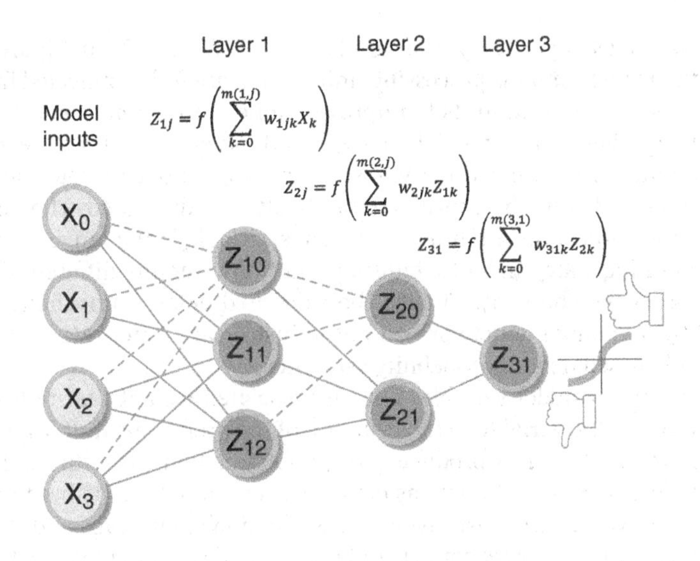

FIGURE 11.9 Illustrative neural network architecture with three basic layers: (a) input layer, (b) intermediate layer, and (c) output layer. Subscript "0" refers to regression constants represented by pseudo-variables $\boldsymbol{X_0} = \boldsymbol{Z_{k0}} = \mathbf{1}$.

to the model fitness *in-sample* and *generalizations* refer to *out-of-sample* predictions.

Neural networks are adaptive nonlinear models composed of layers of artificial neurons (processing units) that interconnect with each other. In simple terms, neural networks are highly stylized representations of brain neurons and their connectivity. Processing units include one or multiple inputs, which are weighted and transformed into an output that reflects a nonlinear response. There are two basic features that characterize a neural network: its connectivity (architecture or topology), and its learning process. For illustration purposes, here we focus primarily on neural networks composed of three basic layers of processing units (see Figure 11.9):

1. an input layer (taking the model variables)
2. a middle -or hidden- layer (composed a single layer or a complex network of connections)
3. an output layer (providing the model outputs).

We assume that there is full connectivity between units in adjacent layers, but none within any given layer, or across more than one layer. The input layer contains processing units whose inputs are the financial ratios and market variables under study (the variables $X_1, X_2, X_3, \ldots, X_m$ in Figure 11.9). The hidden layer contains processing units which are fully connected from all input units to all output units to improve the network's ability to categorize relationships between financial, market and other variables. The output layer contains processing units whose outputs are used to determine if a company or individual borrower is either healthy or distressed. The network outputs are then combined into a continuous variable in the range between 0 (non-defaulting state) and 1 (defaulting state) representing different degrees of distress of the company. Finally, the estimated default probability is calculated by mapping the network output to historical default rates to provide empirically based default probability estimates.

Each input (model variable) has an associated weight that reflects the contribution of the variable to the response of the processing unit. Inputs and weights are combined to produce a response function (or output). Typical response functions (also known as *activation* functions) are (1) hard limiters with a cutoff value, (2) quasi-linear transformations (with caps and floors to reflect saturation of the response), and (3) sigmoid (or S-shape) functions as shown in Figure 11.10.

The most widely used neural network activation function is the sigmoid function based on logistic transformations, which is continuous, differentiable, and easy to use relative to other transformations. Model inputs for each

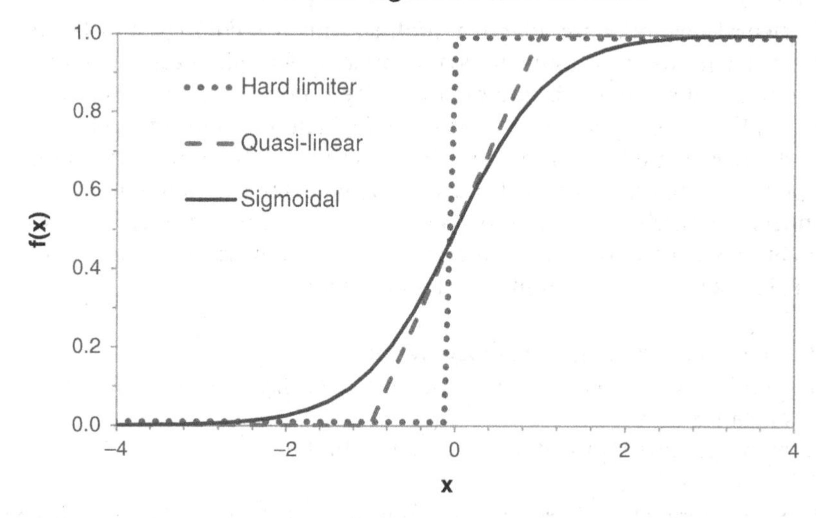

FIGURE 11.10 Typical nonlinear outputs for processing units: (a) hard limiter, (b) quasi-linear transformation, and (c) sigmoid (or S-shape) function.

processing unit (or node) k in the input layer are usually combined linearly and transformed using the activation function:

$$f(Z_k) = \frac{1}{1+e^{-Z_k}} \quad Z_k = \sum_{i=0}^{m} w_{ki}X_i \tag{11.43}$$

Here subscript "0" refers to the model's regression constant represented by pseudo-variables $X_0 = 1$. More generally, each processing unit $0 \leq k \leq n(l)$ in layer $0 \leq l \leq q$ for a q-layer network combines $m(l,k)$ input signals from a previous layer $(l-1)$ into an output signal $f(Z_{lk})$ which feeds another processing unit to reinforce or inhibit observed features. The topology of the neural network could include a wide range of connections across layers and nodes. For illustration purposes, the final output of the neural network is assumed to be a single response function, which combines the contribution of all intermediate layers of processing units:

$$h(X_1, X_2, X_3, \ldots, X_m; \theta) = f\left(\sum_{i=0}^{m(l,1)} w_{l1i}f\left(\sum_{j=0}^{m(l-1,i)} w_{(l-1)ij}f(\ldots)\right)\right) \tag{11.44}$$

Here $h(X_1, X_2, X_3, \ldots, X_m; \theta)$ is the output of the neural network for the inputs $X_1, X_2, X_3, \ldots, X_m$, and $\theta = \bigcup_{\substack{1 \leq l \leq q \\ 0 \leq k \leq n(l)}} \{w_{lk0}, w_{lk1}, w_{lk2}, w_{lk3}, \ldots, w_{lkm(lk)}\}$ represents the set of all variable weights and constants for processing units in all network layers (i.e. all the parameters of the neural network). Subscript "0" refers to the model's regression constants for each network layer represented by pseudo-variables $Z_{l0} = 1, 0 \leq l \leq q$.

Patterns in data are found by mutual reinforcement and feedback from the processing units. This process crudely mimics human inference allowing the model to learn by "example." Network knowledge is stored both in the way the processing elements connect with each other (network topology) and the strength of the interconnections (represented by the coefficients of the nonlinear regressions). A neural network learns by means of changing and reinforcing the internal connections and weights to form associations between inputs and the observed outputs. Depending on how the information is used to adjust the connections, neural network learning is divided into three basic categories:

1. supervised learning
2. associative reinforcement learning
3. unsupervised learning.

In the supervised learning approach, feedback specifies the desired output patterns by explicitly identifying cases for different input variables and changing the connections and weights to minimize the identification errors or the selected cost function. In the associative reinforcement learning, a reinforcement signal is used to indicate whether the actual and desired patterns coincide or disagree. In contrast, in unsupervised learning, the neural network does not receive any external feedback. Learning is self-organized as result of finding seemingly similar clusters that share common patterns. A potential issue with completely unsupervised learning is the identification of spurious patterns that can result in poor model performance.

There are different approaches to propagate errors and determine the parameters of the neural network. For illustration purposes, the learning process of the neural network in Figure 11.9 is the back propagation of errors or, simply, *back propagation*. This learning process is well understood and has proven to be very reliable in a variety of problems. The back-propagation method is a supervised learning process where the network can be trained, for example, on observations of defaulters and non-defaulting companies for which both the inputs and output data are available. The network predictions are then matched up with the true binary outputs of the default events (default or non-default states), and the differences between the calculated and actual values (prediction errors) are recorded and accumulated in the error cost function. In the following we assume that the error cost function is determined by the residual sum of squares for N observations:

$$RSS = \sum_{i=1}^{N} (h(X_{1i}, X_{2i}, X_{3i}, \ldots, X_{mi}; \Theta) - y_i)^2 \tag{11.45}$$

Here $h(X_{1i}, X_{2i}, X_{3i}, \ldots, X_{mi}; \Theta)$ is the final output of the neural network for the inputs $X_{1i}, X_{2i}, X_{3i}, \ldots, X_{mi}$, and y_i is the observed output for each observation $1 \leq i \leq N$.

Supervised learning reduces to the minimization of Equation (11.45), which can be achieved by adjusting the network weights w_{ljk} in each iteration t by an amount proportional to the derivative $\partial RSS/\partial w_{ljk}$ of the residual sum of squares with respect to each weight:

$$w_{ljk}^{t+1} = w_{ljk}^{t} + \Delta w_{ljk} \text{ where } \Delta w_{ljk} = -\eta \frac{\partial RSS}{\partial w_{ljk}} \tag{11.46}$$

The iterative *gradient descent* learning method described above is easy to implement for practical application, although, there are more powerful – but computationally more demanding – algorithms that can also be used.

The back-propagation algorithm minimizes the network's errors by propagating these errors back through the network, adjusting the connection weights (model parameters) between processing units in the direction that reduces errors until ideally the global error minimum is located. Favorable predictions are progressively reinforced, while misclassifications are progressively inhibited. We can envision how the process works by depicting an error landscape composed of hills and valleys of prediction errors. The learning algorithm searches for the lowest feature (the lowest error) on that landscape. For a given set of defaulting and non-defaulting companies given to the network, the landscape is a very complicated function of the many connection weights describing the network. Because the error landscape lies within a high-dimensional space (as many dimensions as network connections) the search for the error minimum is a hard numerical optimization problem. When the error minimum has been found, the network is "trained." That is, it is ready to generalize the default prediction to previously unseen companies based only on the selected model inputs.

VALIDATING STATISTICAL MODELS

The benefits of default risk models cannot be fully realized without a deep understanding of model performance and accuracy. Before we introduce the structural and reduced form frameworks, let's turn our attention to common approaches used to validate and benchmark quantitative credit risk models, including model performance measurement and sampling techniques, as well as other practical considerations associated with performance evaluation, data sparseness and the sensitivity of models to changing economic conditions. Below we present a set of model performance measures and testing approaches that we have found useful for benchmarking default models and validating their performance. The discussion follows closely that of Sobehart, Keenan, and Stein (2000), with additional clarifications and enhancements.

MEASURING MODEL ACCURACY

Model accuracy is often the most prominent dimension in the evaluation of credit risk models. Because these models are used for credit quality assessments that support lending and investment decisions, it is important to understand each model's strengths and weaknesses. When used as classification tools, reliable models should indicate high risk, when the obligor's risk is high (*hit*), and should indicate low risk, when the obligor's risk is low (*correct rejection*). In practice, models can err in one of two ways: (1) the model can indicate low risk, when the risk is high (*miss*); and (2) the model can assign high risk, when the risk is low (*false alarm*). The first situation is referred to as type I error and corresponds to the assignment of low-risk ratings (or low default likelihood) to borrowers who nevertheless default or come close to defaulting on their obligations. The cost for lenders and investors can be the loss of principal and interest, or a loss in the market value of the obligations. The second situation is referred to as type II error and can result in the loss of interest and fees when loans or securities are either turned down or lost through non-competitive bidding. Type II errors may also result in selling obligations at disadvantageous market prices. These situations are illustrated in Table 11.3.

Because models can produce different type I or type II errors, lenders and investors usually seek to select models whose probability of making either type of error is as small as possible. Minimizing one type of error usually comes at the expense of increasing the other type of error. The probability of making a type II error increases as the probability of a type I error is reduced and vice versa.

The issue of *model error cost* is an important one as discussed in previous sections. For example, one model could outperform another model under one set of performance metrics but could be at a disadvantage under a different

TABLE 11.3 Types of classification errors and costs

		Actual	
		Low Credit Quality	**High Credit Quality**
Model	Low credit quality	**Hit** (correct prediction) ■ The model correctly classifies a troubled borrower as a defaulter	**False alarm** (type II error) ■ A financially sound borrower is classified as a defaulter ■ Opportunity costs and lost profits, interest income and origination fees. Sale of asset at disadvantageous prices
	High credit quality	**Miss** (type I error) ■ The model assigns low risk to a defaulter ■ Lost interest and principal through defaults. Recovery costs. Loss in market value.	**Correct rejection** (correct prediction) ■ The model assigns low risk to a financially sound borrower

set of performance metrics. It is difficult to present a single cost function that is appropriate across all model applications. Here we discuss cost functions related primarily to the information content of the models.

A VALIDATION APPROACH FOR QUANTITATIVE MODELS

When evaluating default risk models, we can question whether they are useful in predicting default events rather than just describing them. Validating default prediction can be challenging since many statistical tests of model performance for default risk have extremely low power to reject poorly performing models due to the sparseness of default events. A deeper understanding of the key issues associated with model validation can lead to more useful analysis of a model's benefits and limitations. Irrespective of the validation approach, there are some basic issues to be considered when comparing default risk models (Sobehart, Keenan, and Stein, 2000):

1. Compare models on the same data sets. Performance statistics are often sample-dependent due to the sparseness of credit event data. Differences in sample composition can result in differences in model performance.
2. Avoid using the same data or period to build and test models. It is important to understand how the model will perform in practice, when it is used in populations or periods different from the one in which it was calibrated. Testing using subsequent periods (out-of-time) and using independent data from the one used for building the model (out-of-sample and out-of-universe) can help highlight model limitations. Using out-of-time, out-of-sample, and out-of-universe testing can make overfitted models less appealing:
   - **a.** Out-of-time refers to observations that are not contemporary with the sample used to build the model.
   - **b.** Out-of-sample refers to observations that are not included in the sample used to build the model.
   - **c.** Out-of-universe refers to observations whose properties may differ from the population used to build the model.
3. Avoid relying only on anecdotal cases. Compare performance metrics in aggregate. A common practice to reinforce one's view on a model is to look for anecdotal examples of adverse credit events, such as high visibility defaults, and to examine the model's performance on these events retrospectively. While such analysis may seem reassuring, it can lead to confirmation bias and can obscure the model's true performance since it misses type II errors. The use of aggregate measures such as the *receiver operating characteristics* (ROC), *cumulative accuracy profiles* (CAP curves, also

known as power curves and dubbed curves), and their summary statistics for type I and II errors can provide a more objective and statistically robust assessment of model performance.

4. Understand the variability and robustness of model performance estimates. When observations are sparse, even a handful of records can account for large differences in performance measures. Analysis of the sensitivity of a test's outcome to data sparseness and data outliers can help gain insight in model performance. Techniques such as resampling (or bootstrapping) can be very valuable to understand the variability and robustness of estimates.

Since model performance statistics can be sensitive to the data used for testing, mitigation of sample dependency should include in- and out-of-time, out-of-sample, and out-of-universe testing and resampling on panel datasets (including observations on multiple companies for different time periods) or on cross-sectional datasets (including one observation for each company). However, even this approach could provide a false sense of confidence about a model's reliability if performed incorrectly.

The literature on model testing and validation is quite broad. Here we focus primarily on a few approaches we have found useful in evaluating default and credit risk models across time and across the population of obligors. Figure 11.11 breaks up the model testing procedure along two dimensions: (1) time (along the horizontal axis), and (2) the population of obligors (along the vertical axis). The least restrictive testing procedure is represented by the upper-left quadrant (out-of-sample), and the most stringent by the lower-right quadrant (out-of-time, out-of-sample, out-of-universe). The other two quadrants represent procedures that are more stringent with respect to the time or population dimensions.

Testing approaches across time or changes in population are represented below:

1. The upper left quadrant describes the case where there is only one dataset available for model building and testing. Testing dataset is chosen at random from the available observations in the model fitting dataset. This testing approach relies on the assumption that the properties of the data remain stable over time (stationary process). Because the testing observations are drawn at random from the data used to fit the model, this approach validates the model across the population of obligors preserving its original distribution.
2. The upper right quadrant describes a commonly used testing procedure. Sample observations for model fitting are chosen from any period prior to a certain date and testing observations are selected from periods only after

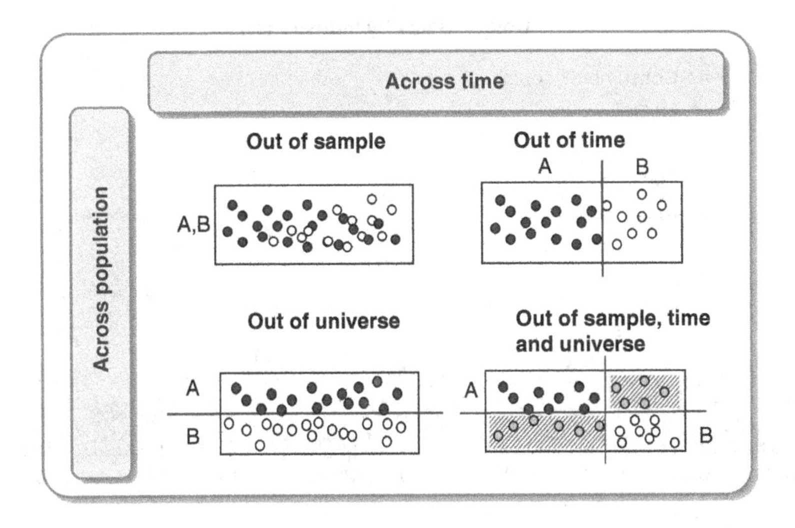

FIGURE 11.11 Testing approaches across time (horizontal axis) and across population (vertical axis). Filled circles represent data used for model fitting (sample A) and open circles represent data used for model testing (sample B).

the selected date. To illustrate, a model built with data from 2010 through 2015 and tested on data from 2016 through 2024 is a simple example of this out-of-time procedure. Because testing is performed with out-of-time observations similarly to how the model is used in practice, the deterioration of the model's forecasting ability over time can be detected.

3. The lower-left quadrant represents the case in which there are two datasets containing no companies in common (out-of-sample testing): one dataset is used for building the model and the other dataset for testing and validation. If the population of the testing set is different from that of the model building set, the dataset is out-of-universe. An example of out-of-universe testing would be a model built using US manufacturing companies but tested on European manufacturing companies. Because the temporal nature of the data is not used, this testing approach does not identify time variability, but it can provide insight into the performance and robustness of the model when applied to a different population.
4. The lower-right quadrant is the preferred testing method since data is segmented across time and across the population of obligors to provide a more comprehensive analysis of model performance. An example of this enhanced testing approach would be a model constructed with data for US manufacturing companies in the period 2010–2015 and tested on

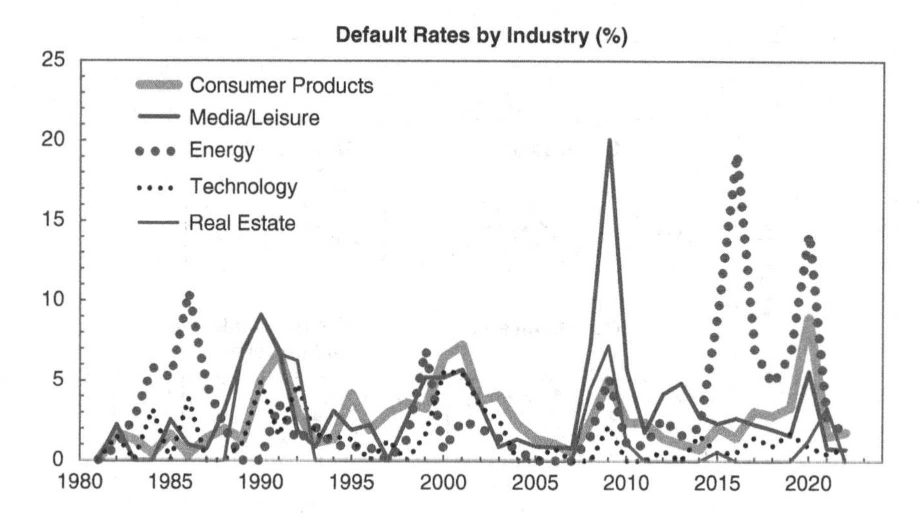

FIGURE 11.12 Default observations for corporate debt issues: (a) shows aggregated default counts for all industries. (b)–(e) show default counts for selected industries: consumer products, media, retail, and technology.

US manufacturing companies in the period 2016–2024 and on European companies in the period 2010–2024 and any sub-period.

The approaches for testing models across time and population differences are relevant for corporate borrowers and other obligors because of the infrequent nature of default events. Sparse default counts and low default rates limit the power of statistical tests to validate models. To illustrate this point, Figure 11.12 shows time series of default events for corporate debt issuers. Figure 11.12(a) shows the aggregated number of defaults for all industries, while Figures 11.12(b)–(e) show the number of defaults for selected industries. Note that data sparseness for different segments could affect the reliability of model testing. Understanding the response of the models across time and population differences can help mitigate the uncertainty of data limitations.

Note that the extension of the observation window to longer time horizons may increase the sample size but may not necessarily translate into more reliable model testing due to the temporal correlation of both default events and model outputs. These correlations cannot be ignored since they impact the confidence of statistical tests. On the other hand, statistical tests based on small samples may incorrectly disqualify relatively good models and accept poor models due to data sparseness.

Because default events are infrequent, and even rare on some segments, and default events and model outputs for consecutive years are often correlated, it is impractical to have large independent hold-out datasets that can be leveraged for out-of-time and out-of-sample model testing. While such tests would be preferred if default data were widely available, this is rarely the case. In practice, model testing needs to balance two conflicting situations:

1. If too many defaults or credit events are left out of the dataset for building models, estimation of model parameters can be impaired.
2. If too many defaults or credit events are left out of the hold-out sample, testing the model performance can be impacted due to poor statistical power.

A practical approach to address this situation combines out-of-time and out-of-sample tests to replicate how models are used in practice. This procedure is usually referred to as *walk-forward* testing (Sobehart, Keenan, and Stein, 2000).

The walk-forward procedure for model testing works as follows:

1. Select a starting year, for example, 2015.
2. Fit the model using all the data available on or before the selected year (e.g. 2010–2015).
3. Once the model is calibrated for the selected period, generate the model outputs for all the companies available during the following year (in this example, 2016). The predicted model outputs for 2016 are out-of-time for companies existing in previous years, and out-of-sample for all the new companies whose data become available after 2016.
4. Save the 2016 predictions and realized defaults as part of a walk-forward testing set.
5. Now move the window up one year (e.g. to 2016), recalibrate the model in the updated period (e.g. 2010–2016) and use data for the following year (2017) for out-of-sample/out-of-time testing.
6. Repeat steps (2) to (5), recalibrating the model and adding the new predictions to the testing set for every year in the walk-forward process until all years are covered.
7. Once the walk-forward testing set containing predictions from different periods has been created, a variety of model performance statistics can be calculated.

The out-of-sample and out-of-time model predictions included in the walk-forward testing set can then be used to analyze the performance of the model across time and across changes in population. Note that this approach mimics how the model could be used in practice. Each year, the model is recalibrated and used to predict the credit quality of known credits, one year hence. The newly calibrated model can be compared to previous versions. The walk-forward process is outlined in Figure 11.13.

The walk-forward approach provides a practical means for testing models, but it is important to recognize that small samples may yield spurious results based on data anomalies. This issue can be mitigated by using different *resampling* techniques.

RESAMPLING

Resampling techniques can be used to leverage the available data and reduce the dependency on specific samples (Cohen, 1988; Efron and Tibshirani, 1993; Sprint, 1998; Cox and Tiao, 1992; Herrity et al., 1999). From the walk-forward results set, a sub-sample is selected at random. The set of performance measures (e.g. number of defaults correctly predicted) are calculated for the sub-sample and recorded. Another sub-sample is drawn at random, and new performance measures are calculated and recorded. The process is repeated until the distribution of performance measures and basic statistics can be determined for each selected performance measure. Figure 11.13 illustrates the resampling process.

Resampling approaches provide two distinct benefits. First, they provide the variability of model performance measures. This variability can be used to determine whether model performance differences are statistically significant. Second, resampling approaches decrease the likelihood that small number of default events, or even individual defaults, will overly influence a particular model's chances of being rated better or worse than another model. For example, if two models (model 1 and model 2) were otherwise identical in their performance, but model 2 erroneously predicted defaults where none occurred, while model 1 did not predict these defaults, we might be tempted to consider model 2 inferior to model 1. However, resampling techniques may be able to show that the models are virtually equivalent over multiple sample draws, which can provide the range of variability of the model performance metrics.

MODEL PERFORMANCE AND BENCHMARKING

Ordinarily, model comparison relies on the analysis of prediction errors for each model. A significant fraction of model validation literature focused on residual error diagnostics (e.g. t-statistics), which are of limited use for model

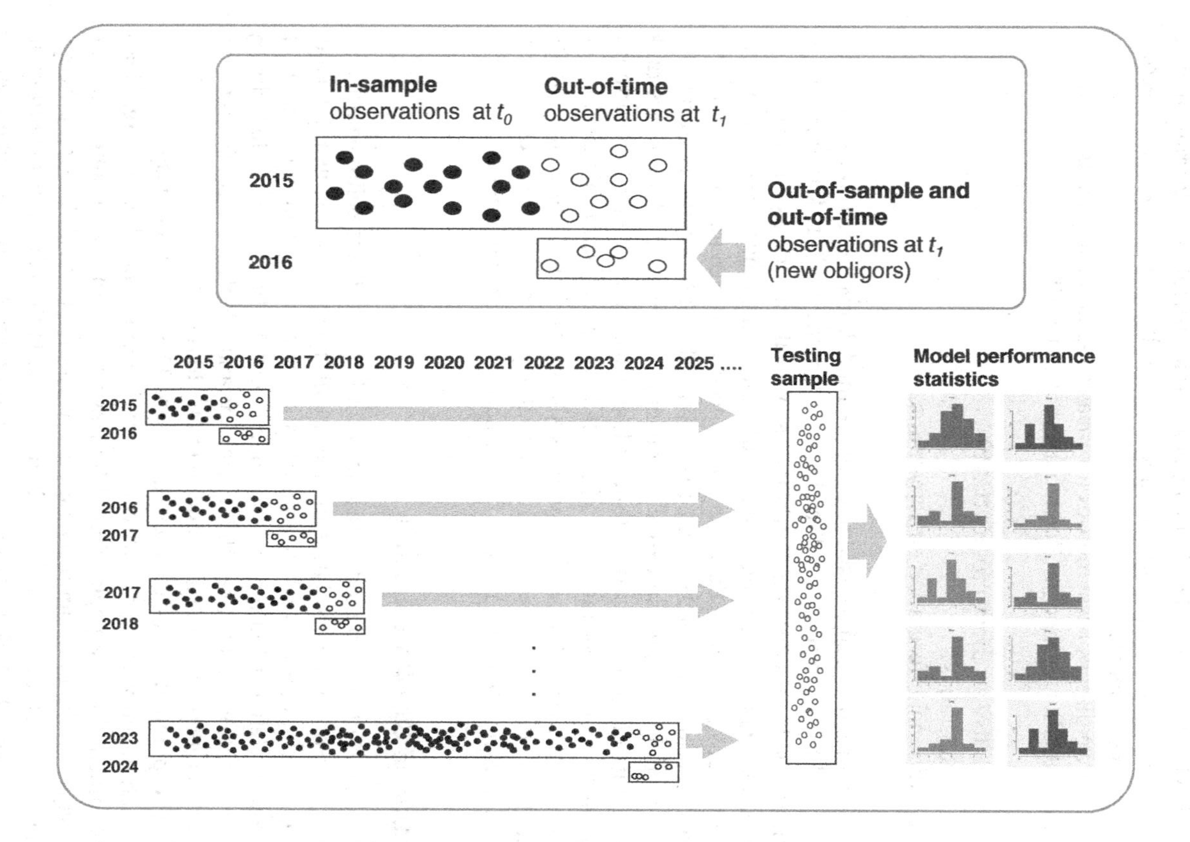

FIGURE 11.13 Walk-forward testing approach based on a sample of historical data. Solid circles represent data used for fitting the model and open circles represent testing data. The walk-forward testing process (bottom left) is performed by fitting the model parameters using data through a selected year, and testing on data from the following year, and then moving the process forward one year. The results of the testing for each year are aggregated (lower left) to calculate model performance statistics.

comparison. Furthermore, key assumptions that underlie residual error diagnostics (e.g. independence of observations or normality of residuals) are frequently violated in practice. In many cases, it is difficult to determine how to correct the model comparison statistics that authors often cite when recommending their models.

Here we discuss basic metrics for measuring and comparing the performance of credit risk models (Keenan and Sobehart, 1999):

1. cumulative accuracy profiles (CAP)
2. accuracy ratios
3. conditional information entropy ratios.

These metrics can be used to compare models even when the model outputs are difficult to compare directly. For example, credit risk ratings can be evaluated side by side with continuous score values such as a probability of default estimate. Comparing models is often a challenging task since the models themselves usually reflect different aspects of the default events and time horizons and may express credit risk using different outputs. For example, some models calculate an explicit probability of default, while others rank risk using ratings that incorporate default and loss expectations. Furthermore, ordinarily agency risk ratings are intended reflect creditworthiness through the economic cycle, that is, they reflect stability over time. This contrasts with models for short-term creditworthiness that rely on the latest market information. However, these models can be compared using the metrics above.

Measure of Cumulative Accuracy

Let's assume that two models (model 1 and model 2) are being tested on a population of borrowers. These models are designed to distinguish the population of defaulters (right distribution) from the population of non-defaulting borrowers (left distribution), as shown in Figure 11.14. Here we follow closely the description in Burton (1972), Hanley and McNeil, 1982; Keenan and Sobehart (2000), and Sobehart and Keenan (2001).

The population of defaulters in Figure 11.14 has been exaggerated for illustration purposes. Ordinarily, the population of defaulters is a relatively small fraction of the population of borrowers, and the two populations can overlap considerably. Figure 11.14(a) shows a hypothetical decision axis that reflects the output of model 1, which represents the borrower's credit quality. In Figure 11.14(a) the criterion C_1 (cutoff point) adopted for classifying a borrower as a defaulter is set at a relatively high value along the model decision axis (i.e. a lax decision in judging credit quality). Borrowers whose risk

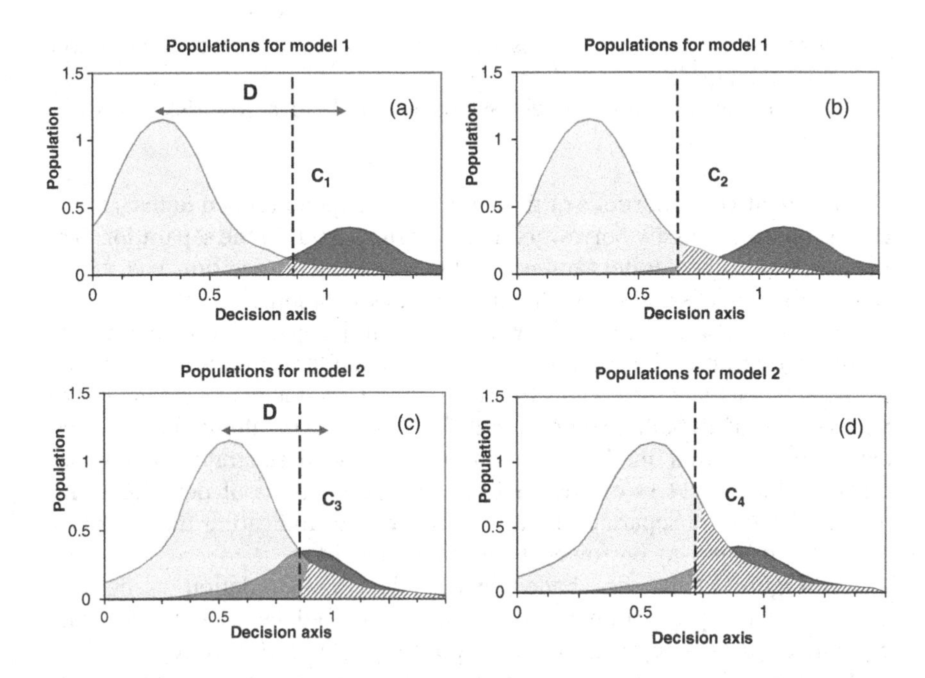

FIGURE 11.14 Separation of populations: non-defaulters on the left and defaulters on the right (exaggerated for illustration purposes). Panels (a) and (b) show the decision axis for model 1. adopting either a lax criterion C_1 or a strict criterion C_2. Panels (c) and (d) show the decision axis for model 2. adopting either a lax criterion C_3 or a strict criterion C_4.

scores are above the cutoff C_1 are classified as likely defaulters. This lax criterion results in a small fraction of actual defaulters to be classified correctly (type I error). In Figure 11.14(b), a lower cutoff C_2 is adopted as a stricter criterion in judging credit quality. Criterion C_2 results in a large fraction of non-defaulters to be classified incorrectly as likely defaulters (type II error). In Figures 11.14(c) and Figure 11.14(d), similar criteria C_3 and C_4 have been adopted for model 2.

There are four possible outcomes when using these models on the population of borrowers:

1. *Hit:* the model assigns high risk to a defaulter.
2. *Miss:* the model assigns low risk to a defaulter.

3. *False alarm:* the model assigns high risk to a low-risk borrower (non-defaulter).
4. *Correct rejection:* the model assigns low risk to a low-risk borrower (non-defaulter).

The relative proportions of these outcomes depend on two items: (1) the cutoff C used to classify borrowers (risk criterion), and (2) the separation D of the population of defaulters and non-defaulters along the decision axis, which reflects the model's ability to differentiate the populations.

Figure 11.14 illustrates how cutoff C and separation D contribute to the proportion of hits and misses for defaulters and non-defaulters. In Figures 11.14(a) and 11.14(b), separation D is relatively small as the populations of defaulters and non-defaulters overlap along the decision axis, reflecting that model 1 has limited ability to separate the populations. In Figures 11.14(c) and 11.14(d), the populations of defaulters and non-defaulters are separated by a greater amount, indicating that model 2 can differentiate risky borrowers better than model 1.

In Figure 11.14, the shaded area under the population of defaulters reflects the proportion of correct *hits* produced by each model. The cross-hatched area represents the proportion of *false alarms* generated by each model. The clear area under the population of defaulters represents *misses* occurring when each model assigns low risk to a defaulter. Finally, the clear area under the non-defaulter population represents *correct rejections.*

The separation of the populations of defaulters and non-defaulters is related to the selected cutoff C (risk criterion) and distance D (model's ability to measure the borrower's creditworthiness). This relation can be visualized in the *receiver operating characteristics* (ROC) and *cumulative accuracy profiles* (CAP) curves (Burton, 1972; Swets, 1988; Provost and Fawcett, 1997; Hoadley, and Oliver, 1998; Keenan and Sobehart, 2000).

Receiver Operating Characteristics

The ROC curve reflects the fraction of hits (correct default predictions) for a given level of false alarms (incorrect default predictions). More precisely, the y-axis of the ROC curve is the *hit rate,* while the x-axis is the *false alarm rate* for a given cutoff value for the model outputs sorted by riskiness. The formula for the hit rate (HR) is

$$HR(C) = \frac{H(C)}{H(C) + M(C)} \tag{11.47}$$

Here $H(C)$ is the number of hits, $M(C)$ is the number of misses for a cutoff C, and $H + M$ is the total number of defaulters. Similarly, the formula for the false alarm rate (FAR) is

$$FAR(C) = \frac{F(C)}{F(C) + R(C)} \tag{11.48}$$

Here $F(C)$ is the number of false alarms, $R(C)$ is the number of correct rejections for a cutoff C, and $F + R$ is the total number of non-defaulter obligors. Note that models may produce different types of scores or outputs, which may require different cutoff values. ROC curves eliminate the dependence on cutoff values C by plotting the hit rate $HR(C)$ versus the false alarm rate $FAR(C)$ as shown in Figure 11.15.

Notice that the upward-sloping diagonal $HR = FAR$ represents the case where the population of defaulters and non-defaulters overlap completely (that is, the non-informative case). The departure of a model's ROC curve from the diagonal $HR = FAR$ measures the ability of the model to separate the populations of defaulters and non-defaulters. Figures 11.14 and 11.15 highlight that there is a tradeoff between type I and type II errors. Minimizing one type of error usually comes at the expense of increasing the other type.

The maximum departure L of the model's ROC curve from the diagonal $HR = FAR$ can be measured along the downward-sloping diagonal $HR = 1 - FAR$ as shown in Figure 11.15.

$$L = \sqrt{\left(HR(C_D) - \frac{1}{2}\right)^2 + \left(FAR(C_D) - \frac{1}{2}\right)^2} \tag{11.49}$$

Here C_D is the cutoff value for which the hit rate $HR(C)$ intersects the diagonal $1 - FAR(C)$ as shown in Figure 11.15. Cutoff C_D represents the point along the decision axis where the shaded area under the normalized population of defaulters equals the unshaded area under the normalized population of non-defaulters in Figure 11.14.

The first term in Equation (11.49) is the difference between the median of the population of defaulters and the cumulative fraction of defaulters correctly classified by the model at cutoff C_D (represented by the shaded area under the distribution of defaulters in Figure 11.14). The second term is the difference between the median of the non-defaulter population and the cumulative fraction of false alarms at cutoff C_D (represented by the shaded area under the distribution of healthy firms in Figure 11.14). Highly predictive models can separate the populations of defaulters and non-defaulters ($HR(C_D) \approx 1$ and $FAR(C_D) \approx 0$). Poorly performing models confuse the two populations ($HR(C_D) \approx 1/2$ and $FAR(C_D) \approx 1/2$). That is, $L \approx 0$ shows the

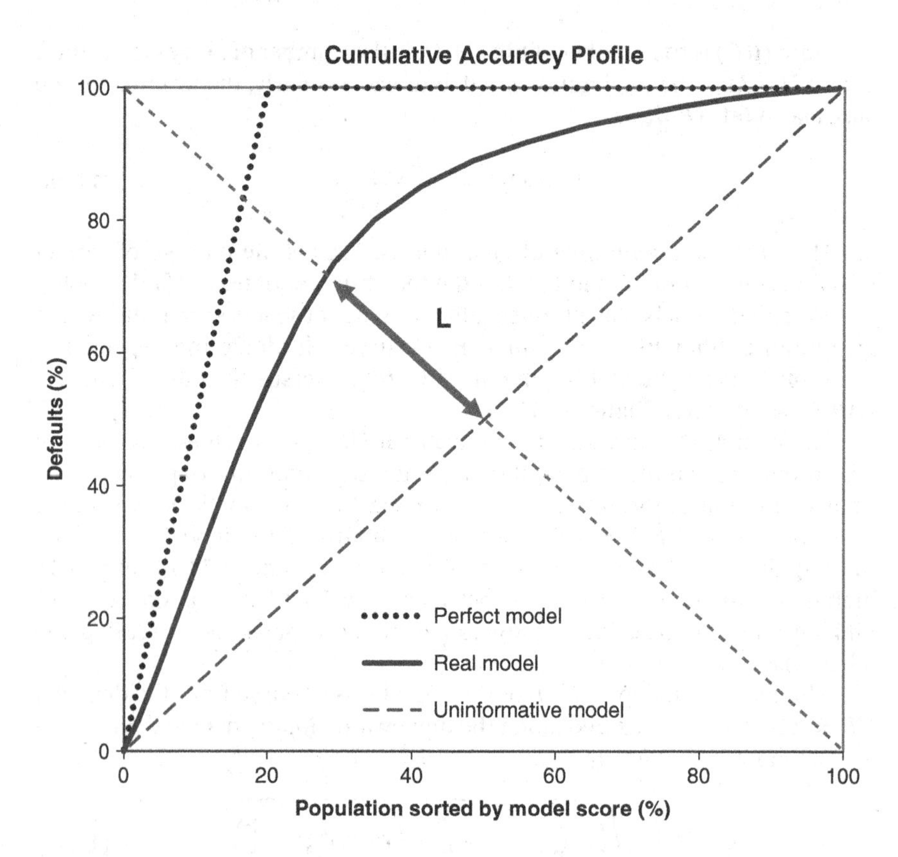

FIGURE 11.15 ROC curves based on the hypothetical decision axes and criteria in Figure 11.14. The distance ***L*** between the model's ROC curve and the uninformative ROC (random assignment of creditworthiness) defines a point (local) measure of the separation between the population of defaulters and non-defaulting companies.

model provides little advantage over a random assignment of model outputs. In contrast, $L \approx 1/\sqrt{2}$ indicates the model provides good separation of the populations of defaulters and non-defaulting companies.

Cumulative Accuracy Profiles

Cumulative accuracy profiles (CAP) represent the cumulative fraction of defaults over the population of defaulters and non-defaulters (as opposed to the non-defaulting population only as in ROC curves). Since defaulters usually represent a small fraction of the population of borrowers, ROC and

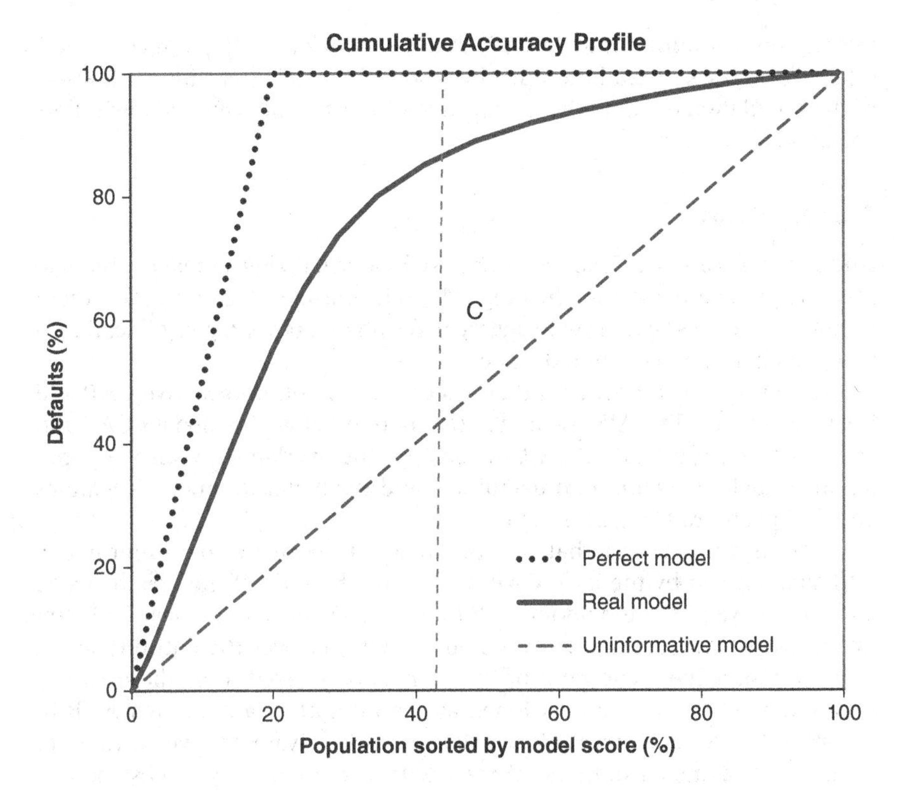

FIGURE 11.16 Type I CAP curve. The solid curved line shows the model performance calculated as the percentage of defaults captured by the model (y-axis) vs. the model score (x-axis). The upper dotted line represents an ideal model that separates defaulters and non-defaulters perfectly. The dashed line represents a random assignment of model outputs (uninformative case).

CAP curves will look very similar. To plot cumulative accuracy profiles, companies are ordered by model score from riskiest to safest. For a given fraction $X\%$ of the total number of companies, a CAP curve is constructed by calculating the percentage $Y(X)$ of the defaulters whose risk score is equal to or lower than the one for fraction X. CAP curves reveal information about the predictive accuracy of the model over its entire range of risk scores for a selected time horizon as shown in Figure 11.16.

A good model assigns the riskiest model scores to the defaults. The percentage of all defaulters identified by the model (y-axis) increases quickly as one moves up the sorted sample (x-axis). In contrast, if the model assigned risk scores randomly (uninformative), we would expect to capture a fraction $X\%$ of the defaulters within about $X\%$ of the observations, generating the diagonal line (random or uninformative CAP). A perfect model would produce an ideal CAP, which is a straight line capturing 100% of the defaults within a fraction of the population equal to the default rate of the sample, when all defaulters are exhausted.

Accuracy Ratios

Although CAP and ROC curves can be used for visualizing a model's ability to separate populations across the range of model outputs, it is more convenient to summarize these curves with aggregated statistics beyond point (local) estimates such as the maximum distance L.

Here we focus on the area that lies *above* the non-informative CAP and is *below* the model's CAP. The more area there is below the model's CAP and above the non-informative CAP, the smaller the overlap between the populations of defaulters and non-defaulters, and the better the model separates populations overall (Figure 11.14).

The maximum area that can be enclosed above the non-informative CAP is identified by the ideal CAP. Therefore, the ratio of the area between a model's CAP and the random CAP to the area between the ideal CAP and the random CAP summarizes the predictive power over the entire range of possible risk values. The ratio of these areas is referred to as the *accuracy ratio* (AR), which can be envisioned as the ratio of area A to area A+B in Figure 11.17. Models with $AR \approx 0$ display little advantage over a random assignment of model outputs. Models with $AR \approx 1$ display almost perfect predictive power.

Technically, the accuracy ratio is defined as

$$AR = \frac{2 \int_0^1 Y(x)\, dx - 1}{1 - f} \tag{11.50}$$

Here $Y(X)$ is the (type I) CAP curve for model output X, and $f = d/(n + d)$ is the fraction of defaults in the population (i.e. the average default rate for the default calculation window), where d is the total number of defaulting obligors and n is the total number of non-defaulting obligors.

Similarly to the definition of AR for CAP curves, we can define a summary statistic for ROC curves by calculating the area that lies *above* the uninformative ROC and is *below* the model's ROC. The maximum area that can be enclosed above the uninformative line is 1/2. Therefore, the ratio of the area between a model's ROC and the uninformative diagonal line to the area between the ideal ROC and the uninformative line defines the ROC accuracy ratio (RAR):

$$RAR = 2 \int_0^1 HR(FAR)\, d(FAR) - 1 \tag{11.51}$$

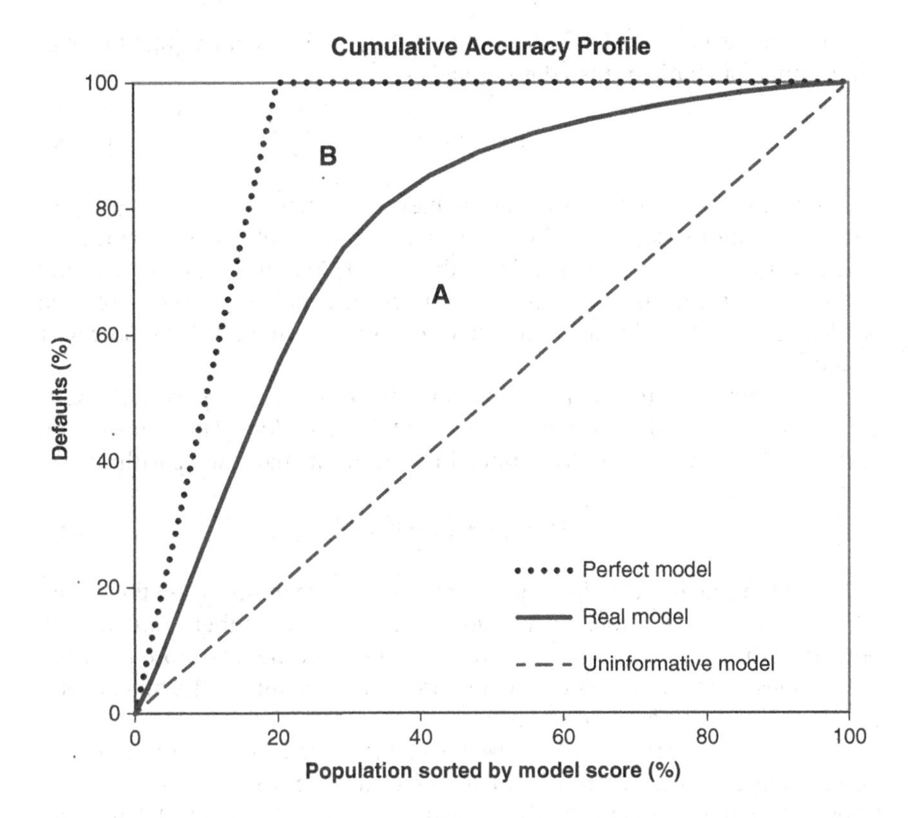

FIGURE 11.17 The accuracy ratio is the ratio of area A under the model's CAP to areas A+B under the ideal CAP.

The accuracy ratio statistic (area under the curve summary) is related to the non-parametric Mann-Whitney U test.

Conditional Information Entropy

Intuitively, the *information entropy* (IE) measures the overall "amount of uncertainty" represented by a probability distribution. The concept of information entropy has its origin in the fields of statistical mechanics and communication theory (Shannon and Weaver, 1949; Jaynes, 1957; Pierce, 1970). Information entropy is conceptually appealing since it can measure objectively how much information is contained in a set of model outputs to predict defaults or credit events.

Following our previous discussion on information entropy, assume the existence of an event with only two possible outcomes: (1) borrower defaults with probability p, and (2) borrower does not default with probability $1 - p$.

The amount of additional information I lenders and investors require to determine which outcome occurred is defined as

$$I = -\log_2(p) \tag{11.52}$$

Here $\log_2(p)$ is the logarithm of p in base 2. When $p = 1$, there is no uncertainty about the outcome and there is no need for more information. However, when the two events are equally likely, then $p = 1/2$ (uninformative case) and the amount of information required to determine which outcome occurred is $-\log_2(p) = 1$ (bit) (i.e. a yes/no answer to the question: did the borrower default?).

The amount of information depends on the selected logarithmic base, which determines the unit of measure of information. The information entropy of the event is usually defined in terms of the natural logarithm:

$$H_0 = -[p\ln(p) + (1-p)\ln(1-p)] \tag{11.53}$$

Note that the information entropy reaches its maximum when the probability is $p = ½$. If the probability of an event is lower or higher than 1/2, one outcome is more likely to occur than the other, and there is more information and less uncertainty. The reduction in the uncertainty of the outcomes is reflected in the reduction of entropy.

When a model is used to describe the likelihood of default events, information entropy can measure the amount of information introduced by the model. Consider again the two mutually exclusive outcomes: borrower defaults (d), and borrower does not default (n). Given a set of m model outputs (risk scores) $S = \{R_1, .., R_m\}$, the conditional entropy $h(R_j)$ measures the information about the possible outcomes d and n for a specific risk score R_j:

$$h(R_j) = -\left(P(d|R_j)\ln\left(P\left(d|R_j\right)\right) + P\left(n|R_j\right)\ln\left(P\left(n|R_j\right)\right)\right) \tag{11.54}$$

Here $P(d|R_j)$ is the probability that the borrower defaults given R_j and $P(n|R_j) = 1 - P(d|R_j)$. Equation (11.54) describes the entropy of events d and n conditional on R_j. The average over all possible risk scores is the *conditional information entropy* for the model scores:

$$H_1(s, \delta) = H_1(R_1, \ldots., R_n, \delta) = \sum_{k=1}^{n} h(R_k)P(R_k) \tag{11.55}$$

The conditional information entropy in Equation (11.55) measures the amount of uncertainty contained in a model with discrete outputs. For models

TABLE 11.4 Comparison of conditional information entropy ratios for multiple models combining financial and market information. The higher the CIER value, the better the model.

Selected Model	In-sample CIER	Validation CIER
Univariate financial ratio: ROA	0.06	0.06
Z'-Score	0.07	0.06
Hazard rate model	0.11	0.11
Modified Merton model	0.14	0.14
Model with market and financial ratios	0.21	0.19

with continuous outputs, the model output can be divided into bins of size δ, which should be big enough to provide meaningful statistics.

Similarly to the reduction of CAP and ROC curves to the AR and RAR statistics, we can use IE to produce another summary statistic for default risk. The *conditional information entropy ratio* (CIER) compares the amount of uncertainty there is about default when we have no model (a state of more uncertainty about the possible outcomes) to the amount of uncertainty remaining after introducing a model (a state of less uncertainty).

First, we measure the uncertainty associated with the event of default $H_0(p)$, where p is the default rate of the sample, when there is no supplemental information about credit quality. Entropy $H_0(p)$ reflects knowledge common to all models. Second, we calculate $H_1(s, \delta)$ based on the model's outputs (scores) s. CIER is the relative different between these two entropies:

$$CIER(S_1, \delta) = \frac{H_0 - H_1(S_1, \delta)}{H_0} \tag{11.56}$$

If the model had no predictive power (random assignment of scores), CIER would be 0. If the model was perfectly predictive (no uncertainty about the outcomes), CIER would be 1. Because the information entropy measures the reduction of uncertainty, a higher CIER indicates a better model. CIER is related to the Kulbach-Leibler entropy statistic discussed earlier (see Burnham and Anderson, 1998).

Table 11.4 shows the CIER results. CIER variability using resampling is of the order of 0.02 and is obtained with a resampling scheme.

REFERENCES

Aggarwal, C.C. (2018). *Neural Networks and Deep Learning*. Springer, eBook, Cham, *Switzerland.*

Altman, E.I. (1968). Financial Ratios, Discriminant Analysis, and the Prediction of Corporate Bankruptcy. *Journal of Finance* September: 589–609.

BCBS (Basel Committee on Banking Supervision) (1999). Credit Risk Modeling Practices and Applications. *Basel Committee on Banking and Supervision* (April).

BCBS (Basel Committee on Banking Supervision) (2000). Supervisory Risk Assessment and Early Warning Systems. *Basel Committee on Banking and Supervision No.* 4, (December).

Black, F. and Scholes, M. (1973). The Pricing of Options and Corporate Liabilities, *Journal of Political Economy* 81: 637–659.

Burnham, K.P. and Anderson, D.R. (1998). *Model Selection and Inference*, Springer, New York.

Burton, G.A. (1972). *Experimental Psychology,* John Wiley, New York, pp. 137–144.

Caouette, J.B., Altman, E.I., and Narayanan, P. (1998). *Managing Credit Risk: The Next Great Financial Challenge*. Wiley, New York, pp. 112–122.

Cohen, J. (1988). *Statistical Power Analysis for the Behavioral Sciences.* Lawrence Erlbaum Associates, Hillsdale, NJ.

Cox, G.E.P. and Tiao, G.C. (1992). *Bayesian Inference in Statistical Analysis*, John Wiley and Sons, New York.

Crouhy, M., Galai, D., and Mark, R. (2001). *Risk Management*. McGraw-Hill, New York, pp. 315–423.

Duffie, D., and Lando, D. (2001). Term Structures of Credit Spreads with Incomplete Accounting Information. *Econometrica* 69: 633–664.

Duffie, D., and Singleton, K. (2003). *Credit Risk: Pricing, Measurement, and Management*. Princeton University Press, Princeton, NJ.

Efron, B. and Tibshirani, R.J. (1993). *An Introduction to the Bootstrap*. Chapman & Hall, New York.

Foster, G. (1986), *Financial Statement Analysis*. Prentice Hall, Englewood Cliffs, NJ, pp. 533–571.

Fridson, M., and Jonsson, J.G. (1997). Contingent Claims Analysis. *Journal of Portfolio Management* (Winter): 31–43.

Goonatilake, S., and Treleaven, P. (1995). *Intelligent Systems for Finance and Business.* John Wiley & Sons, Chichester.

Hanley, A. and McNeil, B. (1982). The Meaning and Use of the Area Under a Receiver Operating Characteristics (ROC) Curve. *Diagnostic Radiology* 143(1): 29–36.

Herrity, J., Keenan, S.C., Sobehart, J.R., Carty, L.V., and Falkenstein, E. (1999). Measuring Private Firm Default Risk. *Moody's Investors Service article* (June).

Hertz, J., Krogh, A., and Palmer, R. (1991). *Introduction to the Theory of Neural Computation*. Santa Fe Institute in the Science of Complexity, Addison-Wesley, Reading, MA.

Hoadley, B., and Oliver, R.M. (1998). Business Measures of Scorecard Benefit. *IMI Journal of Mathematics Applied in Business & Industry* 9: 55–64.

Janessa, T., Jarrow, R.A., and Yildirim, Y. (2001). *Estimating Default Probabilities Implicit in Equity Prices*. Cornell University, Working Paper.

Jarrow, R.A., and Turnbull, S.M. (2000). The Intersection of Market and Credit Risk. *Journal of Banking and Finance* 24(1–2): 271–299.

Jaynes, E.T. (1957). Information Theory and Statistical Mechanics. *Physical Review* 106(4): 62–630.

Keenan, S.C., and Sobehart, J.R. (1999). *Performance Measures for Credit Risk Models*. Moody's Risk Management Services, Research Report 10-10-99.

Keenan, S.C., and Sobehart, J.R. (2000). Credit Risk Catwalk. *Risk Magazine*, July, pp. 84–88.

Kim, J., Ramaswamy, K., and Sundaresan, S. (1993). Does Default Risk in Coupons Affect the Valuation of Corporate Bonds? A Contingent Claims Model. *Financial Management* 45: 117–131.

Kubat, M. (2017). *An Introduction to Machine Learning*. Springer, eBook, Cham, Switzerland.

Levin, J.W., and van Deventer, D.R. (1997). The Simultaneous Analysis of Interest Rate and Credit Risk. In: A.G. Cornyn, R.A. Klein, and J. Lederman (Eds.), *Controlling and Managing Interest Rate Risk*. New York Institute of Finance, New York.

McQuown, J. A. (1993). *A Comment On Market vs.* Accounting Based Measures of Default Risk. KMV Corporation.

Mensah, Y.M. (1984). An Examination of the Stationarity of Multivariate Bankruptcy Prediction Models: A Methodological Study. Journal of Accounting Research 22(1) (Spring).

Merton, R.C. (1973). Theory of Rational Option Pricing. *Bell Journal of Economics and Management Science* 4: 141–183.

Merton, R.C. (1974). On the Pricing of Corporate Debt: The Risk Structure of Interest Rates. -*Journal of Finance* 29: 449–470.

Pierce, J.R. (1970). *Symbols, Signals and Noise: The Nature and Process of Communication*. Harper & Brothers, New York.

Press, J.S. (1977). *Applied Multivariate Analysis*. Holt, Rinehart and Winston, Inc., New York, pp. 263–267.

Provost, F. ,and Fawcett, T. (1997). *Analysis and Visualization of Classifier Performance: Comparison Under Imprecise Class and Cost Distributions*. In: Proceedings of Third International Conference on Knowledge Discovery and Data Mining, Newport Beach, CA, August 14–17.

Refenes, A.P. (1995). *Neural Networks in the Capital Markets*. John Wiley and Sons, Chichester.

Shannon, C. and Weaver, W. (1949). *The Mathematical Theory of Communication*. University of Illinois Press, Urbana.

Shumway, T. (1998). *Forecasting Bankruptcy More Accurately: A Simple Hazard Model*. University of Michigan Business School Working Paper.

Skansi, S. (2018). *Introduction to Deep Learning: From Logical Calculus to Artificial Intelligence*. Springer, eBooks, Cham, Switzerland.

Sobehart, J.R., and Keenan, S. (1999). Equity Market Value and Its Importance for Credit Analysis; Facts and Fiction. *Moody's Investors Service, Risk Management Services, Research Report 10-11-99*.

Sobehart, J.R., and Keenan, S.C. (2001). Measuring Defaults Accurately. *Risk* (March).

Sobehart, J.R., and Keenan, S.C. (2002). Hybrid Default Models. *Risk* (February): 73–77.

Sobehart, J.R., Keenan, S.C., and Stein, R. (2000). Benchmarking Quantitative Default Risk Models: A Validation Methodology, *Moody's Investors Service Rating Methodology* (March).

Sobehart, J.R., Stein, R.M., Mikityanskaya, V., and Li, L (2000). Moody's Public Firm Risk Model: A Hybrid Approach to Modeling Default Risk, *Moody's Investors Service Rating Methodology* (February).

Sprint, P. (1998). *Data Driven Statistical Methods*. Chapman-Hall, London.

Swets, J.A. (1988). Measuring the Accuracy of Diagnostic Systems, *Science* 240 (June): 1285–1293.

Vasicek, O.A. (1984). *Credit Valuation*. KMV Corporation.

Vasicek, O.A. (1995). *EDF and Corporate Bond Pricing*. KMV Corporation.

CHAPTER 12

Pillar 2: Structural Models

THE ROLE OF MARKET INFORMATION IN THE PRICING OF RISKY DEBT

In view of both timeliness and the potential distortions in company financial reporting, academics and practitioners have looked to market information to assess both the true financial leverage of a company and its creditworthiness. Early seminal work on the valuation of financial options led to the insightful observation that the common stock of a company possesses option-like features (Black and Scholes, 1972, 1973; Merton, 1973, 1974). This observation led to an options pricing approach based on the relation between the value of a company's assets, its equity and debt. Central to this view is the idea that equity and debt could be considered as financial *options* (financial claims) on the market value of the company's assets. Furthermore, ownership of common stock is seen as equivalent to being long a "call" option on the company's assets.

This insightful idea of a market-based structural approach that can provide timely information is so appealing that academics, practitioners, and commercial vendors have proposed it for pricing risky debt and for predicting defaults on debt obligations (Black and Cox, 1976; Merton, 1976; Kim et al., 1993; Longstaff and Schwartz, 1995; Babbel and Merrill, 1996; Briys and Varenne, 1997; Fridson and Jonsson, 1997; Levin and van Deventer, 1997; Wei and Guo, 1997; Cathcart and El-Jahel, 1998; Nandi, 1998; Saa-Requjo and Santa Clara, 1999; Kao, 2000; Keenan and Sobehart, 2000; Schwartz and Moon, 2000; Sobehart et al., 2000; Duffie and Lando, 2001; Sobehart and Keenan, 2001; Ergashev, 2002; Kealhofer and Kurbat, 2002; Saunders and Allen, 2002; Giesecke, 2003; Giesecke and Goldberg, 2003; Kealhofer, 2003; Baglioni and Cherubini, 2004; Davydenko, 2004; Benos and Papanastasopoulos, 2005). More precisely, the approach provides the probability of the company's assets being below the value of the promised payment. In this situation, the company is unable to raise additional capital to cover its debts and eventually defaults. This approach does not contemplate cases where companies default on their obligations due, for example, to severe liquidity problems, future legal liabilities due to lawsuits, or catastrophic losses due to climate risk event (Sobehart and Keenan, 2002b).

To illustrate the structural approach, consider the simple case of a company that has equity and one bond issue outstanding. The bond matures in one year, calling for a single payment of principal and interest at maturity T, whose value at par is D_0. There are no interim interest payments during the year. Now consider the position of the stockholders (equity owners). At the end of the period, they have the choice of either repaying the debtholders to retain control of the company or, by not paying the bond, going into default, and surrendering ownership of the company to the debtholders. In a very loose sense, the stockholders own an "option" to buy the company from the debtholders at an exercise price equal to the bond value, with an expiration date equal to the time to maturity of the debt. That is, equity $E(t)$ should be priced as an option $C(A, t)$ on the company's assets $A(t)$, with an exercise price equal to the outstanding bond value D_0 and one year to expiration T.

The total market value of the company's equity and debt must sum the total market value of the assets: $A = E + D$. Note that this value does not represent the book value of the company's assets but the value of all productive assets that account for the potential future cash flows of the company. The fair market value of the debt can be found by subtracting the value of the equity from the (unobservable) market value of the company's assets: $D(t) = A(t) - E(t)$. Because there is a chance that the company could default on its obligations if the market value of the assets decline below the outstanding value D_0, the debt should sell at a price that is less than the present value of its promised payment: $D(t) \leq D_0 e^{-r(T-t)}$, where r is the riskless rate of return. As the total market value of the company's assets increases, the probability of default decreases and the market value of the debt approaches that of the riskless debt.

To gain further insight into the equity-as-an-option analogy we need to know which party owns the company's assets at each point in time and who has a conditional claim against those assets. A simple interpretation of the exact call option analogy $E = C(A, t)$ would imply (counterfactually) that ownership of the assets is transferred temporarily to the debtholders once debt is issued. In this case, the owners of the call options (stockholders) have the right to repurchase the assets at maturity if their value exceeds the exercise price D_0 - the face value of the debt (Figure 12.1(a)). From the call-put parity relationship, the equivalent view is that equity is given by $E = A - D_0 e^{-r(T-t)} + P(A, t)$, where $P(A, t)$ is the value of a put option with exercise price D_0 (Figure 12.1(b)). In this case, stockholders own the assets less the present value of the borrowed amount and have the right to swap the residual value of the assets for the borrowed amount if the value of the assets falls below the payment promised to the bondholders. That is, stockholders

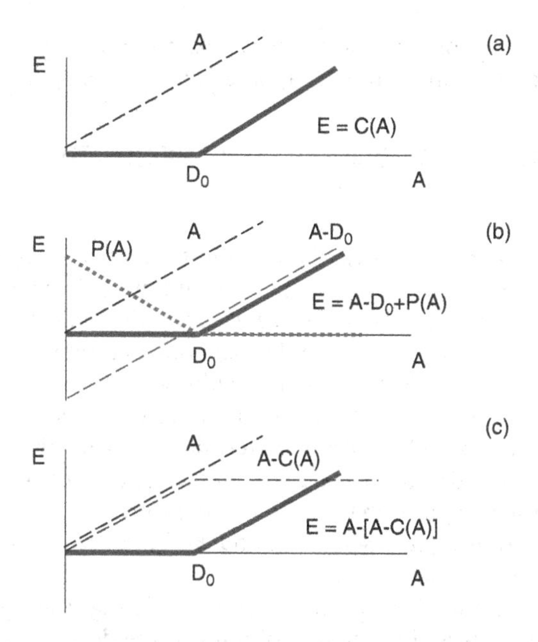

FIGURE 12.1 Option-view of the company's equity E as a function of the company's asset value A at the debt maturity: (a) right to repurchase the company's assets at a fixed price D_0, (b) residual value of the company's assets plus right to sell the assets at fixed price D_0, and (c) ownership of the assets and obligation to repay $min(A, D_0)$ due to limited liability.

can walk away from the company's assets. In reality, the company's assets continue to be owned by the stockholders while debtholders have only a conditional claim against a portion of those assets at the maturity of the debt. This becomes clear if we write $E = A - [A - C(A, t)]$ (Figure 12.1(c)). Stockholders, through the company's management, can dispose of the company's assets as they see fit, and can even liquidate them if it serves their interests and business strategies. Debtholders attempt to protect themselves from these situations by including covenants in the contractual agreement that may limit the amount of dividends paid to stockholders and control some investment and production decisions by specifying the purpose of the debt proceeds. But stockholders and management still have limited liability and can turn over the assets to debtholders for whatever they are worth. This limited liability feature is what makes equity similar to a call option. Because the actual market dynamics of equity and debt may hinge on these underlying relationships, in the following sections we examine the option analogy and the derivation of the options pricing approach more closely.

OPTIONS PRICING, RANDOMNESS, AND THE NOTION OF LIMIT

The popularity of stochastic integrals and infinitesimal (differential) equations in economic and financial models, especially options pricing models, has led to a general disregard for formal derivations as mathematical limit processes. Model derivation shortcuts based on the algebraic manipulation of stochastic differentials directly can yield erroneous results when the nature of the underlying limit is ignored. To illustrate the importance of this issue, we review the derivation of the Black and Scholes and Merton models and discuss the technical implications resulting from *hedging a limit equation for portfolio variations* (the standard pricing approach found in the literature) as opposed to the alternative approach of *hedging the return of the hedge strategy in the limit as the hedging time step approaches to zero.*

In the following we discuss fundamental modeling assumptions, the use of stochastic calculus and the notion of limit in the derivation of credit risk and options pricing models in the context of model design and validation for climate risk applications. Because options pricing and credit models are often used to support investment decisions, it is important not only to understand in detail the strengths and weaknesses of a model's technical assumptions but also the scope of validity of mathematical derivations and their fundamental assumptions.

OPTIONS PRICING AND STOCHASTIC CALCULUS

Here we focus primarily on the derivation of options pricing models based on the no-arbitrage condition (Black and Scholes, 1972, 1973; Merton, 1973, 1974), which leads to the general conclusion that the return on an ideal hedge portfolio must *always* be the riskless rate of return in order to avoid arbitrage opportunities regardless of the curvature of the option as a function of the underlying security.

At first blush for most readers unfamiliar with options pricing, this seems a counter-intuitive result. Here we show that the origin of this counter-intuitive result is the consequence of hedging a limit equation for portfolio variations (the standard pricing approach found in the literature), as opposed to the alternative approach of hedging the return of the hedge strategy in the limit as the hedging time step approaches to zero, which yields a different pricing equation more aligned with intuition. The discrepancy between these two mathematical procedures becomes apparent when we examine the notion of convergence and the nature of the terms being neglected in the limit process (Sobehart and Keenan, 2002c).

Because readers may be skeptical of our opening comments (after all, the application of the no-arbitrage condition to ideal hedge portfolios can be found elsewhere in the literature and has become fundamental to the derivation of options pricing models for over 50 years), we first review briefly the standard derivation of options pricing and the notion of Ito stochastic integrals, and then we present a formal derivation of our arguments.

In the Black-Scholes-Merton (BSM) framework, it is the competitive effort of all market participants to detect and exploit arbitrage opportunities that provides the final constraint required for pricing options. To be able to determine the value of an option from the value of the underlying security Black and Scholes (1972, 1973) introduced a closing condition based on an idealized hedge strategy for a portfolio composed of options and an off-setting position in the underlying security. By construction their ideal hedge portfolio is immune (to first order) to the random changes of its components. They argue that, because such a portfolio should not lead to sustainable arbitrage opportunities, its return must be the risk-free rate (that is, the return on a financial instrument that bears no risk). This condition immediately yields a closed-form differential equation for the value of a financial option as a function of the underlying security. This equation is the foundation for many models for options pricing and the valuation of risky debt.

Central to the derivation of the BSM options pricing model is the use of Ito calculus (Schuss, 1984; Hull, 1993; Neftci, 1996). Following closely early work on this topic (Sobehart and Keenan, 2002c), let us assume that the security's price S follows a log-normal random walk given by the stochastic differential equation:

$$dS = \mu S\, dt + \sigma S\, dZ \tag{12.1}$$

Here the random variable Z follows a Wiener process, μ is the rate of return of the security and σ is its volatility. The standard[1] textbook derivation of the BSM model assumes that, at each time t, investors can construct an ideal hedge portfolio $P(S, t)$ composed of a long position in the option $V(S, t)$ and short a number $\partial V/\partial S$ in the underlying security S

$$P(S,t) = V - \frac{\partial V}{\partial S} S \tag{12.2}$$

A simple Taylor series expansion of the variation of the hedge portfolio $P(S, t)$ in terms of the changes in the security's price (Equation (12.1)) immediately yields the expression:

$$dP = \sigma S \frac{\partial P}{\partial S} dZ + \left(\frac{\partial P}{\partial t} + \mu S \frac{\partial P}{\partial S} + \frac{\sigma^2}{2} S^2 \frac{\partial^2 P}{\partial S^2} \right) dt \tag{12.3}$$

Recognizing that the (tangent) hedge strategy in Equation (12.2) partially removes the stochastic term in dZ in Equation (12.3), the portfolio is claimed to become "riskless". To avoid arbitrage opportunities, the following no-arbitrage condition is imposed on the portfolio change:

$$dP = \left(\frac{\partial P}{\partial t} + \frac{\sigma^2}{2} S^2 \frac{\partial^2 P}{\partial S^2} \right) dt = rPdt \tag{12.4}$$

Here r is the risk-free rate. Equations (12.2) and (12.4) immediately yield the options pricing equation:

$$\frac{\partial V}{\partial t} + rS \frac{\partial V}{\partial S} + \frac{1}{2} \sigma^2 S^2 \frac{\partial^2 V}{\partial S^2} = rV \tag{12.5}$$

The same equation can be obtained using the equivalent self-financing portfolio derivation (Merton, 1973, 1974) and alternative derivations based on the idea that removing the term in dZ yields a system with no residual randomness. Thus, by manipulating stochastic differentials the derivation of the options pricing Equation (12.5) becomes simple and immediate. The above derivation is commonly referred as to the "continuous trading" derivation of options pricing, although it is simply an algebraic manipulation that disregards the meaning of stochastic integrals and differentials.

Unfortunately, due to the nature of stochastic processes the usual rules of the ordinary differential and integral calculus fail, and the algebraic manipulation of stochastic differential equations can lead to incorrect results, the cause of which cannot be seen as long as one looks only at the limit, and not at the limit process. Understanding this limitation is crucial for adequate validation of model assumptions.

Extension to the limit is permitted only when it is the result of a well-defined and well-behaved mathematical limit process. In Ito calculus the differential dZ does not represent the instantaneous value of an infinitesimally small random change that can be hedged away but *the limit of a random realization*, which includes the notion of convergence and the expectation over possible realizations of the random increments of the variable Z_t. Recall that Ito stochastic calculus grew out of the need to assign meaning to differential equations involving continuous stochastic processes such as Brownian motion, providing an alternative description to the standard

Fokker-Planck description (Ito, 1951; Karatzas and Sherve, 1991; Di Paola, 1994). Because some of these random processes cannot be differentiated in the ordinary sense, the concept of *differential*, such as dZ in Equation (12.1), has no meaning apart from that assigned to it when it enters an integral. Technically, the correct procedure for deriving pricing equations would be to evaluate the hedging equation exactly for finite Δt and pass to the limit $\Delta t \to 0$ only afterward. However, since the problem contains a stochastic variable Z_t in addition to t, this procedure is ambiguous until we specify exactly how the limit is approached.

The rest of the discussion is organized as follows. First, we review the concept of stochastic integral and its relation to the notion of convergence. Second, we introduce an alternative derivation of the Black-Scholes-Merton approach taking the limit of a hedge position. Then, we illustrate the main points by showing that the difference between *hedging a limit equation hedging in the limit* yields a persistent skew of the implied volatility. Finally, we introduce an insightful numerical example and summarize key observations on models and assumptions.

ITO STOCHASTIC INTEGRALS, CONVERGENCE, AND THE NOTION OF LIMIT

The fast pace of financial innovation and the proliferation of derivatives models based on stochastic integrals and infinitesimal (differential) equations have led some to disregard formal derivations of models as a limit process for finite intervals and seek for short-cuts based on the algebraic manipulation of stochastic differentials directly. To see the implications of this approach, which contains no indication of which limiting process was used, let us review the concept of the Ito stochastic integral.

Let us consider a continuous, square-integrable martingale $Z = \{Z_t, F_t; 0 \leq t < \infty\}$ on a probability space (Ω, F, P) equipped with a filtration[2] $\{F_t; t \geq 0\}$, i.e. a non-decreasing family of sub-σ-fields of $F_s \subseteq F_t \subseteq F$, for $0 \leq s < t < \infty$.

Given a measurable $\{F_t\}$-adapted process X_t, its Ito stochastic integral can be defined as

$$I_t(X) = \int_a^b X_t(\omega)\, dZ_t = \underset{N\to\infty}{\text{l.i.m}} \sum_{k=1}^{N} X_{t_{k-1}}(Z_{t_k} - Z_{t_{k-1}}) \tag{12.6}$$

Here[3] $\underset{N\to\infty}{l.i.m} Y_N = X$ denotes the following convergence of $\{Y_N\}$ to X in $L^2(\Omega)$: $\lim_{N\to\infty} E[(Y_N - X)^2] = 0$. Note that the integral in Equation (12.6) cannot be defined *path-wise* (i.e. for each $\omega \in \Omega$ separately) as ordinary Lebesgue-Stieltjes integrals. However, it can be constructed in a non-trivial but straightforward manner to any degree of accuracy by a sum with the appropriate notion of convergence (Loeve, 1963; Schuss, 1984; Gardiner, 1985; Karatzas and Shreve, 1991; Lasota and Mackey, 1993; Di Paola, 1994; Lamberton and Lapeyre, 1997).

At a first glance, it is clear what Equation (12.6) is meant to define, however, the formal definition obscures a double dependency. The stochastic integral defined in Equation (12.6) is a mathematical construct describing the properties of the integral itself and the properties of the partition and *convergence scheme* used in its construction. This subtle point is the source of the miscalculation in the derivation of Equations (12.1)–(12.5) and similar model extensions found in the literature, which are based on the algebraic manipulation of stochastic differentials.

The essential idea in the definition of the stochastic integral is refinement and convergence, that is, partitions getting ever finer as the number N of subintervals increases without limit. Letting the number of subintervals N increase without limit does not by itself ensure that the subintervals will all undergo an appropriate contraction.[4] In fact, because the subintervals are stochastic in nature, some of them could have arbitrarily large values for a particular realization of the random variable Z_t. It is here that the notion of ordinary integral and that of stochastic integral diverge in their meaning.

The appropriate convergence may be achieved by demanding that the width of the largest subinterval go to zero in some "probabilistic" sense to remove the uncertainty introduced by the stochastic variations. As indicated in Equation (12.6), in Ito calculus convergence is achieved using the limit of expectation in *mean square sense* (or *l.i.m.*), although useful properties can also be derived using the equally important *almost sure* (*a.s.*) convergence.

Let the expected value of the width of that subinterval be denoted by $\|\Delta Z\| = \max_{0 \leq k \leq N} \sqrt{E[(\Delta Z_{t_k})^2]}$; the stochastic integral is now defined as the limit of well-defined sums as $\|\Delta Z(N)\|$ shrinks into nothingness in mean square sense:

$$\lim_{\|\Delta Z\| \to 0} E\left[\left(\sum_{k=1}^{N} X_{t_{k-1}} \Delta Z_{t_k} - \int_a^b X_t(\omega)\, dZ_t\right)^2\right] = 0 \tag{12.7}$$

With the Ito stochastic integral (12.7) entered into existence, X is said to be integrable (measurable) over the interval $[a,b]$. Its relationship to the concept of an ordinary integral is now formal and conceived at a great distance, so has the concept of the infinitesimal dZ in Equation (12.1). It is now clear that any calculation involving Ito stochastic integrals (or the algebraic manipulation of their differential forms) must take into account the notion of convergence.

HEDGING PORTFOLIO RETURNS IN THE LIMIT VS. HEDGING THE LIMIT OF PORTFOLIO CHANGES

In the following we provide an alternative formal derivation of the options pricing equation deriving simultaneously both Ito's Lemma and the limit of the hedging strategy. Our approach is based on taking the limit of the return on a hedging strategy as the hedging time step approaches to zero, as opposed to the standard derivation based on hedging a limit equation for portfolio variations as discussed earlier (Sobehart and Keenan, 2002c). The difference between these two mathematical procedures becomes apparent when we examine the notion of convergence and the nature of the terms being neglected in the limit process.

Proposition 1. The limit of the variance of the return on the ideal delta-hedged portfolio P_t described in Equation (12.2), composed of options $V(S_t, t)$ and an offsetting position on the underlying security S_t, is $\lim_{\Delta t \to 0} E^{\Delta Z}\left[\left(\frac{\Delta P_t}{\Delta t} - E^{\Delta Z}\left[\frac{\Delta P_t}{\Delta t}\right]\right)^2\right] = \frac{1}{2}\left(\sigma^2 S_t \frac{\partial}{\partial S}\left(S_t \frac{\partial P_t}{\partial S}\right)\right)^2 \geq 0$. Therefore, investors may require an additional risk premium λ above the risk-free rate r as compensation for any residual risk exposure caused by curvature effects.

Proof. Let the option value $V : R^+ \otimes R^+ \to R$ be of class C^2 as a function of the security price[5] $0 \leq S < \infty$ and of class C^1 as a function of time $0 \leq t < \infty$. Also let the security price be $S = S_0 e^X$ where $X = \{X_t, F_t; 0 \leq t < \infty\}$ is a continuous semi-martingale with decomposition [10]

$$X_t = X_0 + \sigma Z_t + \mu t; \qquad 0 \leq t < \infty \tag{12.8}$$

Here $Z = \{Z_t, F_t; 0 \leq t < \infty\} \in M^{c,loc}$, and the security's volatility σ and growth rate μ are constant.

Following an early derivation (Sobehart and Keenan, 2002c), let us fix $t > 0$ and consider the capital gain of the portfolio (12.2) at time $t + \Delta t$. A Taylor series expansion of the portfolio yields the expression:

$$\begin{aligned} P(S_{t+\Delta t}, t + \Delta t) &= P(S_t, t) + \frac{\partial P}{\partial t}(S_t, t)\Delta t + \frac{\partial P}{\partial S}(S_t, t)\Delta S_t \\ &+ \frac{1}{2}\frac{\partial^2 P}{\partial S^2}(\eta_t, t)\Delta S_t^{\,2} + R(S_\zeta, \zeta_t)\Delta t^2 \end{aligned} \tag{12.9}$$

Here $\eta_t(\omega) = S_t(\omega) + \theta_t(\omega)\Delta S_t$ and $\Delta S_t = S_{t+\Delta t} - S_t$ for some appropriate $\theta_t(\omega)$ satisfying $0 \leq \theta_t(\omega) \leq 1$, $\omega \in \Omega$, $R(S_\zeta, \zeta)$ is a residual term, and $\zeta_t = t + \phi_t \Delta t$ for some appropriate ϕ_t satisfying $0 \leq \phi_t \leq 1$. Combining Equations (12.8) and (12.9) yields (Sobehart and Keenan, 2002c):

$$\Delta P_t = P(S_{t+\Delta t}, t + \Delta t) - P(S_t, t) = G_1 + G_2 + G_3 + G_4 + G_5 + O\left(\Delta t^{\frac{3}{2}}\right) \tag{12.10}$$

Here we define:

$$\begin{aligned}
G_1 &:= \sigma S_t \frac{\partial P}{\partial S}(S_t, t)\Delta Z_t \\
G_2 &:= \mu S_t \frac{\partial P}{\partial S}(S_t, t)\Delta t \\
G_3 &:= \left(\frac{\partial P}{\partial t}(S_t, t) + \frac{\sigma^2}{2} S_t^2 \frac{\partial^2 P}{\partial S^2}(S_t, t)\right)\Delta t \\
G_4 &:= \frac{\sigma^2}{2} S_t \frac{\partial}{\partial S}\left(S_t \frac{\partial P}{\partial S}(S_t, t)\right)(\Delta Z_t^2 - \Delta t) \\
G_5 &:= \frac{\sigma^2}{2} S_t^2 \left(\frac{\partial^2 P}{\partial S^2}(\eta_t, t) - \frac{\partial^2 P}{\partial S^2}(S_t, t)\right)\Delta t
\end{aligned} \tag{12.11}$$

Here $\eta_t = S_t + \theta_t \Delta S_t$ where $0 \leq \theta_t \leq 1$. The limit of terms G_1, G_2 and G_3 in Equations (12.10) and (12.11) can be recognized as Ito's lemma. Notice that $|G_4| \sim O(\Delta t)$ for each random realization ΔZ_t but $E^{\Delta Z}[G_4] \sim O(\Delta t^2)$, and $E^{\Delta Z}[G_4^2] \sim O(\Delta t^2)$. In the limit $\Delta t \to 0$, term G_5 vanishes because S_t is continuous. The last term in Equation (12.10) includes higher order contributions in Δt that vanish in the limit $\Delta t \to 0$.

Since the hedge (12.2) removes the leading stochastic term in $\Delta t^{\frac{1}{2}}$, we can take the limit of Equations (12.10) and (12.11) for the rate of change in the continuously hedged portfolio:

$$\lim_{\Delta t \to 0} E^{\Delta Z}\left[\frac{\Delta P_t}{\Delta t}\right] = \hat{L}P_t = \left(\frac{\partial}{\partial t} + \frac{\sigma^2}{2} S_t^2 \frac{\partial^2}{\partial S^2}\right) P_t \tag{12.12}$$

Here $\hat{L}$ is the linear operator defined by the time and price derivatives in parenthesis.

Note that in the BSM framework the return on the ideal hedge portfolio is claimed to be certain and equal to the riskless rate. That is, Equation (12.12) reduces to $\hat{L}P_t = rP_t$. More precisely, since the portfolio is assumed "riskless", the limit of the variance of the portfolio change should vanish identically.

However, from Equations (12.10)–(12.12) the variance of the changes in the hedge portfolio is clearly

$$\lim_{\Delta t\to 0} E^{\Delta Z}\left[\left(\frac{\Delta P_t}{\Delta t} - \hat{L}P_t\right)^2\right] = \lim_{\Delta t\to 0} E^{\Delta Z}\left[\left(\frac{\sigma^2}{2}S_t\frac{\partial}{\partial S}S_t\frac{\partial P_t}{\partial S}(\chi^2 - 1)\right)^2\right]$$

$$= \frac{1}{2}\left(\sigma^2 S_t\frac{\partial}{\partial S}\left(S_t\frac{\partial P_t}{\partial S}\right)\right)^2 \geq 0 \tag{12.13}$$

Here we introduced $\chi^2 = (\Delta Z_t)^2/\Delta t$, where $\chi \sim N(0,1)$ is a normal random variable. Equation (12.13) indicates that the limit of the ideal hedge portfolio is not "riskless" but "risky" to a lower order in $\Delta t \to 0$, which is of the same order as the deterministic component of the portfolio return. That is, the simple tangent replication of the ideal BSM strategy cannot remove the random fluctuations of the portfolio returns related to curvature effects even in the continuous limit $\Delta t \to 0$. More clearly, note from Equation (12.2) that even if $\partial P/\partial S \approx 0$ as result of the hedging strategy, the portfolio is still exposed to additional curvature or gamma risk if $\partial^2 P/\partial S^2 \neq 0$.

Imposing the condition that Equation (12.12) must yield a risk-adjusted return[6] on the portfolio ($\hat{L}P_t = (r + \lambda)P_t$) to compensate for curvature effects that cannot be hedged away, Equations (12.2)–(12.12) immediately yield the alternative options pricing equation:

$$\frac{\partial V}{\partial t} + (r+\lambda)S_t\frac{\partial V}{\partial S} + \frac{\sigma^2}{2}S_t^2\frac{\partial^2 V}{\partial S^2} - (r+\lambda)V = 0 \tag{12.14}$$

This risk-adjusted equation is the same as the BSM Equation (12.5) but with an additional risk premium $\lambda\left(S_t, V_t, t, \lim_{\Delta t\to 0} E\left[\left(\frac{G_4}{P_t}\Delta t\right)^2\right]\right)$ that depends on the risk preferences of the options market participants, who may require additional compensation for any residual risk exposure due to imperfect hedging and curvature effects that may affect the return on their investment. This completes the proof of **proposition 1**.

Finally, note that in order to recover the BSM equation we need to introduce the additional assumption that market participants do not require any compensation for the additional risk exposure G_4 (i.e. $\lambda = 0$). In their seminal paper on options pricing, Black and Scholes (1973, p. 643) argued that this risk could be diversified away if its correlation with a market portfolio that exhibits *normal* random increments vanished.[7] However, the assumption that the market portfolio may exhibit only normal random increments is somewhat restrictive since under general conditions the market portfolio must include all kinds of securities including options and securities with non-normal returns.

The additional assumption that there is no risk premium as compensation for curvature effects is the reason why investors' risk preferences drop from the standard options pricing equation generating a paradox between the BSM riskless result and mathematical intuition (Sobehart and Keenan, 2002c).

RESIDUAL RISK AND VOLATILITY SKEWS

To illustrate the implications of these two different derivations of options pricing equations from the perspective of model design, implementation, and validation, let us discuss an example based on implied volatilities. Let $V(S, t; r + \lambda, \sigma, T, K)$ be a European call option with maturity T and exercise price K obtained from Equation (12.14) for $\lambda \geq 0$. Let us estimate the implied volatility σ_I for the option V inferred from the standard Black-Scholes formula for a call option $V(S, t; r, \sigma_I, T, K)$. That is, we make $V(r + \lambda, \sigma) = V(r, \sigma_I)$ and estimate the implied value σ_I in the presence of a market risk premium as compensation for curvature effects. Let us also assume that both the risk premium, and the discrepancy between the implied volatility σ_I and the security's volatility σ are small. Then, introducing a first-order Taylor series expansion in terms of the implied volatility and risk premium, we obtain

$$V(r + \lambda, \sigma) \approx V(r, \sigma) + \partial_r V \lambda = V(r, \sigma_I) \approx V(r, \sigma) + \partial_\sigma V(\sigma_I - \sigma) \quad (12.15)$$

Let $V(S, t) = SN(d_1) - e^{-r(T-t)}KN(d_2)$ be the solution to the Black-Scholes Equation (12.5) for a European call option at a time t with strike price K and maturity T, where $N(z)$ is the cumulative normal distribution function

$$d_1 = \frac{\log(S/K) + (r + \sigma^2/2)(T - t)}{\sigma\sqrt{T - t}} \quad \text{and} \quad d_2 = d_1 - \sigma\sqrt{T - t} \quad (12.16)$$

Then, Equation (12.15) yields

$$\sigma_I \cong \sigma - \lambda \frac{\partial V/\partial r}{\partial V/\partial \sigma} \cong \sigma + \lambda \frac{KN(d_2)}{S\partial_d N(d_1)} e^{-r(T-t)}\sqrt{T - t} \quad (12.17)$$

Equation (12.17) shows that the implied volatility σ_I contains a persistent skew due to the risk preferences of market participants. Out-of-the-money options will tend to exhibit a smaller implied volatility than in-the-money call options. The effect is stronger for short-term options. Figure 12.2 shows the basic features of Equation (12.17) across the option's "moneyness" (i.e., how far the asset value is from the option's strike price) for different times to expiration. Because we assumed normally distributed returns and constant volatility, interest rates and risk premium for illustration purposes, Equation (12.17) cannot generate volatility smiles or other effects. Note, however, that by introducing small changes in the assumptions above we can obtain a wider range of phenomena. We recognize, of course, that there may be other more relevant

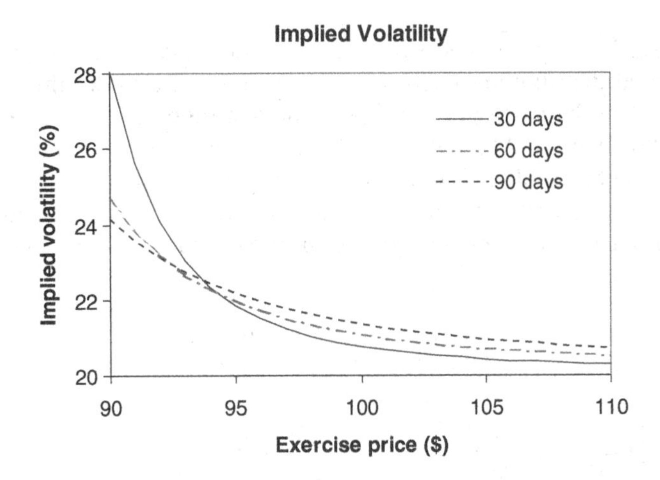

FIGURE 12.2 Implied volatility as a function of the option's moneyness for different times to expiration $T - t =$ 30, 60 and 90 days. The security's price is $S = \$100$, the security's volatility is $\sigma = 20\%$, the riskless rate is $r = 5\%$, and the required risk premium is $\lambda = 1\%$.

factors that can contribute to the actual volatility skew observed in the market such as non-normal distributed returns, stochastic changes in volatility, interest rates or other state variables (Natenberg, 1994; Derman, 1999; Derman and Zou, 2001; Alexander, 2001; Lipton, 2002).

HEDGING STRATEGIES AND RESIDUAL RISK

Here we illustrate the impact of the residual risk using an insightful numerical example (Sobehart and Keenan, 2002c) that shows why hedging away random fluctuations of order $O\left(\Delta t^{\frac{1}{2}}\right)$, which are the leading contributions to the definition of Ito integrals, does not mean that the hedge portfolio is riskless to any order of $\Delta t \to 0$.

Using Equation (12.2) let us construct the ideal BSM hedge portfolio at time t and select a small interval Δt. Keeping the initial value S and time t fixed, generate a sample of n random increments for the stock price and the portfolio during time interval Δt:

$$S_k = Se^{\left(\mu - \frac{\sigma^2}{2}\right)\Delta t + \sigma \Delta Z_k}$$

$$\Delta P_k = (V(S_k, t + \Delta t) - V(S, t)) - \frac{\partial V}{\partial S}(S_k - S) \quad 1 \leq k \leq n \qquad (12.18)$$

Here $\Delta Z_k = X_k\sqrt{\Delta t}$, where $X_1, X_2, \ldots, X_n$ are random draws from a standard normal distribution. Each S_k and ΔP_k represent one of the n random realizations of the stock price and portfolio variation at time $t + \Delta t$. The set of portfolio changes $\{\Delta P_1, \Delta P_2, \ldots, \Delta P_n\}$ provides us with a sample from the distribution of possible portfolio variations at time $t + \Delta t$.

Finally, we calculate the mean value and deviation of the portfolio change as the sample mean M and deviation D for the sample of n random draws:

$$E\left[\frac{\Delta P}{\Delta t}\right] \approx M = \frac{1}{n}\sum_{k=1}^{n}\frac{\Delta P_k}{\Delta t} \tag{12.19}$$

$$E\left[\left(\frac{\Delta P}{\Delta t} - E\left[\frac{\Delta P}{\Delta t}\right]\right)^2\right] \approx D^2 = \frac{1}{(n-1)}\sum_{k=1}^{n}\left(\frac{\Delta P_k}{\Delta t} - M\right)^2 \tag{12.20}$$

According to the standard derivation of the BSM model, the ideal hedge leads to a riskless strategy with a purely deterministic portfolio variation whose return must be the risk-free rate: $dP = rPdt$. We argue that no matter how small Δt, there are always portfolio variations of order $O(\Delta t)$ that are not hedged away and are of the same order than the deterministic part of the portfolio variation. This makes the portfolio stochastic (risky) but to a lower order than the original order $O(\Delta t^{1/2})$, which has been hedged away with the hedging strategy.

Table 12.1 shows the mean and deviation of the portfolio rate of change (Equations 12.19 and (12.20)) using $n = 200{,}000$ random draws for different time intervals Δt. The parameters for the simulations shown in Table 12.1 are $S = \$10$, $K = \$10$, $\sigma = 20\%$, $\mu = 6\%$, $r = 5\%$ and $T - t = 0.5$ years. For this example, the theoretical value for the deviation of the portfolio rate of change is 0.774 (Equation (12.13)), while the theoretical value for the mean portfolio rate of change is −0.264 (Equation (12.12)). Note that we varied the hedging time step 12 orders of magnitude, and the result is always the same: the portfolio rate of change shows random fluctuations with constant deviation around $D = 0.77$ regardless of how small is Δt. Even for a hedging time step of only $\Delta t = 10^{-13}$ years (i.e. an unrealistically small hedging time step of 3 microseconds), the curvature effect is still in reasonable agreement with the theoretical result. The discrepancy is mainly due to numerical rounding errors. The results in Table 12.1 clearly show that after removing the leading stochastic contribution in $\Delta Z \sim O(\Delta t^{1/2})$ the ideal hedge portfolio is still exposed to random variations of order $O(\Delta t)$ due to curvature. This effect does not vanish in the limit $\Delta t \to 0$.

TABLE 12.1 Portfolio rate of change $\Delta P/\Delta t$ for different time steps Δt: mean rate of change M and deviation D for a hedge portfolio using Monte Carlo simulations for Equations (12.19) and (12.20). The theoretical limits are given in Equations (12.12) and (12.13).

Δt (years)	Mean M	Deviation D
1.0E-1	−0.266	0.782
1.0E-3	−0.265	0.773
1.0E-5	−0.266	0.776
1.0E-7	−0.262	0.774
1.0E-9	−0.264	0.777
1.0E-11	−0.264	0.773
1.0E-13	−0.264	0.771
Theoretical value	−0.264	0.774

MODELS AND ASSUMPTIONS

The recognition that alternative mathematical procedures can yield hedge portfolios with residual risk exposure suggests that further research is necessary on pricing models based on limit hedging strategies and highlights the relevance of reviewing technical assumptions and mathematical derivations for adequate model design and validation.

Finally, note that even under the ideal conditions described here, there is still a different source of randomness that can lead – both in theory and in practice – to pricing models significantly different from the BSM model and its extensions. In order to derive Equations (12.5) or (12.14) the option contract value $V(S, t)$ is assumed to be a C^1 differentiable function of time t and a C^2 differentiable function of the security's price S (or, more generally, any set of state variables such as interest rate and volatility). Because in the BSM model the option price is assumed to be a deterministic function of the underlying security's price, changes in the security's price are perfectly correlated with the changes of the option value and the leading random term of the hedge portfolio G_1 in Equation (12.11) can be hedged away. In practice, price changes for options and their underlying securities are not perfectly correlated, which makes the ideal BSM hedging strategy approach questionable. In fact, Black and Scholes (1972, p. 400) recognized early that deficient information and the over- and under-reaction of traders can cause option prices to randomly deviate from their fundamental values temporarily. However, they treated this as an empirical limitation rather than a fundamental reality that needed to be modeled. Since options and their underlying securities are traded in different markets, the additional uncertainty introduced by options market participants (e.g. trading noise or random delays in information flow and orders) can be modeled by making $V(S, t)$ a *stochastic* function of the underlying asset price S and time t (and other state variables). The presence of additional random changes in V unrelated to changes in the underlying state variables (which, in general, cannot be removed with the BSM hedging strategy) can lead to an additional risk premium in Equation (12.14) (Sobehart and Keenan, 2003a, 2003b), as well as the formation of temporary price bubbles and other effects caused by incomplete information and divergence in market expectations (Sobehart, 2005; Sobehart and Keenan, 2005; and Haug and Taleb, 2007).

These issues are relevant for climate risk and ESG due to the market uncertainty generated by changes in policies, regulations, incentives, and restrictions imposed on different business activities to transition to a low-carbon economy, which can lead to additional risk premia required by market participants as compensation for increasing levels of uncertainty in their investment and hedging strategies. Since the (no-arbitrage) options pricing framework is the foundation for many structural models of risky debt and probability of default, understanding the impact of different assumptions is critical to assess the benefits and limitations of these models for use in climate risk and ESG.

ASSET-BASED MODELS AND MARKET UNCERTAINTY

One key fundamental assumption of the standard structural models of a company's equity and debt is that the price of claims on the company's assets is driven solely by changes in underlying state variables such as the market value of the company's assets, and characteristics such as maturity, callability, etc. Here we discuss the situation when participants in the equity and debt markets disagree on their valuation introducing uncertainty (trading noise) in the valuation of the company's equity and debt in addition to uncertainty from changes in the value of the company's assets (Thaler, 1999; Shiller, 2000; Shleifer, 2000; Hirshleifer, 2001). The presence of trading noise in the equity and debt markets interferes with arbitrage and hedging strategies. This situation can result in misalignment of equity and debt prices, creating capital structure arbitrage opportunities (Mitchell et al., 2002; Keenan et al., 2003; Sobehart and Keenan, 2005).

In structural models derived from the Black-Scholes and Merton (BSM) framework (Black and Scholes, 1972, 1973; Merton, 1973, 1974), the position of the equity holders for a company with outstanding debt is equivalent to the position of holders of call options on the company's assets. More precisely, in the same way owners of call options on a stock must pay the strike price to take ownership of the underlying, equity holders must pay bondholders the face value of the bonds (the strike price) to retain ownership of the company and its assets. In this insightful view, driven primarily by the company's leverage and volatility of assets, equity holders have no incentive to repay bondholders unless the value of the company's assets is greater than the amount that must be repaid.

Over the years many enhancements were introduced to address simplifying technical assumptions on the definition of the default boundary (the point at which the company defaults), jumps in the value of assets, taxes, bankruptcy costs, stochastic volatility, interest rates and other characteristics. While many of these model extensions include more realistic descriptions, they still preserve the idealized view of fully informed market participants who agree on their valuation of the company's equity and debt based on the company's assets.

Academics and practitioners have struggled to overcome several practical limitations derived from this idealized view. For example, the predicted default probabilities and credit spreads calculated from these models are often too low to match empirically observed default frequencies and market spread values. Structural models often show relatively poor predictive power for identifying default events once the company's financial difficulties are disclosed. Also, industry, geography and company-specific characteristics are often excluded from these models *a priori* based on the assumption of fully informed market prices and idealized assumptions. Fundamental differences in the dynamics of large and mature companies (blue chip stocks), aggressive research-minded companies (growth stocks), and companies whose earnings are seasonal or follow the business cycle (seasonal and cyclical stocks) are simply ignored under the assumption of fully informed markets. These issues are exacerbated for financial institutions, given their leveraged nature.

The fact that equity and debt markets do not seem to be perfectly informed may impact the ability of these models to accurately forecast default events or estimate credit spreads. The behavior of market participants is often subjected to emotions and under- and over-reaction that can create temporary distortions in prices (Sobehart and Farengo, 2002, 2003). Other significant factors that can distort a company's creditworthiness are the uncertainty introduced by management's ability to refinance, renegotiate, or re-structure a company's obligations, which depends on the company's capacity for borrowing; unanticipated changes in the business, and regulatory environment; and information deficiencies caused by misunderstanding of the business complexity, the company's competitive environment, accounting errors, or fraudulent or misleading disclosures. Market information can be most useful when it is combined with information from the company's balance sheet, income statement, statement of cash flows, and industry and macroeconomic conditions.

Incomplete information and uncertainty in valuation lead to both the possibility of capital structure arbitrage situations and the need for enhanced models that include market, accounting, and other relevant information to compensate for accounting inaccuracies, market uncertainty, and trading noise, and the company's capacity to renegotiate and restructure its debt (Sobehart and Keenan, 2002a, 2002b, 2002c, 2003a, 2003b, 2003c, 2004; Hillegeist et al., 2003; Tudela and Young, 2003a, 2003b).

In the following we discuss the extension of the BSM framework when market participants are neither perfectly informed nor capable of eliminating arbitrage opportunities instantaneously. The introduction of trading noise in the equity and debt market interferes with arbitrage and hedging strategies and, therefore, affects the company's equity and debt prices and can invalidate the assumption that equity and debt prices depend solely on the company's assets (or any set of state variables). Our approach is based on a *forward-looking, singular perturbation analysis* of the joint distribution of the company's assets, equity, and debt using multiple time scales. This powerful mathematical technique provides an alternative valuation approach that can overcome the many limitations of the traditional methods and can provide asymptotic solutions for a wide variety of pricing and valuation problems.

Our approach also shows that information deficiencies and uncertainty lead to a stochastic equity-debt relationship with random fluctuations consistent with the concept of market equilibrium. This relationship reduces to the static equity-debt expression obtained from the BSM framework only on average (expectation). Under certain technical conditions, uncertainty in the equity-debt relationship can lead to capital structure arbitrage or arbitrage based on market timing, and highlights two key issues:

1. The static (no arbitrage) relationship between equity and debt based on perfectly informed market participants often does not hold in practice.
2. The use of structural models with idealized market assumptions for imputing default and recovery values may create arbitrage opportunities in the credit derivatives market, and in the equity and bond markets.

   Addressing the issues above requires a broader class of credit risk models that can include separate sources of uncertainty for the debt and equity markets explicitly.

FORWARD-LOOKING, SINGULAR PERTURBATION ANALYSIS

In their seminal work on options pricing, Black and Scholes (1972, p. 400) recognized that deficient information and the under- and over-reaction of market participants can cause option prices to randomly deviate from their fundamental values temporarily. However, instead of modeling this issue explicitly, they treated it as an empirical limitation of their idealized model. The Merton model exhibits the same limitation.

To overcome this limitation, our approach assumes that participants in the equity and debt markets may disagree on their valuation of the company's assets, equity, and debt, introducing different levels of uncertainty in their prices in addition to the uncertainty generated by intrinsic changes in the company's assets. Furthermore, we recognize that actual tradable instruments may not be the perfectly deterministic functions of the underlying state variables (i.e. the company's assets) alone and may reflect additional sources of randomness due to deficiencies in information and discontinuities in supply and demand. In our model, the company's assets, equity, and debt follow related – but different – stochastic jump processes driven by supply and demand, and the uncertainty in the equity and debt markets. The existence of additional sources of uncertainty caused by information deficiencies makes these markets incomplete and does not allow for the perfect hedging of equity and debt by taking positions in the company's assets and risk-free bonds, as in the traditional BSM approach. This is not a drawback of our approach but the recognition that perfect hedges are no longer determinants of equity and debt prices when there are market information deficiencies.

Because our approach uses general jump processes to describe equity and debt prices, we introduce a forward-looking, singular perturbation analysis of the joint distribution of the company's assets, equity, and debt based on multiple time scales organized as follows. First, we introduce generalized jump processes to describe the evolution of the company's assets. Second, we introduce a singular perturbation asymptotic expansion approach for the probability distribution of assets in terms of the strength of asset jumps (Sobehart, 1999, 2005; Sobehart and Keenan, 2005). Third, we introduce different sources of uncertainty for the participants in the equity and debt markets and show how they affect the valuation of the company's equity and debt. Fourth, we revisit the concept of default point and the impact of the company's debt capacity and business uncertainty in the valuation of the company's assets, equity, and debt and on their joint distribution. Finally, we discuss the effects of trading noise on the relationship between the company's assets, equity, and debt values.

In the following, time is denoted t and the value of the company's assets is denoted A. We assume that the company's assets fluctuate randomly but their logarithmic value $x = \log(A/A_0)$ reverts to a time-dependent value $\phi(t)$, where A_0 is a reference value. The non-stochastic term $\phi(t)$ plays a similar role to the drift term μt in traditional random walk models of the company's assets (where μ is the growth rate of the company's assets). Note, however, that $\phi(t)$ represents a broader set of time-dependent functions to accommodate mature, growth, seasonal, or cyclical companies. The direction and magnitude of the fluctuations in the value of the company's assets are driven by the sensitivity of market participants to detect and exploit discrepancies in asset values above (below) the value $\phi(t)$. Changes in asset values occur at random times t_n that are distributed with a jump process with rate of events (*intensity*) $\lambda(x, t)$-which reflects the intraday frequency of changes in asset values. The company's assets value x is modified by random jumps of size $\varepsilon\ R$, where ε is the noise strength and R is a normalized random variable with a distribution $F(R, t)$.

The evolution equation for the jump process driving the log-value x is:

$$dx = \nu(t)\, dt + \varepsilon \sum_{t_n} R_n\, \delta(t - t_n)\, dt \tag{12.21}$$

Here $\delta(t - t_n)$ is the Dirac delta-function describing an instantaneous price jump occurring at a random time t_n. The parameter $\nu(t)$ represents the actual time-dependent growth rate of the company's assets and can contain seasonal components describing market and business cycles. The relationship between ν and the value ϕ estimated by market participants is described later in this section. Equation (12.21) plays the same role as the differential equation $dS/S = \mu\, dt + \sigma\, dZ$ in asset-based structural models, where μ is the return on the company's assets, σ is the asset volatility, and Z is a Wiener process.

We describe the stochastic jump process in Equation (12.21) in terms of the probability distribution $P(x, t')$ of a random value $X(t')$ in the interval $[x, x + dx]$ at time t'. To obtain the evolution of the probability distribution we compute the stochastic contribution to the leading order in a small interval Δt

$$P(x, t' + \Delta t)\, dx = \{(1 - \lambda(x, t')\Delta t)P(x - \nu\Delta t, t') + \int P(x - \nu\Delta t - \varepsilon R, t')\, F(R, t') \\ \lambda\, (x - \varepsilon R, t')\, dR\, \Delta t\}\, dx + O(\Delta t^2) \tag{12.22}$$

Assuming the limit $\Delta t \to 0$, we obtain the following approximation:

$$P(x - \nu\, \Delta t, t') = P(x, t') - \partial_x P\, \nu\, \Delta t + O(\Delta t^2) \tag{12.23}$$

Combining Equations (12.22) and (12.23) leads to a differential equation for the probability distribution of the company's assets:

$$\partial_t P + \nu\, \partial_x P = \int P(x - \varepsilon R, t')\, \lambda(x - \varepsilon R, t')\, F(R, t')\, dR - \lambda(x, t')\, P(x, t') \\ = \sum_{n=1}^{\infty} \frac{(-\varepsilon)^n}{n!} \partial_x^n (\psi_n P) \tag{12.24}$$

Here $\psi_n(x, t') = \int \lambda(x, t')F(R, t')R^n dR$ for $n > 0$. The series expansion in Equation (12.24) requires the existence of all the moments of the distribution $F(R, t)$, which is a reasonable assumption for a wide range of processes. However, this approach may not be applicable to distributions with divergent moments (e.g. power laws).

As the noise strength ε decreases, the company's assets evolve, reflecting two different time scales: (1) a fast time scale describing the dynamics of the company's assets for short periods of time, and (2) a slower time scale determined by the driving noise, reflecting a diffusive (random walk) process. Thus, we introduce two scales for asymptotic analysis (a fast time scale $t = t'$ and a slow time scale $\tau = \varepsilon\, t'$) and the following singular expansion for the probability distribution in terms of the strength of the jumps (Nayfeh, 1993; Sobehart, 2005):

$$P(x, t') \sim e^{\varphi/\varepsilon}(\Pi + O(\varepsilon)) \tag{12.25}$$

Here $\varphi(x, t, \tau)$ and $\Pi(x, t, \tau)$ are unknown functions to be determined. In the singular perturbation framework, the probability distribution of assets is determined primarily by the value of the singular function $\exp[\varphi(x, t, \tau)/\varepsilon]$, while the function $\Pi(x, t, \tau)$ has a marginal impact on the probability distribution. To provide some insight into this approach, note that if the distribution of assets was log-normal, as in standard structural models, then $\varphi(x, t, \tau)$ would be a quadratic function of x, and $\Pi(x, t, \tau)$ would contain mainly normalization factors and corrections in higher orders of ε.

Substituting Equation (12.25) into (12.24), we obtain a hierarchy of singular perturbation equations in powers of the noise strength ε, which describes the distribution of the company's assets:

$$\frac{1}{\varepsilon}\frac{d\varphi}{dt} + \varepsilon^0\left(\frac{\partial\varphi}{\partial\tau} + \frac{d\Pi}{dt}\right) = \varepsilon^0\left[\sum_{n=1}^{\infty}\frac{(-1)^n}{n!}\psi_n(\partial_x\varphi)^n\right]\Pi + O(\varepsilon) \tag{12.26}$$

The relevant equation of the hierarchy to order $O(\varepsilon^0)$ is:

$$\frac{d\varphi}{dt} = \partial_t\varphi + \nu\partial_x\varphi = 0 \tag{12.27}$$

Equation (12.27) shows that, in the limit $\varepsilon \to 0$, $\varphi(x, t, \tau)$ is a function of the characteristics of Equation (12.21) for the time scale t and can have an arbitrary dependence on the slow time scale τ.

To the next order $O(\varepsilon)$, the singular expansion of Equation (12.24) yields

$$\frac{1}{\Pi}\frac{d\Pi}{dt} = \sum_{n=1}^{\infty}\frac{(-1)^n}{n!}\psi_n(\partial_x\varphi)^n - \partial_\tau\varphi \tag{12.28}$$

In contrast to Equation (12.27), Equation (12.28) includes the first-order correction to the probability distribution introduced by the noise strength ε and depends explicitly on both the fast and the slow time scales t and τ.

Solutions to Equations (12.27) and (12.28) can be obtained using the method of characteristics:

$$x = z + \xi \qquad \xi(t) = \int_{t_0}^{t} v(t')dt' \tag{12.29}$$

Here z is the initial condition at time t_0, and $\xi(t)$, the actual asset growth, generalizes the concept of the drift term μt.

Introducing the transformations $x, t \mapsto z, \tilde{z}$

$$z = x - \xi \qquad \tilde{z} = x + \xi, \tag{12.30}$$

Equation (12.27) reduces to

$$\partial_{\tilde{z}}\phi = 0 \tag{12.31}$$

This condition indicates that $\varphi = \varphi(z, \tau)$ depends only on z and the slow time scale τ.

Because the probability distribution of the company's assets must be bounded, we require the characteristic exponent of Equation (12.28) to vanish over any extended period (stability condition):

$$\lim_{T\to\infty} \frac{1}{T - t_0} \int_{t_0}^{T} \left(\frac{1}{\Pi} \frac{d\Pi}{dt} \right) dt = 0. \tag{12.32}$$

Here we assume that the company's assets do not expire or depreciate completely and, therefore, there is no upper bound on the period used in the long-term average described in Equation (12.32). In practice, there is a technical bound $T >> t_S$, where t_S is the characteristic time scale set by the validity of the two-time scale approach described here. The scale t_S, which is derived later, is the time required to dissipate random fluctuations in the value of the company's assets.

The stability condition (12.32) implies that $\Pi(x, t, \tau)$ cannot grow unbounded and can add only a marginal contribution to the probability distribution $P(x, t)$. From Equations (12.28) and (12.31), we obtain the following condition for the singular expansion function φ

$$\lim_{T\to\infty} \frac{1}{T - t_0} \int_{t_0}^{T} \left(\sum_{n=1}^{\infty} \frac{(-1)^n}{n!} \psi_n (\partial_x \varphi)^n - \partial_\tau \varphi \right) dt = 0. \tag{12.33}$$

Solving Equation (12.33) requires additional simplifications. Let $\psi_1(x,t)$ and $\psi_2(x,t)$ be the leading contributions to the diffusion process followed by the value of the company's assets. Next, we expand the functions $\psi_n(x,t)$ to leading order in the difference between the log-value x and the mean reverting value $\phi(t)$ estimated by market participants. Finally, we simplify the functions $\psi_n(x,t)$ by keeping only the leading terms that describe a standard mean reverting process[8] (Sobehart, 2005; Sobehart and Keenan, 2005):

$$\begin{aligned}\psi_1(x,t) &\approx \psi_1(\phi,t) + \partial_x\psi_1(x-\phi) + \ldots \approx \lambda_0(\omega - \theta(x-\phi))\psi_2(x,t)\\ &\approx \psi_2(\phi,t) + \ldots \approx \lambda_0\chi\psi_n(x,t)/n! \approx 0, \qquad n \geq 3 \qquad (12.34)\end{aligned}$$

Here $\lambda_0, \theta, \omega$, and χ are the coefficients of the Taylor series expansion of $\psi_n(x,t)$. The above approximations assume that market participants react to asset log-values x that differ from the mean reverting value $\phi(t)$ forcing the value to revert to their estimate. The sensitivity of market participants to discrepancies in asset values is $\theta(t)$. The level of uncertainty of the market participants is $\lambda_0\chi$, where $\lambda_0(t)$ is the frequency of changes in assets and $\chi(t)$ is the variance of those changes. Parameter $\omega(t)$ in Equation (12.34) reflects any mispricing of the company's assets due to the difference between the fundamental value of the company's assets $\xi(t)$ and the value $\phi(t)$ estimated by market participants. One should expect any mispricing of the company's assets to vanish over any extended period to avoid sustainable arbitrage opportunities. This issue is discussed further in the context of the relationship between the company's assets, equity, and debt.

Equations (12.30), (12.33), and (12.34) yield the following stability condition for the probability distribution of the company's assets:

$$(-\langle\lambda_0(\omega + \theta(\phi - \xi))\rangle + \langle\lambda_0\theta\rangle z)\partial_z\varphi + \frac{1}{2}\langle\lambda_0\chi\rangle(\partial_z\varphi)^2 - \partial_\tau\varphi = 0 \qquad (12.35)$$

Here $\langle G\rangle \equiv \lim_{T\to\infty} {}^1\!/_{(T-t_0)} \int_{t_0}^{T} G(t)\,dt$ is the forward-looking long-term average of the fast time scale t.

Equation (12.35) is the singular perturbation equivalent to a diffusion equation for the company's assets and determines the leading contribution to the probability distribution of the company's assets $P(x,t) \sim e^{\varphi/\varepsilon}\Pi$. Equation (12.35) indicates that, under long-term stability conditions, the value of the company's assets is determined by a forward-looking estimate of the future uncertainty. This interpretation in terms of future uncertainty does not violate causality because the distribution is not a function of unknown future values, but a function of the long-term time averages described by the time average operator $\langle G \rangle$. This description is consistent with the view that market volatility is a forward-looking estimate of future variability (as shown in Equations (12.38) and (12.43) below).

Inspection of Equation (12.35) suggests that the leading contribution to the probability distribution is normally distributed:

$$\varphi = -\frac{1}{2\gamma}(z^2 + 2\alpha z + \beta) \tag{12.36}$$

Introducing Equation (12.36) into (12.35) yields a polynomial $Q(z)$ of second degree in z whose coefficients are three differential equations for α, β and γ. Since Equation (12.35) indicates that $Q(z) = 0$ for any arbitrary z, each one of the coefficients of $Q(z)$ should vanish identically. This yields to the following equations for α, β and γ:

$$\begin{aligned}
&\partial_\tau \alpha + \langle \lambda_0 \theta \rangle \alpha + \langle \lambda_0 (\omega + \theta(\phi - \xi)) \rangle = 0 \\
&\partial_\tau \beta + \beta \left(2\langle \lambda_0 \theta \rangle - \frac{1}{\gamma} \langle \lambda_0 \chi \rangle \right) + 2\alpha \left(\langle \lambda_0 (\omega + \theta(\phi - \xi)) \rangle + \frac{\alpha}{2\gamma} \langle \lambda_0 \chi \rangle \right) = 0 \\
&\partial_\tau \gamma + 2\gamma \langle \lambda_0 \theta \rangle - \langle \lambda_0 \chi \rangle = 0
\end{aligned} \tag{12.37}$$

Here α and β are related to the drift of the company's assets, while $\varepsilon\gamma$ is the time-dependent level of uncertainty, which is related to the volatility of the assets.

Note that when $\theta = 0$, market participants are insensitive to differences in the value of company's assets above or below the mean reverting value $\phi(t)$ and, therefore, there is no mean reversion of the value of the company's assets. In this case, the uncertainty in the value of the company's assets $\varepsilon\gamma$ grows unbounded as in a typical random walk:

$$\varepsilon\,\gamma(\tau) = \varepsilon\,(\gamma_0 + \langle \lambda_0 \chi \rangle (\tau - \tau_0)) = \varepsilon\,\gamma_0 + \sigma^2 (t - t_0) \tag{12.38}$$

Here γ_0 is the initial value of $\gamma(\tau)$, and $\sigma = \varepsilon \langle \lambda_0 \chi \rangle^{1/2}$ is the volatility of the company's assets. This limit is the standard Brownian motion, where the uncertainty grows as $\sigma\sqrt{t - t_0}$. Note that, in our model, the volatility of the company's assets is equal to the long-term time average of the uncertainty generated by the market, which is not a historical average but a forward-looking average.

When $\theta > 0$, market participants are sensitive to price differences and their reaction leads to mean reversion of the value of the company's assets. The uncertainty and adjustments resulting from Equation (12.37) are:

$$\alpha = -\frac{\langle \lambda_0(\omega + \theta(\phi - \xi))\rangle}{\langle \lambda_0\theta\rangle} + \left(\alpha_0 + \frac{\langle \lambda_0(\omega + \theta(\phi - \xi))\rangle}{\langle \lambda_0\theta\rangle}\right) e^{-\langle \lambda_0\theta\rangle(\tau - \tau_0)}$$

$$\beta = e^{h(\tau)} \int_{\tau_0}^{\tau} e^{-h(s)} 2\alpha(s) \left(\langle \lambda_0(\omega + \theta(\phi - \xi))\rangle + \langle \lambda_0\chi\rangle \frac{\alpha(s)}{2\gamma(s)}\right) ds + \beta_0\, e^{-h(\tau)}$$

$$\gamma = \frac{1}{2}\frac{\langle \lambda_0\chi\rangle}{\langle \lambda_0\theta\rangle} + \left(\gamma_0 - \frac{1}{2}\frac{\langle \lambda_0\chi\rangle}{\langle \lambda_0\theta\rangle}\right) e^{-2\langle \lambda_0\theta\rangle(\tau - \tau_0)}$$

$$h = \int_{\tau_0}^{\tau} \left(2\langle \lambda_0\theta\rangle - \frac{\langle \lambda_0\chi\rangle}{\gamma(s)}\right) ds \tag{12.39}$$

Equation (12.39) indicates that random changes in the company's assets tend to spread out, while the reaction of market participants (damping term) brings values back to the stationary value α/β. The balance between these two opposite tendencies yields an equilibrium distribution with a saturation value of the variance: $\langle \lambda_0\chi\rangle/2\langle \lambda_0\theta\rangle$. The characteristic time scale for reaching equilibrium in asset values is $t_s = 1/\varepsilon\langle \lambda_0\theta\rangle$. Furthermore, when there is no bias in estimating assets values ($\omega \approx 0$), corrections to asset values are small if the market participants' estimate ϕ agrees with the value ξ derived from the actual value of the company's assets.

The probability distribution term $\Pi(x, t, \tau)$ is obtained from Equations (12.28) and (12.39).

$$\log(\Pi(x, t, \tau)) = \int_{t_0}^{t} \Big[-\left(\lambda_0(\omega + \theta(\phi - \xi)) - \langle \lambda_0(\omega + \theta(\phi - \xi))\rangle\right) \partial_z\varphi + (\lambda_0\theta - \langle \lambda_0\theta\rangle) z \partial_z\varphi + \frac{1}{2}(\lambda_0\chi - \langle \lambda_0\chi\rangle)(\partial_z\varphi)^2 \Big] dt \tag{12.40}$$

The probability distribution of the transformed value of the company's assets x at time t is obtained from Equations (12.36), (12.39), and (12.40)

$$P(x, t) = \frac{1}{\sqrt{2\pi\varepsilon\,\gamma(\varepsilon\, t)}} e^{-\frac{(x - \bar{x}(t))^2}{2\varepsilon\,\gamma(\varepsilon\, t)}} \tag{12.41}$$

Here the growth rate of the company's assets reflects the first-order contribution in powers of ε.

$$\hat{x}(t) = \xi + \alpha + \varepsilon \int_{t_0}^{t} [(\lambda_0(\omega + \theta(\phi - \xi)) - \langle \lambda_0(\omega + \theta(\phi - \xi)) \rangle) + \alpha(\lambda_0\theta - \langle \lambda_0\theta \rangle)]dt + O(\varepsilon^2) \tag{12.42}$$

Equations (12.41) and (12.42) indicate that the value of the company's assets is approximately log-normal with volatility $\varepsilon\gamma(\varepsilon t)$, which is a forward-looking estimate based on stability considerations over long periods of time. The width of the distribution depends on the slow time scale $\tau = \varepsilon t$, which gives the relaxation time of changes in the value of the company's asset.

During a small interval Δt the company's assets will grow at an effective rate μ and will have an effective forward-looking volatility σ:

$$\mu = \lim_{\Delta t \to 0} \frac{1}{\Delta t} E[x(t + \Delta t) - x(t)] \sim \frac{d\hat{x}}{dt}$$

$$\sigma^2 = \lim_{\Delta t \to 0} \frac{1}{\Delta t} E[(x(t + \Delta t) - x(t))^2] \sim \varepsilon^2 \langle \lambda_0 \chi \rangle \tag{12.43}$$

Equations (12.41)–(12.43) indicate that the value of the company's assets $A = A_0 e^x$ reflects long-term market expectations that can follow either a standard (random walk) diffusion process ($\theta = 0$) or a mean reverting process ($\theta > 0$), both with an asymptotic log-normal distribution.

MARKET UNCERTAINTY AND THE VALUATION OF EQUITY AND DEBT

Although the equity-as-option analogy is insightful and theoretically appealing, equity holders do not actually relinquish ownership of the firm by issuing debt, and – through the company's management – continue to deploy, dispose of, and even liquidate the company's assets at will, or may refinance and rollover debt to achieve their business goals. Furthermore, equity holders are involved in growing the value of their assets through management, and simply need the cash offered by bondholders to leverage their investment. Bondholders recognize the potential growth and participate by trading off some upside growth potential of equity holders in return for reduced downside potential through the seniority of their claim in bankruptcy. The different objectives and dynamics of the equity and debt markets suggest that the no-arbitrage condition required to close the BSM model does not always hold in practice as the total value of equity and debt can deviate temporarily from the value of the company's assets. This can lead to both "capital structure arbitrage" and "market timing arbitrage."

In our more general approach, equity and debt prices are not necessarily linked in a strict options/underlying type of relationship as proposed in the BSM framework (Sobehart and Keenan, 1999, 2002a, 2002b, 2003). Because we explicitly include additional random fluctuations independent of the changes in the company's assets (and other state variables), the idealized standard equity-debt-asset relationship $E + D = A$ does not hold exactly and shows random deviations around a dynamic equilibrium condition consistent with the concept of market equilibrium.

In the following we assume that equity and debt are European-style options with positive payoffs written on non-dividend-paying underlying assets, the risk-free rate is deterministic and there are no transaction costs, taxes, or problems with the indivisibility of the assets or securities. These simplifying assumptions are made for analytical convenience and can be relaxed to describe more complex situations.

Let's introduce the natural logarithm of the price $y = \log(V(A(x),t)/V_0)$ for a financial claim on the company's assets, where $V(A,t)$ is the value of equity or debt on the company's assets and V_0 is a reference price. We use the notation y_j $1 \le j \le N$, where N is the number of outstanding claims, to describe multiple claims on the company's assets (equity and different types of debt).

Next, let's assume that market participants in the equity and debt markets are neither perfectly informed nor capable of eliminating arbitrage opportunities instantaneously. Because in this situation market information is incomplete, not fully reliable, or contradictory, market prices reflect valuation uncertainty, which emerges whenever information is deficient in some respect. In an ideal world of perfectly informed market participants, who agree on how to price claims on risky assets, the price of the financial claim y and the fundamental value $f(x,t)$ based on the company's assets would be related by the equation $y = f(x,t)$. When market participants disagree on how to price financial claims, y and $f(x,t)$ are different stochastic processes and there is no unique functional relationship between them. The difference between y and $f(x,t)$ is caused by the uncertainty of the participants in the equity and debt markets. Because of deficient information, market participants may believe that the fundamental value derived from the company's assets is some function $g(x,t)$ instead of the correct fundamental value $f(x,t)$. Note, however, that the value $g(x,t)$ determined by market participants and the fundamental value $f(x,t)$ derived from the underlying assets must be closely related to avoid persistent arbitrage opportunities.

The valuation uncertainty of market participants introduces additional changes to the price of the claims on the company's assets, assumed to be random jumps of magnitude $\eta R'$ that occur at random times t_k. Here η is the strength of the price variations due to changes in the supply and demand for the claims on the company's assets, and R' is a random variable with distribution $G(R', t)$. Thus, the value y is modified by two random processes: (1) jumps in the assets of the firm x of magnitude εR, and (2) additional random jumps of size $\eta R'$ due to valuation uncertainty. For simplicity, we assume that the jump processes for R and R' are independent. The jump process for a financial claim on the company's assets has a rate of events $\Lambda(x, y, t)$ and can depend on the transformed asset log-price x, the claim log-price y, and the value $g(x, t)$ estimated by market participants.

As the financial claim contract gets closer to its maturity, the uncertainty for pricing the claim decreases. For an ideal European style contract the relationship between the claim and the value of the company's assets is known with absolute certainty exactly at the contract's maturity $t = T$. This relationship is the boundary condition imposed to solve the pricing equation derived from the BSM approach. In contrast, at the inception of the contract, the uncertainty for pricing the claim may be high. The idealized convergence of opinions as the contract matures can be interpreted as a decreasing level of uncertainty from the inception of the contract to its maturity ($\eta \underset{t \to T}{\to} 0$). Note, however, that supply and demand and other market effects may prevent the valuation uncertainty from vanishing. For analytical convenience, we focus only on the stylized situation where the noise strength η is constant through the life of the contract, which could be a reasonable approximation far from the contract's maturity. A more refined approach would require the inclusion of additional sources of market uncertainty, such as changes in volatility, interest rates, and the preferences of market participants. These additional sources of uncertainty can either preserve or change the ideal convergence of opinion as the claim matures.

The evolution equation for the random process driving value y is:

$$dy = \kappa(t)\, dt + \eta \sum_{t_n} R'_n \delta(t - t_n)\, dt \tag{12.44}$$

Here the growth rate of the claim due to changes in the underlying assets is given by the following expressions:

$$\kappa(x, t) = \varpi(x, t) + \sum_{t_m} s(x, t_m)\, \delta(t - t_m)$$

$$\varpi(x, t) = \partial_t f + v \partial_x f$$

$$s(x, t) = f(x + \varepsilon R, t) - f(x, t) = \sum_{k=1}^{\infty} \frac{1}{k!} (\varepsilon\, R)^k \partial_x^k f \tag{12.45}$$

Equation (12.45) reflects the growth rate of the claim produced by the changes in the assets alone and the uncertainty generated by the market participants.

Next, let us find an approximate solution to the value of a claim on the company's assets in the limit of small market uncertainty $\eta > 0$ and asset volatility $\varepsilon > 0$ using a self-financed hedge portfolio combining the company's assets, equity or debt, and riskless bonds (Merton, 1974; Sobehart and Keenan, 2005). Let $\pi = a\,V + b\,A + c\,B$ dollars be invested in a self-financed portfolio with positions in the claim on the company's assets $V(A,t)$, the underlying assets A, and a riskless bond $B(t)$, whose yield is the riskless rate of return r. Parameters a, b and c represent the numbers of asset claims, company's assets and risk-free bonds that are purchased. This ideal self-financed portfolio requires no initial net investment at time t ($\pi(t) = 0$) and no additional funds at the end of the time increment Δt. That is, the long positions are completely financed by short positions. This imposes a relation between the positions a, b, and c whose values need to be adjusted at the end of the time interval Δt. After a short period Δt the average change in the value of the portfolio due to changes in the riskless bond, the company's assets and claim prices will be given by the following equation (Sobehart, 2005):

$$
\begin{aligned}
\Delta\pi &= a\,\Delta V + b\,\Delta A + c\,\Delta B + O(\Delta t) \\
&= a\left(\partial_t V + r\,\partial_x V + \sum_{k=2}^{\infty} \frac{\varepsilon^k}{k!}\lambda R^k \partial_x^k V - rV + \sum_{k=1}^{\infty} \frac{\eta^k}{k!}\Lambda R'^k \partial_y^k V\right)\Delta t \\
&\quad + \left((\nu - r + \varepsilon\lambda R)(a\ \partial_A V + b) + b\sum_{k=2}^{\infty} \frac{\varepsilon^k}{k!}\lambda R^k\right) A\Delta t
\end{aligned}
\tag{12.46}
$$

Here we used Equations (12.21) and (12.45), and neglected adjustments to a, b and c and higher-order corrections in powers $O(\Delta t^2)$. Here $y = \log(V/V_0)$ and $\partial^n V/\partial y^n = V,\ \ n \geq 0$.

The ideal self-financed portfolio (12.46) can be hedged to partially remove the leading stochastic jump term εR by selecting the hedge ratio $b \approx -a\,\partial_A V$ and neglecting terms in higher powers of parameter ε. However, the hedging strategy does not necessarily remove all stochastic terms and adjustments and, therefore, investors may require an additional risk premium $\rho - r$ as a compensation for holding a hedge portfolio with residual risk. The risk premium $\rho - r$ may depend on the parameters $\nu, \varepsilon, \eta, \lambda$ and Λ. We return to this point when we discuss the equity-debt relationship and capital structure arbitrage opportunities.

The expected value of a claim on the company's assets $E[V|x] = V_0 E[e^{y(x,t)}|x]$ is found by requiring that the expectation of the rate of change of the portfolio $\Delta\pi$ equals the risk-adjusted gain required by investors

$(\rho - r)(\pi - cB)\Delta t$ obtained by continuously hedging the risky part of the portfolio changes described in Equation (12.46).

$$E[\Delta\pi|x] = (\rho - r)\,\mathrm{E}[\pi - cB|x]\,\Delta t \;\Rightarrow \partial_t E[V|x] + \rho\,\partial_x E[V|x] + \sum_{k=1}^{\infty} \frac{\varepsilon^k}{k!} \partial_x^k E[\lambda R^k V|x] + \sum_{k=1}^{\infty} \frac{\eta^k}{k!} \partial_y^k E[\Lambda R'^k V|x] - \rho\,\mathrm{E}[V|x] = 0 \tag{12.47}$$

Equation (12.47) must be solved with the appropriate boundary condition at the maturity of the claim contract. To gain insight into this model extension, note that if we truncated the summation series in Equation (12.47) to second order in powers of the noise strength ε, and neglect the options trading noise η, the resulting equation would be a standard drift-diffusion equation for the asset value $x = \log(A/A_0)$ and the claim's value V. That is, in the drift-diffusion limit, our model describes asymptotically the standard BSM model.

REVISITING THE DEFAULT POINT

In the BSM approach, a company defaults on its obligations when the value of its assets is below the value of the promised debt payment or liabilities (default point). The default event occurs instantaneously with conditional probability (Sobehart and Keenan, 2002c, 2005).

$$p(d = 1\,|\,A_T/D_0) = \begin{cases} 1 & A_T/D_0 \leq 1 \\ 0 & A_T/D_0 > 1 \end{cases} \tag{12.48}$$

Here $d = 1$ indicates a default event occurs at the maturity of the obligations, A_T/D_0 is the company's leverage, and A_T is the value of the company's assets at maturity T and D_0 is the face value of the debt. The probability that no default occurs at maturity ($d = 0$) is $p(d = 0\,|\,A_T/D_0) = 1 - p(d = 1\,|\,A_T/D_0)$. An equivalent condition can be derived for other model extensions.

Note that, in general, the relative value of the company's assets to its liabilities does not necessarily define the default point with absolute certainty. If the company has not exhausted its borrowing capacity, it can refinance its debt, reducing its probability of immediate default. That is, the event of default may be more likely to occur if $A_T/D_0 \leq 1$ but is not necessarily inevitable. This is relevant in the context of climate risk and ESG as changes in public opinion and market perception of the company's strategy and long-term commitments could impact the company's reputation and debt capacity. Furthermore, companies that are solvent according to the condition $A_T/D_0 > 1$ may file for bankruptcy in anticipation of future legal liabilities, lawsuits (e.g. asbestos, tobacco, or product safety issues), severe industry downturns, or unanticipated catastrophic events due to physical risk or transition risk not related to their existing level of liabilities D_0. In this more general situation, the conditional probability of default at time T can be a function of other relevant information ζ

$$p(d = 0 \mid A_T/D_0, \zeta) = p_0(A_T/D_0, \zeta) \text{ and}$$
$$p(d = 1 \mid A_T/D_0, \zeta) = p_1(A_T/D_0, \zeta) \tag{12.49}$$

Here $\int (p_0 + p_1) dA d\zeta = 1$, where ζ represents accounting, market, regulatory geographic, or industry information relevant for determining the company's default, and $p_0(A_T/D_0, \zeta)$ and $p_1(A_T/D_0, \zeta)$ represent the probabilities of default and non-default conditional on information ζ respectively.

The approach presented here covers both stochastic liabilities and situations where the company's liabilities are known with certainty (i.e. they are deterministic) but there is uncertainty as to the actual default point due to exogenous events such as changes in the regulatory environment. This is relevant for climate risk and ESG since changes in the regulatory environment, policies, incentives, and restrictions on business activities caused by the transition to a low-carbon economy can impair a company's ability to operate, impacting its creditworthiness. Furthermore, the volatility of the company's liabilities and the volatility of its revenues (caused by uncertainty in the business strategies and regulatory environment), and the company's capacity to borrow and renegotiate should play a more central role in the development of default risk and pricing models for climate risk and ESG.

THE ROLE OF THE COMPANY'S BORROWING CAPACITY

Although in the BSM approach companies cannot restructure their liabilities, several model extensions have relaxed this limitation. These extensions can cover: (1) multiple liabilities with different time horizons (Geske, 1979; Longstaff, 1990), (2) continuous rollover of debt (Leland, 1999), or (3) stochastic liabilities (Finger et al., 2002). However, these model extensions impose the structure of the company's liabilities exogenously as opposed to endogenously, where the company can reduce its probability of default in the short term by changing its liabilities or maturities, or rolling over its debt. More precisely, a company's liabilities can be operational (e.g. accounts payable) or have debt characteristics for financing business activities. Debt obligations are determined by the company's management in accordance with their strategic goals and opportunities presented by market conditions for capital investment and borrowing. The company can take debt obligations for different time horizons with interest payments (such as loans), coupons (such as corporate bonds) or without coupons (such as commercial paper). These obligations may have call or convertible features, sinking funds, and/or early amortization schedules (Anderson, Sundaresan and Tychon, 1996; Ravid, 1996; Sobehart and Keenan, 2005). Management can roll them over, pay them down, change the mix of debt obligations, or borrow additional capital as needed for investments. Furthermore, some companies keep off their balance sheets significant portions of their debt in the form of lease obligations. These obligations are more difficult to identify because companies can account for them in different ways. The description gets even more complicated when considering the dynamics of financing provided through operations, working capital management, and accounts receivables, which could be monetized through securitization or outright sales.

Key to determining the company's financing strategy is the company's debt capacity for borrowing, which allows management to cancel obligations, refinance old ones, or even add new ones. Debt can be rolled over into multiple types of debt with different features, seniority, and payment schedules. Although the number of different debt refinancing strategies may seem unlimited at first blush, the number of practical alternatives is limited by financing and bankruptcy costs, taxes, and other relevant variables. For illustration, let us assume that the company adopts a simple strategy based on refinancing a fraction F of the debt outstanding at an additional interest charge I. The refinancing fraction F, interest charge I and the seniority of the new claim S will depend on the lender's policy and assessment of the borrower's risk, or on the risk appetite of investors in the capital markets. Variables F, I, and S have a probability distribution $Q(F, I, S)$, which reflects the available refinancing strategies and may depend on the company's industry, business environment, and its competitiveness as well as on the company's credit quality and likelihood of default. Since the company's financing strategy can depend on the company's credit quality and can also be used to determine the company's credit quality, the problem is potentially endogenous and self-referencing.

The adjustment of the original debt outstanding D_0 with the refinanced fraction F and interest charge I has the effect of changing the company's leverage, which can impact the probability of default. To illustrate this point, consider the simple case where Equation (12.49) determines the default event at time T and assume that neither the refinanced fraction F nor the interest charge I depends on the company's assets A. In this situation, Equation (12.47) for a claim V that matures at time T must be solved with the following boundary condition integrated over all possible values of refinancing strategies $Q(F, I, S)$:

$$V(A_T, T|F, I, \zeta) = \begin{cases} A_T(1 - w(S)F) & A_T \leq D_0(1 - F) + I \\ D_0 & A_T > D_0(1 - F) + I \end{cases} \tag{12.50}$$

Note that if the refinanced amount FD_0 is junior to the original debt amount D_0 we have $w(S) = 0$, and if both claims have the same seniority $w(S) = 1$. In the more general case, $0 \leq w \leq 1$ will have a distribution of recovery values.

The difference between the solution obtained from the BSM model and the model described in Equations (12.47)–(12.50) depends on the distribution of refinancing strategies and regulatory and business constraints that can change the effective default boundary. Note that the valuation of equity (the most junior claim) should include boundary conditions for all claims and their maturities. Since here we focus primarily on understanding the impact of market uncertainty, we assume that the solution $y_j = \log(V_j/V_{0j})$ to Equations (12.47)–(12.50) has already been calculated for each one of the claims on the company's assets.

JOINT DISTRIBUTION OF ASSETS, EQUITY, AND DEBT

In the generalized approach discussed here, the transformed asset value $x = \log(A/A_0)$ and financial claims $y_j = \log(V_j/V_{0j})$ $1 \leq j \leq N$ can be described in terms of the joint probability distribution $P(x, y_1, ..., y_N, t')$ of a random transformed asset value $X(t')$ in the interval $[x, x + dx]$, and a random transformed claim price $Y_j(t')$ in the interval $[y_j, y_j + dy_j]$ at time t'. The evolution of the probability distribution of the assets and financial claims is described by the following equation:

$$\partial_t P + \nu \partial_x P + \sum_j \varpi_j \partial_{y_j} P = \int P(x - \varepsilon R, \ldots, y_j - \kappa_j(x - \varepsilon R), .., t') \lambda(x - \varepsilon R, t')$$

$$F(R, t')dR + \sum_j \int P(x, \ldots, y_j - \eta_j R_j', \ldots, t') \Lambda_j(x, y_j - \eta_j R_j', t') G_j(R_j', t') dR_j'$$

$$-\left(\lambda(x, t') + \sum_j \Lambda_j(x, y_j, t') \right) P(x, y_1, \ldots, y_N, t') \tag{12.51}$$

As the trading noise η_j for each financial claim decreases, the evolution of the claim's price becomes a two-time scale problem involving the actual time t'

and its product with the noise strength $\eta_j t'$. Therefore, we introduce a fast time scale $t = t'$ and slow time scales $\tau = \varepsilon t'$ and $\upsilon_j = \eta_j t'$ $1 \leq j \leq N$ related to the trading noise in the company's assets (x), equity, and debt markets (y_j). This also motivates the use of the following singular expansion for the probability of the company's assets and their financial claims (equity and debt):

$$P(x, y_1, \ldots, y_N, t') \sim e^{\varphi/\varepsilon + \sum_j \Phi_j/\eta_j}(\Pi + O(\varepsilon, \eta_1, \ldots, \eta_N)) \tag{12.52}$$

Here $\varphi(x, t, \tau)$, $\Phi_j(x, y_j, t, \tau, \upsilon_j)$ and $\Pi(x, y_1, \ldots y_N, t, \tau, \upsilon_1, \ldots, \upsilon_N)$ are unknown functions to be determined.

To illustrate the nature of the joint probability distribution of the company's assets and claim prices $P(x, y_1, \ldots, y_N, t')$, let's consider the limit case $\varepsilon > 0$ and $\eta_j = 0$ $1 < j < N$. In this case, the solution to Equation (12.51) is of the generic form:

$$P(x, y_1, \ldots y_N, t') = P(x, t') \prod_j \delta(y_j - f_j(x, t')) \tag{12.53}$$

Here $\delta(y_j - f_j(x, t'))$ is the (singular) Dirac delta-function. Note that in this limit, the only source of uncertainty in the value of the financial claim is the stochastic diffusion of the underlying assets, that is, there is no uncertainty in the relationship $y_j = f_j(x, t')$. In this limit we recover the idealized BSM framework where market participants agree on *how* to price financial claims on the company's assets at each point in time because they firmly believe that equity and debt are completely determined by the value of the company's assets.

In practice, participants in the equity and debt markets often disagree on their valuation methodologies due to differences in information, investment goals, and other factors that can lead to random deviations between the observed prices of financial claims on the company's assets and the values derived from pricing models (Sobehart and Keenan, 1999, 2002a, 2002b, 2002c, 2003). As a result, the joint distribution of assets, equity, and debt prices will not exhibit the singular characteristics described in Equation (12.53). This situation is represented by the general condition: $\varepsilon > 0$ and $\eta_j > 0$ $1 < j < N$.

The relative magnitude of the noise parameters ε and η_j can help us determine an asymptotic solution to Equation (12.51) by noting that $\varepsilon \sim q/A$ and $\eta_j \sim q/V_j(A, t')$, where q is the smallest meaningful price change observed across markets, and $| V_j(A, t') | \leq A$ for a variety of financial claims.

Combining Equations (12.51) into (12.52) and considering the condition $\varepsilon \leq \eta_j < 1$ for the relative order of the perturbation parameters, we obtain a hierarchy of equations for the probability distribution corresponding to different orders in ε and η_j. In the limit $\varepsilon/\eta_j << 1$, the relevant equations of the hierarchy to orders $O(\varepsilon^0, \eta^{-1})$, $O(\varepsilon^0, \eta^0)$ and $O(\varepsilon, \eta)$ are respectively:

$$\partial_t \varphi + \nu \partial_x \varphi = 0$$

$$\partial_t \Phi_j + \nu\, \partial_x \Phi_j + \sum_j \varpi_j \partial_{y_j} \Phi_j = 0$$

$$\frac{1}{\Pi}\frac{d\Pi}{dt} = \left(\sum_{n=0}^{\infty} \frac{(-1)^n}{n!} \psi_n (\partial_x \varphi)^n - \lambda - \partial_\tau \varphi \right)$$

$$+ \sum_j \left(\sum_{m=0}^{\infty} \frac{(-1)^m}{m!} \Psi_{jm} (\partial_{y_j} \Phi_j)^m - \Lambda_j - \partial_{v_j} \Phi_j \right) \tag{12.54}$$

Here $d\Pi/dt = \partial_t \Pi + \nu\, \partial_x \Pi + \sum_j \varpi_j\, \partial_{y_j} \Pi$ is the total derivative of Π in the fast time scale t, and $\Psi_{jm} = \int \Lambda_j(y_j, x, t)\, G_j(R, t)\, R^m\, dR$ for $m > 0$.

Let's define the following transformations:

$$z = x - \xi$$

$$\tilde{z} = x + \xi$$

$$w_j = y_j - f_j(x, t) \tag{12.55}$$

Introducing the reduced variables $z, \tilde{z}$, and w_j into Equation (12.54), the first two expressions in Equations (12.54) reduce to:

$$\partial_{\tilde{z}} \varphi = 0, \qquad \partial_{\tilde{z}} \Phi = 0 \tag{12.56}$$

Equation (12.56) implies that $\varphi = \varphi(z, \tau)$ and $\Phi_j = \Phi_j(z, w_j, v_j)$.

In the following we extend the approximation in Equation (12.34) for the functions $\psi_n(x, t)$ under the assumption that the price adjustment process for the financial claims is driven only by discrepancies between the actual log-value of the claim y_j and the value $g_j(x, t)$ determined by market participants. The expansion of $\Psi_n(x, y, t)$ in Equation (12.54) to leading order in powers of the difference between the transformed claim price y_j and the mean value $g_j(x, t)$ leads to the following mean-reverting process for financial claims on the company's assets:

$$\Psi_{j0} \approx \Lambda_0$$

$$\Psi_{j1}(x, y_j, t) \approx \Lambda_0(\Omega_j - \Theta_j(y_j - g_j))$$

$$\Psi_{j2}(x, y_j, t) \approx \Lambda_{j0} X_j$$

$$\Psi_{jk}(x, y_j, t)/n! \approx 0 \quad k \geq 3 \quad 1 \leq j \leq N \tag{12.57}$$

Here Λ_0, Θ, X and Ω are the coefficients of the Taylor series expansion of the functions ψ_n. Note that one should expect the fundamental value estimated by market participants to be the same as the value derived from the company's assets $g_j(x,t) \approx f_j(x,t)$ to avoid sustainable arbitrage opportunities. Equations (12.57) indicate that market participants react to prices that differ from their estimates $g_j(x,t)$ to generate a mean-reverting process for the claim on the company's assets with fluctuations around its fundamental value $f_j(x,t)$. The sensitivity of the market participants to pricing errors is $\Theta_j(t)$. The level of uncertainty of the market participants is $\Lambda_{j0}X$, where $\Lambda_{j0}(t)$ is the frequency of price changes for the claims and $X_j(t)$ is the variance of those changes. The parameter $\Omega_j(t)$ represents any possible mispricing of claim y_j, which should vanish over any extended period to avoid sustainable arbitrage opportunities.

Assuming that the distributions of prices for assets and claims are stable over time, a stability condition can be imposed on the solution to Equations (12.54) using the long-term time average $\langle f \rangle$. In principle, the distribution of debt prices should remain stable, at least, until the financial claim y_j expires at time T_j, while the distribution of prices for the company's assets should remain stable for an arbitrarily long period of time independent of the maturity of the claims if the assets do not expire (Sobehart, 1999, 2005; Sobehart and Keenan, 2005). However, if new claim contracts are generated and traded regularly with a minimum level of activity that can justify a stable market condition for any claim contract, one can assume that the price distributions for both the company's assets and claims remain stable for any extended period independent of the claims' maturity. This condition may not necessarily apply if the financial claims are offered opportunistically in response to changing market conditions, for example, in booming stock markets or decreasing interest rate environments.

Combining Equations (12.54), (12.55), and (12.57) yields the following stability condition for the company's assets and their financial claims:

$$[(-\langle \lambda_0(\omega + \theta(\phi - \xi)) \rangle + \langle \lambda_0 \theta \rangle z)\, \partial_z \varphi + 1/2 \langle \lambda_0 \chi \rangle (\partial_z \varphi)^2 - \partial_\tau \varphi]$$

$$+ \sum_j [(-\langle \Lambda_{0j}(\Omega_j + \Theta_j(g_j - f_j)) \rangle + \langle \Lambda_{0j}\Theta_j \rangle w_j)\, \partial_w \Phi_j$$

$$+ \frac{1}{2} \langle \Lambda_{0j} X_j \rangle (\partial_w \Phi_j)^2 - \partial_v \Phi_j] = 0 \qquad (12.58)$$

Because $\varepsilon/\eta_j << 1$, and we assume that there is no correlation between the stochastic jump processes describing changes in the company's assets, equity and various types of debt, the functions φ and Φ_j are not coupled. In the following, we assume that the uncertainty in the equity and debt markets does not affect any of the other markets. That is, each term in brackets in Equation (12.58) can be set to zero independently.

Inspection of the nonlinear Equations (12.35)–(12.58) suggests that φ can be described by Equation (12.36) and Φ_j can be described by the following expression:

$$\Phi_j = -\frac{1}{2\tilde{\gamma}_j}(w_j^2 + 2\tilde{\alpha}_j w_j + \tilde{\beta}_j) \tag{12.59}$$

Combining Equation (12.35), (12.36), (12.58), and (12.59) yields Equations (12.37) for α, β, and γ, whose solutions are described by Equations (12.39). Three similar equations are obtained for $\tilde{\alpha}_j, \tilde{\beta}_j$, and $\tilde{\gamma}_j$, whose solutions are obtained using a zero-order expansion in terms of the discrepancy $g_j - f_j$ that leads to the following approximate expressions in the limit $|g_j - f_j| << f_j$ (quasi-equilibrium with low levels of uncertainty)

$$\tilde{\alpha}_j \approx -\frac{\langle \Lambda_{0j}(\Omega_j + \Theta_j(g_j - f_j))\rangle}{\langle \Lambda_{0j}\Theta_j\rangle} + \left(\tilde{\alpha}_{0j} + \frac{\langle \Lambda_{0j}(\Omega_j + \Theta_j(g_j - f_j))\rangle}{\langle \Lambda_{0j}\Theta_j\rangle}\right) e^{-\langle \Lambda_{0j}\Theta_j\rangle(\upsilon_j - \upsilon_{0j})}$$

$$\tilde{\beta}_j \approx e^{\tilde{h}_j(\upsilon)} \int_{\upsilon_{0j}}^{\upsilon_j} e^{-\tilde{h}_j(s)} 2\tilde{\alpha}_j(s)$$

$$\left(\langle \Lambda_{0j}(\Omega_j + \Theta_j(g_j - f_j))\rangle + \langle \Lambda_{0j}X_j\rangle \frac{\tilde{\alpha}_j(s)}{2\tilde{\gamma}_j(s)}\right) ds + \beta_{0j}\, e^{-\tilde{h}_j(\upsilon_j)}$$

$$\tilde{\gamma}_j = \frac{1}{2}\frac{\langle \Lambda_{0j}X_j\rangle}{\langle \Lambda_{0j}\Theta_j\rangle} + \left(\tilde{\gamma}_{0j} - \frac{1}{2}\frac{\langle \Lambda_{0j}X_j\rangle}{\langle \Lambda_{0j}\Theta_j\rangle}\right) e^{-2\langle \Lambda_{0j}\Theta_j\rangle(\upsilon_j - \upsilon_{0j})}$$

$$\tilde{h}_j = \int_{\upsilon_{0j}}^{\upsilon_j} \left(2\langle \Lambda_{0j}\Theta_j\rangle - \frac{\langle \Lambda_{0j}X_j\rangle}{\tilde{\gamma}_j(s)}\right) ds \qquad 1 \le j \le N \tag{12.60}$$

Here $\tilde{\alpha}_{0j}, \tilde{\beta}_{0j}$, and $\tilde{\gamma}_{0j}$ are the initial conditions. Equations (12.60) have a characteristic time scale $t_\nu = 1/\eta_j\langle \Lambda_{0j}\Theta_j\rangle$ for transient effects. To avoid mispricing and eliminate arbitrage opportunities in the case $\Omega_j \approx 0$, the value estimated by market participants should be the same as the fair value derived from the company's assets: $g_j(x,t) \approx f_j(x,t)$. In this situation the dependence of $\tilde{\alpha}_j$ and $\tilde{\beta}_j$ as functions of $z = x - \xi$ can be considered negligible or small. The correct description of the situation when the difference between $g_j(x,t)$ and $f_j(x,t)$ is significant would require a revision of Equations (12.54) and (12.58) to include higher-order corrections in terms of the relative noise strength: ε/η. This analysis would correspond to far-from-equilibrium markets, which is beyond the scope of our analysis.

Equations (12.60) show that at the inception of the financial claim contract on the company's assets the uncertainty generated by the market participants is of the order $\tilde{\gamma}_0$ and increases or decreases as the claim matures. The direction and magnitude of the change depend on the relationship between the initial uncertainty $\tilde{\gamma}_{0j}$ and the final value $\langle \Lambda_{0j} X_j \rangle / 2 \langle \Lambda_{0j} \Theta_j \rangle$. Note that, if there is no risk of default by the counterparty offering the contracts or by exchange mediating the trading of the company's assets and their claims (equity or debt), there should be no uncertainty on the value of the claim exactly at maturity. That is, $\eta_j \tilde{\gamma}_j(t = T_j) = 0$. The simple model of market uncertainty described in Equations (12.57) cannot accommodate the ideal situation where market participants become completely certain exactly at the maturity of the claim unless the condition $\eta_j \underset{t \to T_j}{\to} 0$ is satisfied. In practice, this model limitation is not severe for pricing risky debt or calculating the likelihood of default for small level of market uncertainty far from the exact expiration of the claim.

After determining φ and Φ_j, the term Π_j can be obtained from Equations (12.54) and (12.60) to the leading order in ε and η_j and, therefore, the joint probability distribution of the transformed asset log-price x and claim log-price y_j at time t can be obtained from Equation (12.52). Because we derived the model using only functions $\psi_1, \psi_2, \Psi_{1j}$ and Ψ_{2j} (drift-diffusion approximation), the probability distribution of the transformed company's assets and their claims is asymptotically joint normal.

$$P(x, y_1, .., y_N, t) = \frac{1}{\sqrt{2\pi\varepsilon\gamma(\varepsilon t)}} \prod_j \frac{1}{\sqrt{2\pi\eta_j \tilde{\gamma}_j(\eta_j t)}} e^{-\left[\frac{(x-\hat{x}(t))^2}{2\varepsilon\gamma(\varepsilon t)} + \sum_j \frac{(y_j - \hat{y}_j(x,t))^2}{2\eta_j \tilde{\gamma}_j(\eta_j t)}\right]} \tag{12.61}$$

Here

$$\hat{y}_j(x,t) = f_j + \tilde{\alpha}_j + \eta_j \int_{t_0}^{t} \{\Lambda_{0j}(\Omega_j + \Theta_j(g_j - f_j)) - \langle \Lambda_{0j}(\Omega_j + \Theta_j(g_j - f_j)) \rangle$$
$$+ A(\Lambda_{0j}\Theta_j - \langle \Lambda_{0j}\Theta_j \rangle)\} dt + \ldots \tag{12.62}$$

Note that the growth rate in Equation (12.62) includes only a first-order contribution $O(\varepsilon, \eta)$ in the strength of the trading noise. Equations (12.61) and (12.62) indicate that value of the company's assets is log-normal with volatility $\varepsilon\gamma$, and that the value of individual claims on these assets can deviate temporarily from the fundamental value due to trading noise. Random deviations for the financial claim (equity or debt) y_j around their fundamental value are normal and their magnitude is $\eta\tilde{\gamma}_j$.

UNCERTAINTY, ARBITRAGE, AND EQUITY-DEBT RELATIONSHIP

In the framework discussed here, equity and debt prices have additional sources of randomness independent of the company's assets. As a result, random changes in the price of equity and debt cannot be replicated using the price changes of the underlying assets alone. In this scenario, equity and debt may require supplemental risk premia $\rho - r \geq 0$ due to random variations that cannot be hedged away completely. That is, hedge portfolios are always risky, and investors need to be compensated for the residual risk. The risk premia for equity and debt are related since portfolios that combine equity, debt, and the company's assets should not create sustainable arbitrage opportunities for extended periods of time. However, the idealized standard equity-debt relationship, $E + D = A$ in the BSM model does not hold exactly. Furthermore, the relationship between equity, debt and the company's assets shows random fluctuations around a dynamic equilibrium condition consistent with the concept of market equilibrium. These short-lived fluctuations may lead to unsustainable capital structure arbitrage opportunities.

To illustrate, let's discuss the impact on a company with a simple capital structure: equity E and a bullet bond with face value D_0 that matures at time T. The volatility of the company's assets and interest rates are constant, and the company pays no dividends and the risk premium $\rho - r$ is also constant. Let's assume that the company has negligible debt capacity for borrowing, and the default boundary is determined by the standard BSM constraint (12.48). We also neglect the sensitivity of market participants to price differences and take the limit of small noise $\eta \to 0$. Truncating Equation (12.47) to second order in powers of ε we obtain a modified BSM equation for equity $E = e^w C_0(A, t; \sigma, \rho)$, where $w = y - \log(E_T[e^y])$. Introducing the boundary condition $C_0(A, T) = \max(A - D_0, 0)$ at maturity T (European-type contract), and approximating the solution with the BSM call-option formula $C_{BSM}(A, t; \sigma, \rho)$ for a European-type claim contract on the company's assets, the equity value results

$$E(A, t) = e^w C_{BSM}(A, t; \sigma, \rho) \tag{12.63}$$

Here

$$C_{BSM}(A, t; \sigma, \rho) = AN(d_1^\rho) - e^{-\rho(T-t)} D_0\, N(d_2^\rho) \tag{12.64}$$

where $N(z)$ is the cumulative normal distribution function, and

$$d_1^\rho = \frac{\log(A/D_0) + (\rho + \sigma^2/2)(T-t)}{\sigma\sqrt{T-t}} \text{ and } d_2^\rho = d_1^\rho - \sigma\sqrt{T-t} \tag{12.65}$$

Note from Equations (12.63) and (12.64) that the equity value resembles the BSM solution but includes a risk premium $\rho - r$ to compensate for the trading noise e^w and the non-ideal hedge Equation (12.46). A similar expression can be derived for the company's debt $D(A,t)$ using the BSM put-option formula $P_{BSM}(A,t;\sigma,\rho)$ for a European-style claim on the company's assets, with an additional term in the trading noise w'. For simplicity, in the following we consider that both trading noises w and w' are independent but have the same value for the noise parameter η and trading noise volatility $\tilde{\gamma}(t)$.

Now, consider the following idealized portfolios (Sobehart and Keenan, 2005):

Portfolio 1: long one riskless bond $D_0\, e^{-r(T-t)}$.

Portfolio 2: one share of the company's assets A, short one share of equity $E(A,t)$ and long one synthetic European put option on the company's assets $P(A,t) = D_0\, e^{-r(T-t)} - D(A,t)$.

If we consider that the additional uncertainty for the claims on the company's assets must vanish exactly at the bond's maturity ($\eta \rightarrow_{t\rightarrow T} 0$), portfolios 1 and 2 will be worth D_0 at expiration with certainty. If there were no additional sources of uncertainty, these portfolios would have identical expected values at time $t \leq T$. This would mean a static equilibrium condition:

$$A = E(A,t) + D_0\, e^{-r(T-t)} - P(A,t) = E + D \tag{12.66}$$

The expression above is the standard call-put relationship between equity, debt, and the company's assets. However, Equations (12.63)–(12.65) for equity and debt show that Equation (12.66) does not hold. The reason is that, at time t, there is no uncertainty in the construction of portfolio 1 but, given any arbitrary value of the company's assets A, there is uncertainty in how much portfolio 2 costs due to the additional noise in equity and debt prices. For portfolio 1 to be equivalent to portfolio 2, the riskless bond in portfolio 1 needs to be discounted at the extra risk premium $\rho - r$ required as compensation for the additional uncertainty created by equity and debt price fluctuations. These random fluctuations lead to instantaneous random profits/losses:

$$\begin{aligned} \Xi &= E + D - A \\ &= [e^w - 1]\, C_{BSM}(A,t;\sigma,\rho) - [e^{w'} - 1]\, P_{BSM}(A,t;\sigma,\rho) \neq 0 \end{aligned} \tag{12.67}$$

Equation (12.67) describes an equilibrium condition between equity, debt, and the company's assets with random fluctuations caused by market uncertainty and trading noise. As a result of these fluctuations, the total value of equity and debt can temporarily deviate from the value of the company's assets. The relative magnitude of the random fluctuations is determined by the trading noise $\eta\Gamma$. The time scale t_ν determines the relaxation time for equity and debt price deviations from to the fundamental value derived from the company's assets.

Note that if the parameters in Equation (12.60) were such that $E_t^w[e^w] = 1$ for an arbitrary time t, and transient effects were neglected, the random profits/losses in Equation (12.67) would vanish on average (that is, $E[\Xi] = 0$). In this scenario, there would be temporary random fluctuations but no "sustainable" arbitrage opportunities. In the general case $E_t^w[e^w] \neq 1$, Equation (12.67) can lead to potential "capital structure arbitrage" opportunities (that is, $E[\Xi] \neq 0$) caused by the difference between the fundamental value of the claims $f(x, t)$ and the value estimated by market participants $g(x, t)$.

At first blush, one might argue that, even in the case when there are no sustainable arbitrage opportunities, a strategy could be devised to profit from the "market timing" opportunities caused by trading noise by simply selecting over- and under-priced equity and debt claims at each point in time if transaction costs are not excessive. Note, however, that this would require executing orders simultaneously in two (or more) markets as soon as equity and debt prices are disclosed to market participants. This is unrealistic because for any small delay in the execution of the trading strategy the expected profits/losses from Equation (12.67) would vanish. Also note that the additional risk premium $\rho - r$ compensates investors for the characteristic fluctuations of equity and debt prices around the BSM-type values, and for the uncertainty in the profits/losses from Equation (12.67) (measured by the standard deviation of the equity-debt discrepancy Ξ).

In this approach, the relationship between equity, debt, and the company's assets reflects the situation where market participants are neither perfectly informed nor capable of eliminating arbitrage opportunities instantaneously. This scenario highlights the impact of deficient information and trading noise in the pricing of risky debt (Sobehart and Keenan, 1999, 2005). The model discussed above includes naturally the uncertainty of the equity and debt market and the concept of volatility not as a historical average, but as a forward-looking average based on market stability expectations. This approach characterizes the dynamics of the company's assets, equity, and debt using assumptions more consistent with markets driven by different degrees of uncertainty, expectations, and beliefs. These issues are critical in the evaluation of transition risk and physical risk for climate risk and ESG, where uncertainty, trading noise, and differences between fundamental values and market perceived values can impact a company's credit assessment and reputation.

NOTES

1. Black and Scholes assumed a portfolio composed of a long position in the security and a short position in the option. In contrast, Merton assumed a self-financing portfolio composed of the underlying assets, options, and risk-free bonds.
2. Given a stochastic process, the simplest choice of a filtration is that generated by the process itself, i.e. $F_t(X) = \sigma(X_s; 0 \leq s \leq t)$. In simple terms, at each point in time, the process history is known.
3. An older notation for the limit in quadratic mean l.i.m is q.m. lim.
4. This is also true in the deterministic case. Suppose, for example, that the interval between $a = 0$ and $b = 1$ is partitioned according to the following scheme: $0 < 1/2^N < 1/2^{N-1} < ... < 1/8 < 1/4 < 1/2 < 1$. The largest subinterval in this partition has width 1/2 no matter how N is increased.
5. That is, the stock price is the only source of uncertainty for the option.
6. Here we introduced a risk premium adjustment proportional to the portfolio value. Note, however, that more general adjustments are also possible (Sobehart, 2005; Sobehart and Keenan, 2005).
7. Black and Scholes (1973, p. 643) discussed the correlation between G_4 and a stylized market portfolio with *normal* increments only. Under the CAPM framework, this situation yields $\lambda = 0$.
8. Similar expansions can be derived from a combination of additive and multiplicative noises.

REFERENCES

Alexander, C. (2001). Principles of the Skew. *Risk (January)*, 29–32.

Anderson, R.W., Sundaresan, S., and Tychon, P. (1996). Strategic Analysis of Contingent Claims. *European Economic Review* 40: 871–881.

Babbel, D.F., and Merrill, C.B. (1996). Valuation of Interest Sensitive Financial Instruments. SOA Monograph M-FI96-1.

Baglioni, A., and Cherubini, U. (2004). *Accounting Fraud and The Pricing of Corporate Liabilities*. Catholic University of Milan, Working Paper.

Benos, A., and Papanastasopoulos, G. (2005). *Extending the Merton Model: A Hybrid Approach to Assessing Credit Quality*. Department of Banking and Financial Management, University of Piraeus, Greece, Working Paper.

Black, F., and Cox, J.C. (1976). Valuing Corporate Securities: Some Effects on Bond Indenture Provisions. *Journal of Finance* 31: 351–367.

Black, F. and Scholes, M. (1972). The Valuation of Option Contracts and a Test of Market Efficiency. *Journal of Finance* 27: 399–418.

Black, F., and Scholes, M. (1973). The Pricing of Options and Corporate Liabilities. *Journal of Political Economy* 81: 637–659.

Briys, E., and Varenne, F. (1997). Valuing Risky Fixed Rate Debt: An Extension. *Journal of Financial and Quantitative Analysis* 32(2): 239–248.

Cathcart, L., and El-Jahel, L. (1998). Valuation of Defaultable Debt. *Journal of Fixed Income* 8(1): 65–78.

Davydenko, S.A. (2004). *When Do Firms Default? A Study of the Default Boundary.* London Business School, Working Paper

Derman, E. (1999). Regimes of Volatility. Risk (April): 55–59.

Derman, E., and Zou, J. (2001). A Fair Value for the Skew. Risk (January): 111–113.

Di Paola, M. (1994). Stochastic Differential Calculus. In: F. Casciati (Ed.), *Dynamics Motion: Chaotic and Stochastic Behavior. International Center for Mechanical Sciences Courses and Lectures 340.* Springer-Verlag, Berlin.

Duffie, D., and Lando, D. (2001). Term Structures of Credit Spreads with Incomplete Accounting Information. *Econometrica* 69: 633–64.

Ergashev, B. (2002). *On Valuing Corporate Debt with The Volatility of Corporate Assets According to an Ornstein-Uhlenbeck Process.* Washington State University, Working Paper.

Finger, C., Flinkestein, V., Pan, G., Lardy, J.P., and Ta, T. (2002). Credit Grades Technical Document. *RiskMetrics Group*, pp. 23–30.

Fridson, M., and Jonsson, J.G. (1997). Contingent Claims Analysis. *Journal of Portfolio Management* (Winter): 31–43.

Gardiner, C.W. (1985). *Handbook of Stochastic Methods.* Springer-Verlag, Berlin.

Geske, R. (1979). The Valuation of Compound Options. *Journal of Financial Economics* 12: 211–235.

Giesecke, K. (2003). *Default and Information.* Cornell University, Working Paper.

Giesecke, K., and Goldberg, L. (2003). *Forecasting Defaults in the Face of Uncertainty.* Cornell University, Working Paper.

Haug, E.G., and Taleb, N.N. (2007). *Why We Have Never Used the Black-Scholes-Merton Option Pricing Formula*, Working Paper.

Hillegeist, S.A., Keating, E.K., Cram, D.P., and Lundstedt, K.G. (2002). *Assessing the Probability of Bankruptcy.* Kellogg School of Management, Northwestern University, Working Paper.

Hirshleifer, D. (2001). *Investor Psychology and Asset Pricing.* Ohio State University, Working Paper.

Hull, J.C. (1993). *Options, Futures and Other Derivatives.* Prentice Hall, Englewood Cliffs, NJ, pp. 218–222.

Ito, K. (1951). On the Stochastic Differential Equations. *Memoirs of the American Mathematical Society* 4: 1–51.

Kao, D.L. (2000). Estimating and Pricing Credit Risk: An Overview. *Financial Analysts Journal* 56(4): 50–66.

Karatzas, I., and Shreve, S.E. (1991). *Brownian Motion and Stochastic Calculus.* Springer, New York.

Kealhofer, S. (2003). Quantifying Credit Risk I: Default Prediction. *Financial Analyst Journal* (January/February): 30–44.

Kealhofer, S., and Kurbat, M. (2002). Predictive Merton Models. *Risk* (February): 67–72.

Keenan, S.C., and Sobehart, J.R. (2000). Credit Risk Catwalk. *Risk* (July): 84–88.

Keenan, S.C., Sobehart, J.R., and Benzschawel, T.L. (2003). *The Debt and Equity Linkage and the Valuation of Credit Derivatives.* Credit Derivatives: *The Definitive Guide.* Ed. J. Gregory, Risk Books, London.

Kim, J., Ramaswamy, K., and Sunderasan, S. (1993). Does Default Risk in Coupons Affect the Valuation of Corporate Bonds? *A Contingent Claims Model. Financial Management* 22(3): 117–131.

Lamberton, D., and Lapeyre, B. (1997). *Introduction to Stochastic Calculus*. Chapman & Hall, London.

Lasota, A., and Mackey, M.C. (1993). *Chaos, Fractals and Noise: Stochastic Aspects of Dynamics*. Springer, New York.

Leland, H.E. (1999). *The Structural Approach to Credit Risk. Frontiers in Credit-Risk Analysis*, AIMR Conference Proceedings, pp. 36–46.

Levin, J.W., and van Deventer ,D.R. (1997). The Simultaneous Analysis of Interest Rate and Credit Risk. In: A.G. Cornyn, R.A. Klein, and J. Lederman (Eds.), *Controlling and Managing Interest Rate Risk*. New York Institute of Finance, New York, pp. 494–506.

Lipton, A. (2002). The Vol Smile Problem. Risk (February): 61–65.

Loeve, M. (1963). *Probability Theory*. van Nostrand Co., Princeton, NJ, pp. 455–482.

Longstaff, F.A. (1990). Pricing Options with Extendible Maturities: Analysis and Applications. *Journal of Finance* 45(3): 935–957.

Longstaff, F.A., and Schwartz, E.S. (1995). A Simple Approach to Valuing Risky Fixed and Floating Rate Debt. *Journal of Finance* 50: 789–819.

Merton, R.C. (1973). The Theory of Rational Options Pricing. *Bell Journal of Economics and Management Science* 4: 141–183.

Merton, R.C. (1974). On the Pricing of Corporate Debt: The Risk Structure of Interest Rates. *Journal of Finance* 29: 449–470.

Merton, R.C. (1976). The Impact on Option Pricing of specification Error in the Underlying Stock Price Returns. *Journal of Finance* 31 (2): 333–349.

Mitchell, M., Pulvino, T., and Stafford E. (2002). Limited Arbitrage in Equity Markets. *Journal of Finance* 57(2): 551–584.

Nandi, S. (1998). *Valuation Models for Default-Risky Securities: An Overview*. Federal Reserve Bank of Atlanta, Economic Review (Fourth Quarter): 22–35.

Natenberg, S. (1994). *Options Volatility and Pricing*. McGraw-Hill, New York, pp. 385–418.

Nayfeh, A.H. (1993). *Perturbation Methods*. Wiley-Interscience, New York.

Neftci, S.H. (1996). *An Introduction to the Mathematics of Financial Derivatives*. Academic Press, San Diego.

Ravid, S.A. (1996). Debt Maturity - A Survey. *Financial Markets, Institutions, and Instruments* 5(3): 1–69.

Saa-Requejo, J., and Santa-Clara, P. (1999). *Bond Pricing with Default Risk*. University of California, Working Paper.

Saunders, A., and Allen, L. (2002). *Credit Risk Measures*. 2nd edition. John Wiley and Sons, New York, pp. 9–22, 48–66.

Schuss, Z. (1984). *Theory and Applications of Stochastic Differential Equations*. John Wiley and Sons, New York.

Schwartz, E.S., and Moon M. (2000). Rational Pricing of Internet Companies. *Financial Analyst Journal* (May–June): 62–75.

Shiller, R.J. (2000). *Irrational Exuberance*. Princeton University Press, Princeton, NJ, pp. 135–168.

Shleifer, A. (2000). *Inefficient Market, An Introduction to Behavioral Finance*. Oxford University Press, Oxford, pp. 28–52.

Sobehart, J.R. (1999). Beyond Black-Scholes: Pricing Options under Market Uncertainty and Trading Noise. Moody's Risk Management Services Research Report 2-3/25/99.

Sobehart, J.R. (2005). A Forward-Looking Singular Perturbation Approach to Pricing Options under Market Uncertainty and Trading Noise. *International Journal of Theoretical and Applied Finance* 8(5): 635–658.

Sobehart, J.R., and Farengo, R. (2002). Fat Tails Bulls and Bears. *Risk (December),* S20–S24.

Sobehart, J.R., and Farengo, R. (2003). A Dynamic Model of Irrational Exuberance. *Journal of Risk Finance*, 5(4): 91–116.

Sobehart, J.R., and Keenan, S.C. (1999). *Equity Market Value and its Importance for Credit Analysis: Facts and Fiction.* Moody's Investors Service, Working Paper.

Sobehart J.R., and Keenan S.C. (2001). A Practical Review and Test of Default Prediction Models. *RMA Journal* (November): 54–59.

Sobehart, J.R., and Keenan, S.C. (2002a). The Need for Hybrid Models. Risk (February): 73–77.

Sobehart, J.R., and Keenan, S.C. (2002b). Hybrid Contingent Claims Models: A Practical Approach to Modeling Default Risk. In: M. Ong (Ed.), *Credit Risk.* 1st edition. Risk Books, London.

Sobehart, J.R., and Keenan, S.C. (2002c). A Paradox of Intuition: Hedging the Limit or Hedging in the Limit? *International Journal of Theoretical and Applied Finance* 5(7): 729–736.

Sobehart, J.R., and Keenan, S.C. (2003a). Uncertainty in Pricing Tradable Options. *International Journal of Theoretical and Applied Finance* 6(2): 103–117.

Sobehart, J.R., and Keenan, S.C. (2003b). The Role of Uncertainty in the Pricing of Risky Debt. *Journal of Risk Finance* 4(2): 56–67.

Sobehart, J.R., and Keenan, S.C. (2003c). Hybrid Probability of Default Models: A Practical Approach to Modeling Default Risk. *Citigroup Global Markets, Quantitative Credit Analyst* 3 (October): 5–29.

Sobehart, J.R., and Keenan, S.C. (2004). The Score for Credit. Risk (February): 54–58.

Sobehart, J.R., and Keenan, S.C. (2005). Capital Structure Arbitrage and Market Timing Under Uncertainty and Trading Noise. *Journal of Credit Risk* 1(4): 1–27.

Sobehart, J.R., Stein, R., Mikityanskaya, V., and Li, L. (2000). *Moody's Public Firm Risk Model: A Hybrid Approach to Modeling Short Term Default Risk.* Moody's Investor Service, Rating Methodology.

Thaler, R.H. (1999). The End of Behavioral Finance. *Financial Analyst Journal* 55(6): 12–17.

Tudela, M. and Young, G. (2003a). A Merton Model Approach to Assessing the Default Risk of UK Public Companies. *Bank of England, Working Paper* 194.

Tudela, M., and Young, G. (2003b). Predicting Default Among UK Companies: A Merton Approach. *Financial Stability Review* (June): 104–114.

Wei, D.G., and Guo, D. (1997). Pricing Risky Debt: An Empirical Comparison of the Longstaff and Schwartz, and Merton Models. *Journal of Fixed Income* (Fall): 8–28.

CHAPTER 13

Pillar 3: Supply of Credit: Modeling Lender's Behavior and Business Strategies

PORTFOLIO MANAGEMENT

Banking and financial investments are primarily risk-taking activities, which are successful as long as risks are controlled within reasonable bounds through sound risk management practices (Matten, 1996; Bessis, 2002; van Deventer et al., 2005). Financial institutions manage their assets, primarily loans for commercial banks, to maintain the liquidity needed to meet investors' demands or deposit withdrawals, cover expenses including credit, market, or operational losses, while earning a profit to provide a reasonable return to their stockholders. These business goals must be aligned to the institution's risk appetite, and internal policies and guidelines for sound risk taking and risk management, where safety of principal and assurance of repayment in a timeline manner should take priority over profits.

Commercial banks and other financial institutions lend to different borrowers for a variety of reasons, under different contractual terms, periods, and borrowed amounts. As a result, loans and other credit products can vary widely in their liquidity and risk. Furthermore, credit concentrations must be limited so that no single occurrence can impact the portfolio significantly, leading to large losses. Institutions manage the growth of their portfolios and asset diversification in relation to the capital needed to support that growth. As financial institutions turn their attention to climate risk and ESG issues to define their own low-carbon emission and net zero strategies, investments and risk-asset growth become increasingly selective, and the risk-reward characteristics of high-carbon emission assets undergo closer scrutiny.

In analyzing the amount of risk that the institution is willing to take, consideration is usually given to the obligor's character and experience, capabilities, net profitability, liquidity, market presence, net worth, leverage and debt obligations, and financial commitments. Additional consideration is also given to the obligor's climate and environmental impact and its carbon-emission strategies and commitments. Some obligors and industries must be recognized as intrinsically riskier than others due to their risk characteristics (e.g. high-tech start-ups, or high-leverage obligors), while other industries are riskier due to the potential implications of transition risk (e.g. coal, oil and gas, or power generation) or physical risk (e.g. real estate developers). Whenever a significant change in an obligor's operations is expected (such as a new business undertaking, product, factory, new process, changes in regulations or business incentives), the risk should be heightened beyond normal conditions. Furthermore, fundamental changes in industry, production, or technology in response to climate risk can uncover vulnerabilities in the obligor's creditworthiness. In many instances, banking and financial investment problems can be traced to poor credit decisions or lax terms of lending and risk taking, or weak risk management practices. Risk criteria must be precisely specified for each segment:

1. Determine what type of borrowers, obligors, or counterparties are acceptable.
2. Determine the risk-acceptance criteria and credit terms: form of borrowing, price, tenor, collateral, support, seniority of claims, and documentation requirements.
3. Understand the risk-reward characteristics and their measuring, monitoring, and management in alignment with the institution's risk appetite and business strategies.
4. Specify what professional and technical skills are needed for the credit assessment, measurement, monitoring, and risk management of the obligor or counterparty relationship.

While loan agreements and financial contracts are usually drafted on a case-by-case basis for different products, segments, or situations, an alignment of the terms and conditions for high-carbon-emission situations and the institution's own low-emission strategies can provide the institution with an element of control to protect itself against loss or reputation damage. In today's volatile and uncertain world, it is difficult to imagine writing debt covenants that would cover every possible contingency that could impact the obligor or its business environment. However, strong risk practices and sound risk management require understanding how changes in the business environment, policies, regulations, or business restrictions could adversely affect obligors. It is quite possible that an obligor's ability to repay may have deteriorated significantly because of substantial realized losses or potential losses due to pending litigation or reputation damage related to climate risk or ESG issues. In other cases, a material adverse change clause may have to be extended or amended to include guarantors or support.

Previous economic recessions and crises (e.g. the 1990s real estate crisis, the 2000s dot-com bubble, the 2008–2009 Great Recession) highlighted the dangers of excessive risk taking and concentrations on specific sectors or asset classes. Those lessons should be borne in mind not only in the current phase of the business cycle but in the context of climate risk and ESG, where transitions to a low carbon-emission economy or an increase in the frequency or severity of storms, hurricanes, floods, and other physical perils could affect obligors' creditworthiness.

Loans and other banking products can vary considerably across markets, obligors, and product segments. Some are short-term, self-liquidating, and seasonal, while others are longer-term (often limited to a maximum period) and amortizing with protective debt agreements. These products can be unsecured, or secured, and can reflect different types of collateral, support, or seniority.

In general, the purpose of the loans and other credit products should be the basis of their repayment. Short-term, seasonal liquidating loans to support current assets or to take advantage of attractive supplier terms are usually the most appealing in terms of liquidity. Loans to fund non-current assets or long-term investments carry greater risk. As liquidity diminishes, debt repayment must come from long-term sources, which may extend the repayment timeline, adding repayment uncertainty.

For banks, typical loans and related credit products usually fall within the following broad categories:

1. **Commercial and industrial (C&I) loans** could include short-term, long-term, secured (with collateral) and unsecured loans. When lines of credit are used to finance seasonal or cyclical working capital needs, these needs should be self-liquidating in the normal course of business (e.g. through the conversion of receivables and inventory into cash). Strong borrowers with access to long-term debt and equity markets (capital markets) can use credit lines to bridge the financing of their permanent capital requirements. Their ability to repay their debt may depend on economic and industry conditions, which can be affected by climate risk and ESG issues. Usually, financial institutions will require a *second way out*, based on the obligor's balance sheet and claims on the available assets.

2. **Real estate loans** could be amortizing and must be based on a thorough appraisal and analysis of supply and demand for development projects. Prudent lending and diversification practices require reducing geographic concentrations. Construction loans usually have takeouts (firm commitments from acceptable lenders) and must be monitored closely to ensure compliance with the takeout requirements.
3. **Consumer loans** enable borrowers to spend money (in the form of credit or cash) before earning it. In contrast, commercial credit (e.g. C&I loans) is used primarily to generate income by funding business activities or investments. Furthermore, consumer lending is a volume activity that requires efficient administration, sound market segment positioning, and adequate diversification.
4. **Financial institution lending** is loans to the securities industry, finance companies, insurance companies, banks, and thrift institutions. A bank may be able to collect more detailed information about its commercial, industrial, consumer retail and government borrowers than it can get about other banks. Usually, information about their management, the quality of their portfolios, asset concentrations, liquidity, asset-liability management, and contingent liabilities is limited and can change significantly in a short period of time. In some countries, banks may be more regulated or managed than in others, leading to questions about their solvency and liquidity. For example, a bank could be solvent if it may be able to liquidate assets over a reasonable period of time without realizing any loss, while being illiquid in the short run. Liquidity problems for a bank usually begin on the asset side of the balance sheet. The bank can become insolvent if enough of the assets are impaired, regardless of how much liquidity or capital the bank has available.
5. **Specialized loans** cover a variety of industries, disciplines, and areas such as airlines, aerospace, asset-based lending, banks and finance institutions, hospitals and healthcare, energy, chemicals, mining, farming and agriculture, nonprofit, information systems and communications, insurance, contractors, petroleum, shipping, project finance, public utilities, and real estate. Several of these specialized segments are exposed to climate risk and ESG issues due to the nature of the industries or due to changes in the regulatory environment and business incentives for the transition to a low carbon economy (e.g. energy and power generation).
6. **Project finance** typically involves loans put together by banking groups to develop projects, whose repayment depends upon the success of these projects. Projects usually have long development periods and have different economic assumptions with respect to costs, pricing, and market conditions. Because this type of financing usually involves high leverage,

it may require some guarantees for the selected market and sufficient revenues to successfully achieve the completion of the projects.

7. **Loans to foreign governments and related entities** cover a range of financial activities by central and local governments and related entities and must fall within established country limits. Loans to governments often finance viable projects or are used to bridge temporary shortfalls in their budgetary processes. This is relevant given the significant investments needed to transition to a low-carbon economy (e.g. infrastructure, adopting new technologies or retrofitting existing technologies), or to mitigate the impact of physical risk (e.g. building dams and levees for flood control). Due to the nature of sovereigns and related entities, loans should meet all risk management criteria and include loan agreement provisions to provide appropriate protection in case of default or significant credit deterioration. This includes the stipulation of the applicable laws and courts of a recognized commercial jurisdiction outside the jurisdiction of the sovereign to resolve disputes.
8. **Export-import trade financing** includes the use of commercial letters of credit for facilitating import and export trade activities between different parties. A letter of credit provides a means of paying the beneficiary of the letter (e.g. exporter) on behalf of the applicant (e.g. importer), whose credit standing is replaced by that of the bank or institution issuing the letter of credit.
9. **Standby letters of credit** are frequently used as a form of payment guarantee to a beneficiary in the case of nonperformance of a contractual obligation by the applicant of the letter. Similar to the case for commercial letters of credit, the issuing bank provides its creditworthiness to support that of the obligor requesting the letter.
10. **Commercial paper backstop lines** are used to bridge periods when an otherwise creditworthy borrower is unable to access the paper market because of external factors beyond its control, or when there is a cost advantage. These lines are not intended for bailing out commercial paper investors who hold paper of deteriorating borrowers.
11. **Loans secured by shares or securities** have varied credit and market risks and require exhaustive appraisal of the risks and rewards involved. Obligors should have the financial resources to maintain margin requirements, which should consider the downside potential to permit action in case of market deterioration.
12. **Multi-bank credit and fronting activities** involve the institution as an agent bank or as a participating institution in a group that collectively extends credit to an obligor, where the fronting bank provides funding on behalf of the other participants.

The broad activities described above can be affected by climate risk and ESG issues. The transition to a low-carbon economy to mitigate the effects of climate risk will affect countries differently, which will impact the portfolio risk characteristics of many financial institutions. While energy-producing countries (coal, oil, gas) will see their revenues affected by changes in supply and demand due to the global transition to a low-carbon economy, non-energy-producing countries will face the problems of adjusting to the new energy economics and their socio-economic costs. Furthermore, expansionary economics and the re-alignment of business activities in countries that seek to maintain a high rate of consumption and growth during their transition to a low-carbon economy can result in fiscal deficits and balance-of-payment problems. At some point, refinancing their deficits and debt may be unsustainable for these countries, which will confront country governments, lenders, and investors with the possibility of credit defaults and workout problems. In this situation, local economies, companies, and consumers can be adversely affected and face significant economic hardship when a country manages down and squeezes its internal economy in the hope of maintaining its credit standing. These effects can worsen as the global competitive landscape shifts and some countries take advantage of the situation while other countries assume the burden of increasing climate-related regulations and business restrictions that can impair their economies and competitiveness.

ESTIMATING PORTFOLIO LOSSES

Here we discuss a framework for credit loss estimation including multiple scenarios designed to reflect the uncertainty in economic environments, including scenarios driven by climate risk and ESG issues. The framework leverages default likelihood and severity information and their relation to economic activity and provides a more robust approach to the estimation of portfolio losses during periods of significant economic and market uncertainty. This framework can also help assess the impact of different transition and physical risk scenarios aligned to the institution's business strategies and low-carbon emission commitments.

Historically, credit conditions and actual portfolio losses for financial institutions have been volatile over time, and quite cyclical in response to business cycles, and monetary and socio-economic policies (Sobehart and Keenan, 2004; Sobehart, 2009; Sobehart and Giacone, 2012). When financial institutions evaluate the risk inherent in their credit portfolios, they do so against a backdrop of macroeconomic and market uncertainty. As discussed in previous chapters, the relationship between the *economic cycle* (as measured by economic indicators) and the *credit cycle* (as measured by default rates or recovery rates) shifts over time. But since expected losses may be highly dependent on these macroeconomic trends, even with lending practices held constant, it is important to define a framework that can provide potential credit losses under different levels of economic uncertainty. When the estimation is performed over extended periods of time as required for climate risk forecasts, the uncertainty includes both borrowers' behavior and demand for credit and the lender's investment strategy and ability and willingness to provide funds.

Simple portfolio loss models found in the literature often assume that rating transition probabilities, probabilities of default, loss given default, and credit spreads representing borrowers' characteristics are stable across obligor types and across the business cycle. These models assume that a single set of the average historic rating transition, default rates, and credit spreads, and simple assumptions about cumulative effects of losses are good enough to characterize potential future losses in an institution's portfolios. However, credit migration, loss likelihood, and loss severity depend on the state of the economy and a variety of other factors. For example, the probability of downgrades and credit default are often greater in business downturns than in upturns. Furthermore, historic credit losses are often volatile and driven by the heterogeneity of the portfolios (including large and small loans). Therefore, models based on the average rating transition or average default probability may understate potential tail losses – exactly the quantity most institutions want to measure with some degree of precision. Also note that credit losses calculated using scenarios based on average loss statistics usually refer to possible losses of hypothetical scenarios that might never materialize as business and economic activities rarely exhibit "average" conditions. This contrasts with the estimation of losses for a wide range of possible scenarios that can reflect economic and market uncertainty as well as the impact of transition and physical risks at higher levels of confidence.

A more comprehensive approach is to directly model the relationship between rating transition probabilities, default probabilities, loss severity and macroeconomic factors and climate risk drivers, and then analyze their impact on plausible credit loss patterns over time by generating different conditions (Sobehart and Keenan, 2004; Sobehart, 2009; Sobehart and Giacone, 2012). This alternative macroeconomic and climate risk approach provides a sound means of constructing a wide range of possible loss likelihood and loss severity estimates for credit reserves and credit risk capital calculations that can preserve the basic relationships observed in historical rating transitions and observed default and loss rates. The approach can also be used for advanced stress testing across different markets and economies as required for short- and long-term climate risk. This approach is important for analyzing both credit stress conditions and the tail end of the distribution of credit losses under different scenarios, which is often driven by the irregularities and small sample size effects found in published default and loss data.

The approach described here provides credit loss estimates that are generally aligned to the US Current Expected Credit Loss (CECL) and International Financial Reporting Standard 9 (IFRS9), which are accounting standards for financial instruments used for recognizing and measuring financial assets, financial liabilities, and certain nonfinancial items (IFRS, 2014; FASB, 2016). Under both the CECL and IFRS9 accounting standards, the expected credit loss (ECL) over the life of an asset is a critical concept for the estimation of impairments of assets such as loans held for investment, or securities held to maturity. The shift to the ECL accounting framework helped reduce weaknesses identified during the 2008–2009 global financial crisis around the timely recognition of credit losses. In order to implement the CECL and IFRS9 accounting standards, banks must use sound judgment, analytics, and robust considerations for providing forward-looking information that can reflect a wide range of possible benign and downturn economic conditions.

Although most credit downturns and financial crises may result from unique circumstances, they usually exhibit common patterns associated with economic expansions and contractions, and periods of divergence between fundamental risk assessment and the market's perception of risk. Here, we present a multiple-scenario ECL approach for capturing uncertainty in economic environments that leverages default likelihood and severity information and their relation to credit cycles, climate risk drivers and other factors that can determine credit losses in an institution's portfolio.

Cash Shortfalls and Credit Losses

Both CECL and IFRS9 Standards were introduced in response to the shortcomings of the incurred loss approach used by financial institutions, whose impact was reflected during the 2008–2009 financial crisis (Great Recession). These two accounting standards share many points in common for loss and impairment estimation, but they also differ in several fundamental aspects (IFRS, 2014; FASB, 2016; OCC, 2020, 2021). While CECL provides general guidelines on the estimation of ECL through the life of the assets in an institution's portfolio, IFRS9 is more prescriptive and provides specific guidance for the estimation of ECL for a 1-year period and through the life of the assets in scope. Under the IFRS9 Standard, the determination of which one of the ECL estimates should be used for reporting the asset's allowance for credit losses is determined by the changes in the asset's credit quality (significant increase in credit risk). ECL estimates depend on forward-looking forecasts for the assets, which could be driven by different economic and business scenarios. This is relevant in the context of climate risk and ESG since both transition and physical risks can have profound implications for changes in asset quality and asset impairment.

The estimation of credit losses depends on a variety of factors: the credit quality of the obligors, the level of subordination of its obligations and collateral, the economic environment and other risk drivers. Thus, the estimation of expected credit losses can leverage several of the fundamental risk concepts discussed in previous chapters, including obligor and debt concepts such as (Sobehart, 2017, 2019, 2020, 2021):

1. Probability of Default (PD) and risk rating migration.
2. Exposure at Default (EAD) (for outstanding balances and incremental drawdowns).
3. Loss Given Default (LGD).

Also note that, under IFRS9, the allowance for credit losses is based on discounted cash shortfalls, while under CECL, the allowance for credit losses is defined as a valuation account that is deducted from the amortized cost basis of the financial assets to present the net amount expected to be collected on the financial asset. Note, however, that the IFRS9 and CECL definitions can be aligned under specific technical conditions as described below (Sobehart, 2021).

Given the uncertainty generated by different economic and climate risk scenarios, our description focuses on probability-weighted estimates of the present value of all cash shortfalls over the expected life of the financial instruments in an institution's portfolio. More precisely, cash shortfalls are the difference between the cash flows that are due from the obligors to the financial entity in accordance with the contractual obligation (*contractual cash flows*) and the cash flows that the entity expects to receive in practice (*expected cash flows*) due to default risk and other factors. Because the estimation of financial losses needs to consider both the amount and timing of payments, a credit loss could arise even if the financial institution expects to be paid in full but later than the contractual payment due date. More precisely, the *contractual* cash flow CF_t provides the expected cash flow to be received in each period t as agreed with the obligor. If the *expected* cash flow CF_t^E reflected the situation where the entity may not be paid as contractually agreed (in terms of the amount or timing of payments), the cash shortfall would be $S_t = CF_t - CF_t^E$. For assets eligible under CECL and IFRS9, the cash shortfall needs to be recognized as a credit loss.

Let's assume we have a simple debt obligation (e.g. loan or bond) with a principal amount V_0 outstanding at the time $t = 0$. The debt obligation requires interest and amortization payments at regular time intervals $t = 0, 1, 2, \ldots, T$, where T is the remaining tenor of the debt measured in payment periods (e.g. monthly, quarterly, semi-annual, or annual interest payments).

Let's define the contractual cash flows CF_t that are paid at regular time periods t as:

$$CF_t = \text{contractual cash flow due at time } t \quad 0 \le t \le T \tag{13.1}$$

Cash flow CF_t can include a combination of interest payments (e.g. coupons, interest, fees) $C_1, C_2, \ldots, C_T$, and amortization payments $W_1, W_2, \ldots, W_T$ of the total principal amount:

$$C_t = \text{coupon, interest, or fee due at time } t \tag{13.2}$$

$$W_t = \text{Principal amortization payments at time } t \tag{13.3}$$

As a result of the amortization of the principal amount, the principal amount value V_t outstanding at the beginning of each payment period t is:

$$V_t = V_0 - \sum_{s=0}^{t-1} W_s = V_{t-1} - W_{t-1} \tag{13.4}$$

Notice that the cash flows can be described as follows:

$$CF_t = C_t + W_t \quad 0 \le t \le T \tag{13.5}$$

Each cash flow CF_t paid at time t needs to be present valued at time $t = 0$ to be able to estimate the fair value of the obligation. The appropriate discounting factor D_t is determined by the effective interest rate (EIR) R, which depends on the contractual interest and amortization payments, a reference interest rate (e.g. for floating rate loans) and the term structure of interest rates. For a flat term structure with constant effective interest rate R, the discounted factor is simply:

$$D_t = \frac{1}{(1+R)^t} = \text{discount factor at time } t \tag{13.6}$$

Table 13.1 describes the expected contractual cash flows described above during the remaining life of the obligation for a generic debt instrument (e.g. loan or bond).

To illustrate our approach, for a simple loan (or bond) paying fixed interest and fees (or coupons) with a single balloon payment of principal $W_T = V_0$ at expiration T, the cash flows are simply:

$$CF_t = \begin{cases} C_t & 0 \le t \le T \\ C_T + W_T & t = T \end{cases} \tag{13.7}$$

TABLE 13.1 Contractual cash flows through the life of the asset

Period t	Interest payment	Amortization schedule	Principal outstanding	Cash flow	Discount factor	Discounted cash flow
0	C_0	W_0	V_0	$CF_0 = C_0 + W_0$	D_0	$D_0\ CF_0$
1	C_1	W_1	$V_1 = V_0 - W_0$	$CF_1 = C_1 + W_1$	D_1	$D_1\ CF_1$
2	C_2	W_2	$V_2 = V_1 - W_1$	$CF_2 = C_2 + W_2$	D_2	$D_2\ CF_2$
3	C_3	W_3	$V_3 = V_2 - W_2$	$CF_3 = C_3 + W_3$	D_3	$D_3\ CF_3$
4	C_4	W_4	$V_4 = V_3 - W_3$	$CF_4 = C_4 + W_4$	D_4	$D_4\ CF_4$
...	...	...	...	....	...	....
T	C_T	W_T	$V_T = V_{T-1} - W_{T-1}$	$CF_T = C_T + W_T$	D_T	$D_T\ CF_T$
PV						$\sum_{t=0}^{T} D_t CF_t$

Table 13.2 describes the expected contractual cash flows during the remaining life of the debt obligation for a fixed interest rate loan (or bond) with principal payment $W_T = V_0$ at expiration T, and interest payments $C_t = RV_0 \; 0 < t \leq T$ with no initial interest payment at $t = 0$ (i.e. $C_0 = \$0$), and a constant term structure of interest rates.

Now let's turn our attention to the estimation of cash flows when there is a risk of default on the contractual obligations. The contractual cash flow CF_t describes the expected cash flow to be received in each period as agreed with the obligor. That is, when there is no risk of default. However, if the cash flow CF_t^E is not expected to be paid as agreed in terms of the amount or timing of payments, there would be a cash shortfall $S_t = CF_t - CF_t^E$ that needs to be recognized as a loss.

Let's assume that cash shortfalls only occur as a result of a default on the contractual obligation as opposed to a more general case, where credit losses could be due to partial payments of interest and/or principal. In this situation, default events can happen at any time t with a (cumulative) probability of default:

$$P_t = Prob(Default, 0 \leq t' \leq t) \tag{13.8}$$

In principle, the cumulative probability P_t could be determined by the obligor's credit quality (i.e. its risk rating), the nature and risk characteristics of the contractual obligation, and expectations of the economic environment. In practice, obligors may have multiple contractual obligations and can default due to a variety of reasons. Therefore, the probability of default on any debt obligation is assumed independent of the characteristics of the obligations but it still depends on the obligor's characteristics, economic forecasts, and climate risk drivers.

Note that the cumulative probability of default P_t can be expressed as the probability of default P_{t-1} in a previous period $t-1$ plus the probability of survival until time $t-1$ and then defaulting during the period $[t-1, t]$ with (marginal) probability $p_{t-1,t}$:

$$P_t = P_{t-1} + (1 - P_{t-1})\, p_{t-1,t} \tag{13.9}$$

Equation (13.9) provides a means for constructing the cumulative probability of default for any period by compounding the effects of marginal probabilities period after period. This property can will be used in the impairment staging process to align the cumulative probabilities estimated at the obligation's origination with the current projections for the remaining life of the debt instrument.

Note that in the approach discussed here the obligor can default on the debt obligation within each payment period t (that is, the period between the

TABLE 13.2 Contractual cash flows for a simple loan (or bond)

Period t	Interest payment	Amortization schedule	Principal outstanding	Cash flow	Discount factor	Discounted cash flow
0	C_0	0	V_0	$CF_0 = C_0$	$D_0 = \frac{1}{(1+R)^0}$	$\frac{0}{(1+R)^0}$
1	C_1	0	$V_1 = V_0$	$CF_1 = C_1$	$D_1 = \frac{1}{(1+R)^1}$	$\frac{R\,V_0}{(1+R)^1}$
2	C_2	0	$V_2 = V_0$	$CF_2 = C_2$	$D_2 = \frac{1}{(1+R)^2}$	$\frac{R\,V_0}{(1+R)^2}$
3	C_3	0	$V_3 = V_0$	$CF_3 = C_3$	$D_3 = \frac{1}{(1+R)^3}$	$\frac{R\,V_0}{(1+R)^3}$
4	C_4	0	$V_4 = V_0$	$CF_4 = C_4$	$D_4 = \frac{1}{(1+R)^4}$	$\frac{R\,V_0}{(1+R)^4}$
...	...	...		....		
T	C_T	$W_T = V_0$	$V_T = V_0$	$CF_T = C_T + W_T$	$D_T = \frac{1}{(1+R)^T}$	$\frac{(1+R)V_0}{(1+R)^T}$
PV						V_0

previous payment due at time $t-1$ and following payment due at time t). The probability of defaulting during time interval $[t-1, t]$ (before payment due at time t) is:

$$Prob(t-1 < t' \leq t) = P_t - P_{t-1} = (1 - P_{t-1})\, p_{t-1,t} \tag{13.10}$$

For a given period t, if the obligor did not default previously, the obligor either pays all the cash flows (interest and amortization payments) as scheduled (with survival probability $(1 - P_t)$) or pays all cash flows due before time t but fails to pay the contractual interests and amortization payment due at time t. That is, the cash shortfalls for a default in the time interval $[t-1, t]$ are:

$$S_{t,s} = \begin{cases} 0 & 0 \leq s \leq t \text{ with } (1 - P_s) \\ CF_s - CF_s^E & t \leq s \leq T \text{ with } (P_t - P_{t-1}) \end{cases} \quad \text{for } 0 \leq t \leq T \tag{13.11}$$

Furthermore, once the obligor defaults during the interval $[t-1, t]$, there are no further interest or principal amortization payments. However, there is an immediate recovery payment amount Q_t defined in terms of: (1) the Exposure At Default $EAD_t = V_t$, which is the principal amount outstanding at the time of default t, and (2) the Loss Given Default LGD_t, which is discounted to the time of default. More precisely,

$$CF_s^E = \begin{cases} Q_t & s = t \\ 0 & t < s \leq T \end{cases} \quad \text{for } 0 \leq t \leq T \tag{13.12}$$

Here the recovery payment Q_t is:

$$Q_t = EAD_t\,(1 - LGD_t) \tag{13.13}$$

The outstanding principal amount at default (at time t) is simply:

$$EAD_t = V_t = V_0 - \sum_{s=0}^{t-1} W_s \tag{13.14}$$

The case of additional drawdowns for unfunded exposures (unused commitments and contingent exposure) is discussed later below. Table 13.3 describes the cash flow payments in period s as determined by the timing of the default event that occurs in period t.

TABLE 13.3 Expected cash flows for a simple loan (or bond) with default risk

Period s	Default in t=0	Default in t=1	Default in t=2	Default in t=n	Default in t=T
0	Q_0	CF_0	CF_0	CF_0	CF_0
1		Q_1	CF_1	CF_1	CF_1
2			Q_2	CF_2	CF_2
3				CF_3	CF_3
4				CF_4	CF_4
...			...	Q_n	....
T					Q_T
Probability of default	P_0	$P_1 - P_0$	$P_2 - P_1$	P_n - P_{n-1}	P_T - P_{T-1}

Probability Weighted Cash Shortfalls

The loss given default and recovery amount described above provide the potential credit losses in the event of default. Note that even if there is a full recovery of the principal amount at the time of default (i.e. $LGD_t = 0$), there are cash shortfalls in each period that reflect the time value of money, since there are no additional contractual payments after the obligor defaults. The pattern of cash shortfalls for all payment periods $0 \le s \le T$ until the expiration of the obligations at time T depends on the scenario for the obligor default. If the obligor defaults in period t (with probability of default $P_t - P_{t-1}$), the cash shortfalls for any payment period s with a default in period t are defined as follows (Sobehart, 2021):

$$S_{t,s} = \begin{cases} 0 & 0 \le s \le t \quad \text{with} \quad (1 - P_t) \\ CF_t - Q_t & s = t \quad \text{with} \quad (P_t - P_{t-1}) \\ CF_s & t < s \le T \quad \text{with} \quad (P_t - P_{t-1}) \end{cases} \quad \text{for } 0 \le t \le T \quad (13.15)$$

As a result, the probability weighted discounted cash shortfalls DS_t for a default event that occur at time t (i.e. in the time interval $[t-1, t]$) are simply:

$$DS_{t,s} = (P_t - P_{t-1}) \times \begin{cases} D_t(CF_t - Q_t) & s = t \\ D_s\, CF_s & t < s \le T \end{cases} \quad \text{for } 0 \le t \le T \quad (13.16)$$

Although CECL requires only lifetime credit loss estimates, IFRS9 requires two different estimates: (1) lifetime expected credit losses, and

(2) 1-year expected credit losses. Since T is the remaining tenor of the obligations and represents the lifetime period, let's introduce T_{1Y} as the 1-year period. Both expected credit losses can be calculated as follows:

Lifetime expected credit loss:

$$ECL_L = \sum_{t=0}^{T}\sum_{s=t}^{T} DS_{t,s} = \sum_{t=0}^{T}\left(\sum_{s=t}^{T} D_s CF_s - D_t Q_t\right)(P_t - P_{t-1})$$

$$= \sum_{t=0}^{T}\left(\left(\sum_{s=t}^{T} D_s CF_s - D_t EAD_t\right) + D_t EAD_t LGD_t\right)(P_t - P_{t-1}) \quad (13.17)$$

1-year expected credit loss:

$$ECL_{1Y} = \sum_{t=0}^{T_{1Y}}\sum_{s=t}^{T} DS_{t,s} = \sum_{t=0}^{T_{1Y}}\left(\sum_{s=t}^{T} D_s CF_s - D_t Q_t\right)(P_t - P_{t-1})$$

$$= \sum_{t=0}^{T_{1Y}}\left(\left(\sum_{s=t}^{T} D_s CF_s - D_t EAD_t\right) + D_t EAD_t LGD_t\right)(P_t - P_{t-1}) \quad (13.18)$$

Here $P_{-1} = 0$, which reflects that the obligor does not start the period in default. Table 13.4 illustrates the calculation of the lifetime ECL using the discounted cash shortfall estimates for each default period.

For a simple fixed interest rate loan (or bond) with principal payment $W_T = V_0 = EAD_0$ at expiration T, and interest payments $C_t = RV_0$ $0 < t \leq T$ with no initial interest payment at $t = 0$, Equations (13.17) and (13.18) reduce to:

$$ECL_L = \sum_{t=0}^{T} D_t LGD_t EAD_0 (P_t - P_{t-1}) \quad (13.19)$$

$$ECL_{1Y} = \sum_{t=0}^{T_{1Y}} D_t LGD_t\, EAD_0 (P_t - P_{t-1}) \quad (13.20)$$

For this simple example, the expected credit loss is the cumulative default loss described by the LGD_t at time t but discounted (present-valued) to the initial period $t = 0$ (reporting period) by the factor D_t since the projected defaults occur at different points in time in the future and their losses need to be present-valued to the reporting period. The calculation of cash shortfalls and expected credit losses described above applies to a wide range of banking products and debt securities under the *amortizing cost* (AC) and *fair value through other comprehensive income* (FVOCI) required under IFRS9. In principle, these calculations could be applied more generally to assets not covered under IFRS9 and CECL.

TABLE 13.4 Discounted cash shortfall and expected credit losses

Period s	Discounted cash shortfall t=0	Discounted cash shortfall t=1	Discounted cash shortfall t=2	Discounted cash shortfall t=n	Discounted cash shortfall t=T	Discounted cash flow
0	DS_{00}	0	0	0	0	DS_{00}
1	DS_{01}	DS_{11}	0	0	0	$DS_{01}+DS_{11}$
2	DS_{02}	DS_{12}	DS_{22}	0	0	$DS_{02}+DS_{12}+DS_{22}$
3	DS_{03}	DS_{13}	DS_{23}	DS_{n3}	0	$DS_{03}+\ldots+DS_{33}$
4	DS_{04}	DS_{14}	DS_{24}	DS_{n4}	0	$DS_{02}+\ldots+DS_{44}$
...	...	...	...	...	...	.....
T	DS_{0T}	DS_{1T}	DS_{2T}	DS_{nT}	DS_{TT}	$DS_{0T}+..+DS_{TT}$
PV						$\sum_{t=0}^{T}\sum_{s=t}^{T} DS_{t,s}$

Cash Shortfalls for Funded and Unfunded Obligations

The lifetime and 1yr ECLs for unfunded loans with unused commitments (e.g. revolving lines) with principal amortization schedules are described by a modified version of Equations (13.17) and (13.18). For products with bullet payment of principal (i.e. no principal amortization schedule), the lifetime and 1yr ECL reduce to Equations (13.19) and (13.20).

The contractual cash flows and the effective interest rate *EIR* used for discounting cash flow are determined by whether the loans are fixed rate or floating rate instruments. The former requires the calculation of *EIR* at the loan origination, while the latter requires the calculation of (1) the original credit spreads and (2) the current reference interest rate (and discounting rate), which are combined to determine *EIR*.

The contractual cash flows for unused commitments include two separate fees/interest rate components:

1. interest rate payments on the drawn amount V_t (with interest rate r_v)
2. interest fees on the undrawn amount U_t available for future drawdowns (with fee rate r_u)

That is, the cash flows can be described as follows:

$$CF_t = r_v V_t + r_u U_t \quad 0 \le t \le T \tag{13.21}$$

As a result of the amortization of the principal amount W_t and incremental drawdowns d_t, the principal amount value V_t outstanding at the beginning of each payment period t is simply:

$$V_t = V_0 + \sum_{s=0}^{t-1} d_s - \sum_{s=0}^{t-1} W_s = V_{t-1} + d_{t-1} - W_{t-1} \tag{13.22}$$

Similarly, the unused commitment amount available is:

$$U_t = U_0 - \sum_{s=0}^{t-1} d_s + \sum_{s=0}^{t-1} W_s = U_{t-1} - d_{t-1} + W_{t-1} \tag{13.23}$$

Here the incremental drawdown is determined by the difference in incremental utilization between periods $t-1$ and t:

$$d_t = IU_t - IU_{t-1} \tag{13.24}$$

The incremental utilization IU_t represents the expected drawdown of the unused commitment in each period t.

Each contractual cash flow CF_t is based on the amount outstanding at that point in time (as opposed to an interest rate payment on a fixed initial principal). Thus, the cash shortfall before default is zero since the expected contractual cash flows are projected in the same way as the expected cash flow before default. (Note that in our approach only the event of default creates a cash shortfall.) That is, there is no difference in facility utilization between the contractual cash flows and the expected cash flows and, therefore, Equations (13.17) to (13.19) used for funded exposures can be extended to the case where EAD_t includes both the funded amount and the expected funded amount (expected drawdown) at default from the available unused committed exposure represented by the *credit conversion factor* (CCF):

$$EAD_t = V_t + CCF_t U_t \tag{13.25}$$

Note that the effects of incremental use of unused commitments are included in the definitions of V_t and U_t. Also, for an unused commitment with an amortizing schedule, the cash shortfall after default includes the drawdown at default as reflected in EAD (see Table 13.4).

Portfolio Loss Projections

Since different sectors of the economy could be affected differently by physical and transition risks, financial institutions should adopt a comprehensive sectoral approach tailored to the institution's business strategy for understanding climate risks across industries and geographies and providing adequate risk identification. To illustrate, real estate could be impacted by an increased severity of extreme weather resulting from physical risk, while the energy and power generation sectors could be impacted by the transition to a low-carbon-emission economy with new business restrictions, changes in regulations, increased carbon taxes and costs, or migration to alternative sources of energy. Key to assessing the impact of different business strategies on the institution's portfolios is to be able to estimate future credit losses for different scenarios and portfolio industry mix over a time horizon aligned to the institution's commitments on carbon emission reductions.

The expected credit loss $\overline{ECL_{0t}}$ at a future time t measured at the starting point $t = 0$ is determined by the projections of PD, LGD, and EAD for different scenarios and the changes in the portfolio composition due to: (1) rating migration and changes in credit quality, and (2) the institution's strategy for providing funding to different market segments, refinancing existing clients, and supporting new origination. This is related to the institution's business strategy and commitments on the transition to a low-carbon-emission economy.

To illustrate this point, under the assumption that the institution will keep the same portfolio composition over time, the expected $\overline{ECL_{0t}}$ at a future time t for an obligor that has an initial risk rating R at $t = 0$ is simply:

$$\overline{ECL_{0t}}(R) = \sum_{R'=1}^{M} TM_t(R, R')\, ECL_t(R') \tag{13.26}$$

Here $TM_t(R, R')$ is the rating migration matrix from an initial rating R to any rating $1 \leq R' \leq M$ (including default), and ECL_t is the expected credit loss discounted appropriately with the discounting curve D_{tk} at time t for different periods $t \leq k \leq T$.

$$ECL_t(R') = \sum_{k=t}^{t+T} D_{t,k} LGD_k EAD_k (P_k - P_{k-1}) \tag{13.27}$$

The expected credit loss $ECL_t(R')$ reflects the cumulative default loss described by the number of incremental defaults events in each payment period $t \leq k \leq t + T$ for loan maturity T, discounted (present-valued) to period t using the discounting factor D_{tk} since defaults occur at a future period k and need to be discounted to time t (which represents the future ECL reporting date). Equations (13.26) and (13.27), combined with cumulative default loss estimates and supplemental assumptions about debt refinancing and new origination, provide a simple framework for estimating default losses and changes in allowance for credit losses that can help assess different portfolio strategies and scenarios required to understand the impact of climate risk on the institution.

REFERENCES

Bessis. J. (2002). *Risk Management in Banking*. Wiley Finance, New York.

FASB (Financial Accounting Stability Board) (2016). Financial Instruments – Credit Losses (Topic 326) Measurement of Credit Losses on Financial Instruments (Current Credit Expected Loss). *FASB Financial Accounting Series* No. 2016-13 (June). https://www.fasb.org/jsp/FASI3/FASI3Content_C/CompletedProjectPage&cid=1176168232014

IFRS (International Financial Reporting Standard) (2014). IFRS 9 Financial Instruments, *International Financial Reporting Standard* (July). https://www.ifrs.org/issued-standards/list-of-standards/ifrs-9-financial-instruments/

Matten, C. (1996). *Managing Bank Capital*. John Wiley & Sons, Chichester.

OCC (Office of the Comptroller of the Currency) (2020). Interagency Policy Statement on Allowances for Credit Losses. Rules and Regulations. Federal Register 85(105) (June). https://www.federalregister.gov/documents/2020/06/01/2020-10291/interagency-policy-statement-on-allowances-for-credit-losses

OCC (Office of the Comptroller of the Currency) (2021). Allowances for Credit Losses. *OCC Comptroller's Handbook* (April). https://www.occ.treas.gov/publications-and-resources/publications/comptrollers-handbook/files/allowances-for-credit-losses/index-allowance-credit-losses.html

Sobehart, J.R. (2009). Uncertainty, Credit Migration, Stress Scenarios and Portfolio Losses. In: D. Roesch, and H. Scheule (Eds.), *Stress Testing for Financial Institutions: Applications, Regulations and Techniques*. Risk Books, London, pp. 197–235.

Sobehart, J.R. (2017). Forward Looking Expected Credit Losses, Model Risk and Uncertainty - A Foundational Approach for IFRS9 and CECL. Presented at RiskMinds International 2017. Amsterdam, the Netherlands, December 4–8, 2017.

Sobehart, J.R. (2019). Leveraging IFRS 9 Implementation Lessons to Solve CECL Requirements – IFRS9 and Beyond. Presented at RiskMinds International 2019. Amsterdam, the Netherlands, December 3–6, 2019.

Sobehart, J.R. (2020). CECL Forecasting. Presented at RiskMinds International 2020. Boston, September 21–23, 2020.

Sobehart, J.R. (2021). CECL and IFRS9 Expected Credit Loss Estimation in Uncertain Economic Environments. *Journal of Risk Management for Financial Institutions* 14(4): 367–380.

Sobehart, J.R., and Giacone, R. (2012). Addressing Procyclicality: Credit Cycles, Stress Scenarios and Portfolio Losses. In M. Ong (Ed.), *Measuring and Managing Capital for Banks and Financial Institutions*. Risk Books, London, pp. 388–433.

Sobehart, J.R., and Keenan, S.C. (2004). Modeling Rating Migration for Credit Risk Capital and Loss Provisioning Calculations. *RMA Journal* (October): 30–37.

van Deventer, D.R., Imai, K., and Mesler, M. (2005). *Advanced Financial Risk Management*. John Wiley and Sons, Singapore, pp. 5–63.

Acknowledgments

I am very grateful to Valeria, Nadia, and Lionel who provided me with encouragement along the way; to Sean Keenan for teaching me about market behavior and credit risk, and for his invaluable friendship and support over the years; and Susan Dunsmore and Alice Hadaway for their contributions, dedication and effort to make this book a reality. All remaining errors and omissions are my own.

About the Author

Jorge Sobehart has over 35 years' experience in advanced quantitative modeling in industry and government, having worked for prestigious institutions, including the Atomic Energy Commission of Argentina (Nuclear Fusion Division), the Center for Nonlinear Studies at the US Los Alamos National Laboratory, Moody's Investors Service, and the Center for Adaptive Systems Applications (CASA), a cutting-edge financial consulting start-up, making contributions in the fields of risk management, behavioral finance, theoretical and applied physics, computation, and mathematical modeling. He also has acted as a technical reviewer for several book editors and over a dozen professional journals in these fields. He is currently a Managing Director of a large financial institution, leading analytics for wholesale credit and climate risk and risk ratings globally. During his career, he designed and developed advanced frameworks for wholesale credit risk capital and allocation, credit and climate risk stress testing (CCAR, ICAAP, Climate Risk), credit reserves (CECL, IFRS9, FAS5), risk ratings, probability of default, and various early warning tools of credit deterioration.

About the Author

Index

Please note that page references to Tables are in italics, and references to Figures are in bold. Note references are indicated by page and number of Note

B

F

M